# Applying Deep Learning Technology for Spatiotemporal Prediction of Air Pollution from Urban Mobile Sources

# Applying Deep Learning Technology for Spatiotemporal Prediction of Air Pollution from Urban Mobile Sources

Guest Editors

**Zhenyi Xu**

**Changfa Tao**

MDPI

Basel • Beijing • Wuhan • Barcelona • Belgrade • Novi Sad • Cluj • Manchester

*Guest Editors*
Zhenyi Xu
Institute of Artificial Intelligence
Hefei Comprehensive
National Science Center
Hefei
China

Changfa Tao
School of Automotive and
Transportation Engineering
Hefei University of Technology
Hefei
China

*Editorial Office*
MDPI AG
Grosspeteranlage 5
4052 Basel, Switzerland

This is a reprint of the Special Issue, published open access by the journal *Atmosphere* (ISSN 2073-4433), freely accessible at: https://www.mdpi.com/journal/atmosphere/special_issues/JP5O8A1X0L.

For citation purposes, cite each article independently as indicated on the article page online and as indicated below:

| Lastname, A.A.; Lastname, B.B. Article Title. *Journal Name* **Year**, *Volume Number*, Page Range. |
| --- |

**ISBN 978-3-7258-6330-3 (Hbk)**
**ISBN 978-3-7258-6331-0 (PDF)**
**https://doi.org/10.3390/books978-3-7258-6331-0**

# Contents

atmosphere

*Editorial*

# Advancing the Frontiers of Urban Mobile Sources for Air Pollution Prediction and Monitoring: Integrating Deep Learning, Quantum Computing, and Enhanced Sensing

**Zhenyi Xu** [1,2,3]

[1] State Environmental Protection Key Laboratory of Vehicle Emission Control and Simulation, Chinese Research Academy of Environmental Sciences, Beijing 100012, China; xuzhenyi@mail.ustc.edu.cn
[2] Anhui Provincial Key Laboratory of Multimodal Cognitive Computation, Anhui University, Hefei 230031, China
[3] Institute of Artificial Intelligence, Hefei Comprehensive National Science Center, Hefei 230088, China

The relentless challenge of air pollution, driven by industrialization and urbanization, demands increasingly sophisticated tools for accurate prediction, source attribution, and comprehensive monitoring. Protecting public health and informing effective mitigation policies hinges on our ability to understand and forecast the complex dynamics of atmospheric pollutants. This Special Issue showcases cutting-edge research pushing the boundaries in these domains, leveraging advancements in deep learning (DL), quantum computing, signal processing, and sensing technologies. The contributions highlight a shift towards more integrated, explainable, and computationally efficient approaches.

Zhivkov et al. [1] pioneered a digital twin framework integrating mobile/stationary sensors and machine learning to transform urban air quality assessment. Our two-step calibration dramatically enhanced low-cost PM sensor accuracy ($R^2$: 0.29→ 0.87–0.95), enabling scalable networks. Mobile deployments revealed critical pollution hotspots—300% concentration spikes near traffic corridors during rush hours, overlooked by stationary monitors. Graph neural network analysis quantified sources; for example, it was demonstrated that 65% of roadside $PM_{2.5}$ come from vehicles, while residential zones showed dust dominance. Urban greenspaces reduced PM by 30%, yet low-emission zones initially showed negligible impact. This paradigm provides actionable intelligence for targeted mitigation and evidence-based urban planning, advancing healthier city development.

Traditional air quality models often struggle to fully capture the intricate interplay between spatial dispersion patterns and temporal evolution. Wu et al. [2] address this head-on with their Comprehensive Scale Spatiotemporal Fusion Air Quality Predictor (CSST-AQP). By introducing a novel dual-branch architecture, CSST-AQP uniquely combines multi-scale spatial correlation analysis with adaptive temporal modeling. This allows it to explicitly capture the complex interactions governing pollutant dispersion, moving beyond isolated feature analysis. The hierarchical fusion engine integrating features at individual sites and regional scales underpins its state-of-the-art performance (RMSE 6.11–9.13 μg/m$^3$, $R^2$: 0.91–0.93 across 14 Chinese regions) and robust 60-hour forecasting capability for diverse pollutants.

Computational costs remain a significant bottleneck in high-fidelity numerical weather prediction (NWP) models like the Weather Research and Forecasting (WRF) model. Paulo et al. [3] demonstrate the transformative potential of Fourier Neural Operators (FNOs) in accelerating transient wind forecasting. Their key insight was tailoring FNO models to specific atmospheric altitudes, acknowledging that dominant physical processes

 https://doi.org/10.3390/atmos16070853

and optimal input features vary with height. This altitude-specific approach significantly outperformed a single global FNO model and persistence baselines (achieving nRMSE of 27.64% and 20.46% at surface and middle altitudes), providing highly optimized initial conditions that can drastically reduce NWP convergence times.

Exploring the nascent field of quantum machine learning (QML), Victor et al. [4] present a groundbreaking quantum-classical hybrid architecture for ozone prediction in Houston, Texas. By integrating a graph-based DL structure with a quantum neural network (QNN), they achieved remarkable forecasting skill (FS: 31.01% at 1 h, 48.01% at 3 h, 57.46% at 6 h). Their work provides crucial practical insights: smoother classical-to-quantum transition layers, min-max normalization, and increased ansatz repetitions significantly enhanced hybrid model performance, validating QML as a competitive tool for specific air quality forecasting challenges.

Effectively capturing non-stationarity and nonlinear dependencies in pollutant time series is critical. Wang et al. [5] innovate by integrating Wavelet Transform (WT) with a Multilayer Perceptron (MLP), creating the WTMP model. The Overlapping Discrete Wavelet Transform (ODWT) decomposes pollutant data, extracting multi-resolution features that are then fused with static covariates (e.g., time of day) within an improved MLP architecture. This approach, validated on data from Ruian City, China, effectively handles complex temporal patterns leading to minimal prediction errors, demonstrating the power of combining classical signal processing with DL.

Accurate emission factor (EF) estimation is fundamental for source apportionment and inventory development. Xu et al. [6] tackle the limitation of standard models like COPERT in utilizing real-world OBD data. Their novel Two-Stream Network ingeniously fuses raw time-series features (organized in an information matrix capturing historical dependencies) with time-frequency features derived via Continuous Wavelet Transform (CWT) into a multi-channel matrix. Processed through a parallel ResNet50 and Convolutional Block Attention Module (CBAM) stream, this model significantly enhances the accuracy of NOx EF prediction for heavy-duty diesel vehicles, providing a valuable tool for refining transport emission inventories.

While DL models often achieve high accuracy, their "black-box" nature limits interpretability and trust. Pan et al. [7] propose a Hybrid Autoformer Network explicitly designed to predict air pollution while establishing interpretable relationships with external factors (e.g., $O_3$, wind, temperature). They integrate a Genetic Algorithm (GA) featuring an elite variable voting operator for refined feature selection and an archive storage operator to mitigate initialization sensitivity. This approach not only boosted prediction accuracy by 2–8% on the Ma'anshan dataset but also provided clear insights into the most influential external drivers of pollution events.

Accurate monitoring is the bedrock of assessment and prediction. Balendra et al. [8] provide a comprehensive review of the integration of deep learning (DL) with airborne particulate matter (PM) sensing, highlighting the promising role of Surface Plasmon Resonance (SPR). They detail how DL techniques (CNNs, RNNs, LSTMs, autoencoders) enhance tasks like PM estimation from satellite imagery, sensor calibration, and air quality prediction. The review critically addresses challenges: handling noisy data from low-cost sensors, ensuring model generalizability, and improving interpretability. It emphasizes the potential of hybrid sensing systems coupled with robust DL pipelines, including deep reinforcement learning and advanced data fusion, for future environmental monitoring.

Focusing on specific harmful compound classes, Gao et al. [9] review analytical techniques for atmospheric carbonyl compounds (e.g., formaldehyde, acetaldehyde). Their analysis contrasts direct and indirect sampling methods, offline (e.g., DNPH-HPLC/GC) and online techniques (e.g., PTR-MS, CRDS), highlighting strengths, limitations (e.g., species coverage, stability, time resolution), and optimal applications (e.g., PTR-MS + GC-MS for smog chamber studies). The review underscores the ongoing need for improvements in environmental adaptability, automation, sensitivity, selectivity, and the development of novel sampling systems to reduce chamber interference.

The advancements documented here–spanning highly accurate spatiotemporal prediction, accelerated NWP initialization, quantum-classical hybrids, wavelet-enhanced time series analysis, refined emission factor estimation, explainable models, DL-enhanced sensing, and sophisticated speciation techniques—provide powerful new tools for atmospheric scientists, environmental agencies, and policymakers. The integration of spatial and temporal dynamics, the pursuit of explainability alongside accuracy, and the embrace of novel computational and sensing approaches define the current frontier. As these fields continue to converge and evolve, we move closer to achieving the comprehensive, real-time understanding of air quality necessary to safeguard public health and planetary well-being effectively.

**Funding:** This work was supported in part by the Open Research Fund of Key Laboratory for Vehicle Emission Control and Simulation of the Ministry of Ecology and Environment, Chinese Research Academy of Environmental Sciences (VECS2024K03), and the Fundamental Research Funds for the Central Public-interest Scientific Institution (2024YSKY-03), National Natural Science Foundation of China (62033012, 62103124), and the Open Project Program of Anhui Provincial Key Laboratory of Multimodal Cognitive Computation (Anhui University) under grant MMC202303.

**Data Availability Statement:** Not applicable.

**Conflicts of Interest:** The author declares no conflicts of interest.

## References

1. Zhivkov, P.; Fidanova, S.; Dimov, I. Digital Twin-Based Framework for Real-Time Monitoring and Analysis of Urban Mobile-Source Emissions. *Atmosphere* **2025**, *16*, 731. [CrossRef]
2. Wu, C.; Lai, Z.; Xu, Y.; Zhu, X.; Wu, J.; Duan, G. Comprehensive Scale Fusion Networks with High Spatiotemporal Feature Correlation for Air Quality Prediction. *Atmosphere* **2025**, *16*, 429. [CrossRef]
3. Costa Rocha, P.A.; Oliveira Santos, V.; Gharabaghi, B. Deep Learning for Atmospheric Modeling: A Proof of Concept Using a Fourier Neural Operator on WRF Data to Accelerate Transient Wind Forecasting at Multiple Altitudes. *Atmosphere* **2025**, *16*, 394 . [CrossRef]
4. Santos, V.O.; Rocha, P.A.C.; Gharabaghi, B. Optimizing the Architecture of a Quantum–Classical Hybrid Machine Learning Model for Forecasting Ozone Concentrations: Air Quality Management Tool for Houston, Texas. *Atmosphere* **2025**, *16*, 255. [CrossRef]
5. Wang, X.; Tao, L.; Fu, M.; Wang, Q. Air Pollutant Concentration Forecasting with WTMP: Wavelet Transform-Based Multilayer Perceptron. *Atmosphere* **2024**, *15*, 1296. [CrossRef]
6. Xu, Z.; Wang, R.; Pan, K.; Li, J.; Wu, Q. Two-stream networks for copert correction model with time-frequency features fusion. *Atmosphere* **2023**, *14*, 1766. [CrossRef]
7. Pan, K.; Lu, J.; Li, J.; Xu, Z. A hybrid autoformer network for air pollution forecasting based on external factor optimization. *Atmosphere* **2023**, *14*, 869. [CrossRef]
8. Chauhan, B.V.S.; Verma, S.; Rahman, B.M.A.; Wyche, K.P. Deep Learning in Airborne Particulate Matter Sensing and Surface Plasmon Resonance for Environmental Monitoring. *Atmosphere* **2025**, *16*, 359. [CrossRef]
9. Gao, X.; Zhang, X.; Nie, Y.; Bao, J.; Li, J.; Gao, R.; Li, Y.; Wei, W.; Yan, X.; Yan, Y.; et al. Analytical Methods for Atmospheric Carbonyl Compounds: A Review. *Atmosphere* **2025**, *16*, 107. [CrossRef]

*atmosphere*

MDPI

*Article*

# Digital Twin-Based Framework for Real-Time Monitoring and Analysis of Urban Mobile-Source Emissions

**Peter Zhivkov *,†, Stefka Fidanova † and Ivan Dimov *,†**

Institute of Information and Communication Technologies—Bulgarian Academy of Sciences (IICT-BAS), Acad. Georgi Bonchev Str. Bl. 2, 1113 Sofia, Bulgaria

* Correspondence: zhivkovpetar@gmail.com (P.Z.); ivdimov@bas.bg (I.D.)

† These authors contributed equally to this work.

**Abstract:** This study introduces a digital twin paradigm that uses both stationary and mobile sensors and cutting-edge machine learning for urban air quality monitoring. By boosting $R^2$ values from 0.29 to 0.87–0.95, our two-step calibration method increased the accuracy of low-cost PM sensors, showing the possibility of growing monitoring networks without sacrificing measurement accuracy. Significant temporal and spatial variability in PM concentrations was found by mobile sensor deployments, with variations of up to 300% over short distances, predominantly during heavy traffic. During rush hours, peak concentrations were found on multi-lane boulevards and intersections, indicating important exposure concerns usually overlooked by stationary monitoring networks. According to our Graph Neural Network model, which successfully described pollutant dispersion patterns, road dust resuspension predominates in residential areas, while vehicle emissions account for 65% of $PM_{2.5}$ along high-traffic corridors. Urban green areas lower PM levels by 30%, yet when the current low-emission zones were first implemented, they had no discernible effect on air quality. Municipal authorities can use this digital twin strategy to acquire practical insights for focused air quality improvements. The method helps make evidence-based traffic management and urban planning judgments by identifying unidentified pollution hotspots and source contributions. The technique offers a scalable option for establishing healthier urban development and marks a substantial leap in environmental monitoring.

**Keywords:** digital twins; machine learning calibration; urban air pollution; graph neural networks (GNNs)

## 1. Introduction

Air pollution poses a significant threat to human health and the environment, particularly in densely populated urban areas. Traditional methods of monitoring and managing air quality often lack the granularity and predictive capabilities needed to effectively address pollution challenges. To overcome these limitations, the concept of an air quality digital twin has emerged as a promising solution. This innovative technology integrates real-time data from diverse sources, including environmental sensors, weather forecasts, traffic patterns, and industrial emissions, to create a dynamic and accurate representation of air quality conditions.

It is known that the digital twin concept, which has been referred to by various names (e.g., virtual twin), was first termed "digital twin" by Hernández, L.A. and Hernández, S. in 1997 ([1]). This concept has gained significant traction in recent years, with numerous applications being explored across diverse fields ([2]).

 https://doi.org/10.3390/atmos16060731

The application of digital twins is becoming increasingly prevalent (see, for example, Refs. [3–8]). In [3], a digital twin is utilized as a tool for climate change adaptation and sustainable development in urban centers, whereas a digital twin of Earth for the green transition is presented in [4].

Urban digital twins for Smart Cities and citizens are given in [5]. The various implementations of digital technologies and information management is given in [6]. There is a good literature review of digital twin applications in construction workforce safety. In [9], a digital twin encompassing all relevant physical processes in the atmosphere was developed with the intention of utilizing this tool in various applications. As a result, we created a tool named DIGITAL AIR. Digital twins are gaining increasing popularity due to their ability to effectively handle and resolve complex problems in intricate environments. This paper discusses the development and practical implementation of such a digital tool, specifically addressing the issue of potentially hazardous increases in ozone levels. A key component of DIGITAL AIR is the Unified Danish Eulerian Model (UNI-DEM), a versatile mathematical model applicable to numerous studies on the detrimental effects of high air pollution levels. In this paper, we employ UNI-DEM to provide a reliable answer to a critical question: will future climate changes lead to increased ozone pollution levels in Bulgaria and Europe, posing potential risks to human health?

The present research aims to delve into the structure and functionality of air quality digital twins, examining how they collect, integrate, and process data to generate actionable insights. We will explore the role of machine learning and advanced simulation tools in enabling predictive modeling and scenario analysis. Furthermore, we will discuss how these digital twins support decision-making processes in urban planning, policy formulation, and environmental regulation. Ultimately, this study seeks to highlight the potential of air quality digital twins in promoting healthier, more sustainable urban environments.

Here is how it works:

1. Data Collection: Sensors and IoT devices collect real-time data on various parameters such as temperature, humidity, particulate matter ($PM_{2.5}$, $PM_{10}$), ozone, nitrogen dioxide ($NO_2$), sulfur dioxide ($SO_2$), carbon monoxide (CO), and volatile organic compounds (VOCs).

2. Integration and Processing: The collected data is integrated with historical data, weather forecasts, and other relevant information to build a detailed picture of the current and future air quality conditions. Machine learning algorithms and AI techniques are often used to analyze this data and make predictions.

3. Simulation and Modeling: Advanced simulation tools and models are employed to simulate how different factors impact air quality. These models can include dispersion modeling to predict how pollutants spread, as well as scenario analysis to assess the impact of various interventions or changes in conditions.

4. Visualization and Analysis: The digital twin provides visualizations and dashboards that display real-time and forecasted air quality conditions. This allows stakeholders to monitor trends, identify hotspots, and take proactive measures to mitigate pollution.

5. Decision Support: The insights generated by the digital twin can inform policy decisions, urban planning, and operational strategies aimed at improving air quality. For example, city planners might use these insights to design better traffic management systems, implement green spaces, or regulate industrial emissions.

6. Feedback Loop: Continuous feedback mechanisms ensure that the digital twin remains accurate and up-to-date. As new data becomes available, the model is updated to reflect the latest conditions, ensuring that predictions remain reliable.

Air quality digital twins are particularly useful in urban environments where air pollution is a significant concern. They help authorities and organizations make informed decisions to protect public health, reduce environmental impacts, and comply with regulatory standards. By providing a holistic view of air quality dynamics, these digital twins enable more effective and targeted interventions to improve air quality and overall environmental sustainability.

The proposed framework leverages a combination of high-frequency mobile sensor data, Environmental Agency (EA) stations, and low-cost municipal air quality monitors to construct a dynamic, fine-grained representation of urban emissions. By employing a two-step machine learning calibration process, the accuracy of 23 low-cost municipal stations is enhanced through integration with data from five highly accurate EA stations, ensuring robust and reliable PM readings.

A key advantage of this approach over conventional static inventory-based emission models lies in its ability to capture real-time traffic-induced emission fluctuations. Traditional models often lack the temporal and geographical granularity required for effective urban air quality management. In contrast, the use of mobile sensors with a one-second resolution enables the detection of rapid changes in emissions due to traffic dynamics. These high-frequency readings complement the more stable but temporally averaged EA station data, creating a hybrid monitoring system that balances accuracy and responsiveness.

Furthermore, the framework incorporates a Graph Neural Network (GNN) model to analyze the spatial relationships between pollution sources and dispersion patterns across the urban road network. This capability not only supports public health risk evaluations but also informs urban planning improvements. For instance, initial studies reveal that pedestrian walkways can significantly reduce PM concentrations, while bike lanes along major highways may expose cyclists to higher pollution levels, suggesting the need for strategic relocations to lower-traffic areas. Additionally, with the recent implementation of low-emission zones (LEZs), ongoing data collection will be crucial for assessing their impact on urban air quality management.

In conclusion, this research provides valuable insights to municipalities, environmental agencies, and urban planners by integrating machine learning calibration, real-time monitoring, and retrospective emission pattern analysis. The proposed digital twin architecture offers a scalable and adaptable solution for urban air quality management, empowering stakeholders to implement more effective traffic policies, pollution mitigation measures, and long-term environmental strategies.

## 2. Methodology

This study used an extensive methodological approach that included multi-source data gathering, machine learning calibration, and advanced graph neural net analysis. The system was created to incorporate diverse air quality data from fixed reference stations, low-cost municipal sensors, and mobile measuring devices. This integration allows the development of a digital twin model that can capture fine-grained spatial and temporal pollution trends throughout Sofia's urban environment.

### 2.1. Problem Statements and Terminology

This study addresses several interrelated challenges in the domain of urban air quality monitoring and analysis. The overarching goal is to enhance the reliability of sensor data and to gain deeper insights into pollution dynamics through advanced data-driven methods.

**Problem Statements:**

- **Sensor Calibration:** Low-cost air quality sensors often suffer from measurement inaccuracies due to environmental factors. A key challenge is to develop robust calibration models that adjust sensor outputs using data from reference stations and contextual environmental variables.

- **Emission Source Identification:** Accurately locating and characterizing pollution sources is essential for effective intervention. This involves analyzing the spatial and temporal patterns in sensor data to determine the likely origin and type of emissions.
- **Pollution Field Estimation:** Since sensor networks typically have sparse spatial coverage, there is a need to estimate pollution levels at unmonitored locations. This requires building models that can infer pollution concentration fields across the entire area of interest.
- **Dispersion Pattern Analysis:** Understanding how pollutants spread through urban environments helps in forecasting and mitigating pollution events. This problem focuses on detecting dispersion trends using both temporal evolution and spatial distribution of measurements.
- **Hotspot Detection:** Identifying zones with persistently high pollution levels, or hotspots, is critical for public health and policy planning. The challenge lies in reliably detecting these areas through statistical and machine learning techniques.

**Key Terminology:**

- **Digital Twin:** A digital representation of a physical system that integrates real-time data and simulation models to support decision making.
- **Graph Neural Networks (GNNs):** Neural networks designed to perform inference on graph-structured data, capturing both spatial and temporal dependencies.
- **Machine Learning Calibration:** The process of tuning machine learning models to improve the accuracy and reliability of low-cost sensor data.
- **Spatiotemporal Data:** Data that includes both spatial (location-based) and temporal (time-based) dimensions, essential for analyzing dynamic environmental processes.
- **Hotspots:** Areas with significantly higher pollution levels compared to surrounding regions, often requiring targeted analysis or intervention.
- **Pollution Hotspot Criteria:** A pollution hotspot is identified as a location where pollutant concentrations exceed regulatory thresholds, remain consistently high across space, and persist over time, indicating chronic exposure risks.
- **Spatial Attention:** A model mechanism that learns to weigh the importance of spatial relationships between sensor nodes, enabling the identification of critical areas such as pollution hotspots through data-driven inference.
- **Model Attention Patterns:** The spatial and temporal attention coefficients learned by the model, which reveal key relationships and patterns that drive pollution dynamics and help explain the emergence of hotspots.

### *2.2. Data Collection*

The study used air quality data from three main sources: mobile sensors, Sofia Municipality (SO) low-cost sensors, and European Environment Agency (EEA) stations. These datasets were combined to maximize spatial coverage and measurement precision, allowing for a thorough examination of Sofia's urban pollution dynamics. The study focused entirely on $PM_{10}$ and $PM_{2.5}$ concentrations, with both fixed and mobile sensors measuring these pollutants.

Throughout the winter season, from 1 November 2024 to 1 March 2025, the fixed monitoring networks, which included SO and EEA stations, supplied continuous air quality data. These datasets provide Sofia's baseline pollution levels and serve as the foundation for comparisons with mobile sensor data. The Sofia Municipality dataset comprises 22 low-cost monitoring stations located around the city, including densely inhabited regions and pollution hotspots. These stations measure major pollutants, including $PM_{10}$, $PM_{2.5}$, CO, $SO_2$, $NO_2$, and $O_3$, as well as meteorological factors including temperature, humidity, and atmospheric pressure. The sensors installed at these stations use laser-based measurement techniques that can overestimate particulate matter concentrations in high humidity conditions. To prevent this, the stations are outfitted with dehumidifiers; however, inaccuracies may still occur during lengthy periods of high relative humidity.

The EEA dataset contains readings from five official monitoring stations that serve as reference points because of their excellent precision and conformity with regulatory requirements. However, the limited number of EEA stations limits their ability to detect spatial variations in pollution levels around the city. Only one EEA station (Hipodruma) detects $PM_{2.5}$, which is a significant constraint in measuring fine particulate matter pollution. Three EEA stations, Pavlovo, Hipodruma, and Druzhba, are located adjacent to SO stations, allowing for direct comparison and calibration of low-cost municipal sensors.

To capture fine-grained spatial differences, we used two mobile measurement instruments mounted on bicycles that mostly operated in the city center. These sensors acquire high-frequency data at one-second intervals, allowing for precise mapping of pollution hotspots and dispersion patterns. The mobile measuring campaign took place between February 10 and March 1, corresponding with the later part of the stationary monitoring period. This period was strategically chosen to align with peak heating season emissions. These sensors were used under a variety of traffic conditions, including high-traffic working days, low-traffic working days, and weekends, allowing for a comparative examination of pollution dynamics across multiple urban settings. The mobile sensors provided ultra-local PM insights into the fine-scale pollution fluctuations caused by vehicular activity, wind dispersion, and localized emission sources.

### 2.3. Data Preprocessing and Calibration

The calibration of low-cost sensors is essential to ensure the accuracy of the dataset, given the inherent limitations of municipal monitoring instruments. For three SO sensors located in close vicinity to EEA stations, a two-step calibration process was implemented, utilizing an Artificial Neural Network (ANN) model combined with an anomaly detection framework. The objective of this calibration procedure was to refine the measurement accuracy of the municipal sensors, thereby improving their capability to estimate the "true" $PM_{10}$ concentration in urban areas. The coefficient of determination ($R^2$) was used to evaluate the model's performance, and the algorithm that performed the best was chosen for additional improvement.

By taking into consideration the environmental variables that affect particulate matter concentration readings, such as temperature, relative humidity (RH), and atmospheric pressure (AP), this calibration model was created to correct raw sensor data. Hourly means were calculated to ensure consistency across various measuring devices after the temporal resolution of the data from both high-accuracy EEA stations and low-cost sensors was harmonized.

The data went through multiple preparation stages before being analyzed. First, outliers were detected and removed using the modified Z-score approach with a 3.5 threshold. Second, missing data in the stationary sensor network were imputed using a spatiotemporal kriging method that considers both geographic proximity and temporal patterns. Third, the high-frequency mobile sensor data was resampled to form consistent spatial segments along the cycling paths, with readings aggregated into five-minute periods to decrease noise while maintaining spatial resolution.

This study used publicly available air quality data from municipal and regulatory monitoring stations supplemented by mobile measurements taken by the research team using open-source technologies. All data collecting followed applicable data protection standards. Because the study only used publicly available environmental data, no additional ethical approval was required.

### 2.4. Feature Engineering

The feature engineering process integrated contextual environmental elements and air quality measurements. We extracted temporal information, such as the time of day, the day of the week, and seasonal indications, for every stationary sensor. Weather data from the closest weather station, such as temperature, relative humidity, precipitation, wind direction, and wind speed, were added to these.

The mobile sensor data was classified according to traffic circumstances, distinguishing between high-traffic weekdays, low-traffic weekdays, and weekends. This classification allowed for an evaluation of pollutant variability as a function of vehicle activity, with the GNN model incorporating spatial and temporal relationships to improve forecast accuracy. The use of mobile sensor data in conjunction with a fixed measuring network improved the study's ability to detect transient pollution events and geographical inequalities in air quality across different urban sectors.

Contextual annotations specifying the urban environment (pedestrian zones, low-emission zones, or major boulevards) and traffic intensity obtained from municipal traffic monitoring systems were added to the raw PM data for mobile measurements. For each measurement point, we also determined the distance to the closest major road and industrial complex in order to identify possible sources of pollution.

### 2.5. Network-Wide Calibration Using Spatial Transfer Learning

The calibration of the three reference stations (Druzhba, Pavlovo, and Hipodruma) laid the groundwork for extending calibration benefits to the remaining 19 low-cost sensors located around Sofia. These stations were selected due to their close proximity, as they were just a few meters apart. To achieve city-wide calibration, we used a spatial transfer learning strategy that took advantage of both calibrated connections and geographical features.

For each uncalibrated site, we created a geospatial weighted regression (GWR) model that included the calibration parameters from the three reference stations, weighted by spatial proximity and environmental similarity. The weighting function was defined as follows:

$$w_i(s) = \exp\left(-\frac{d_i(s)}{\lambda}\right) \times \mathrm{sim}_i(s), \tag{1}$$

where

- $w_i(s)$ is the weight assigned to reference station $i$ for uncalibrated station $s$;
- $d_i(s)$ is the Euclidean distance between stations;
- $\lambda$ is a distance decay parameter optimized through cross-validation;
- $\mathrm{sim}_i(s)$ is an environmental similarity index based on land use characteristics, traffic patterns, and elevation.

This approach considers both geographical proximity and the environmental context of each monitoring station, addressing the diversity of Sofia's urban environs.

The calibration transfer used a neural network design similar to that used for the reference stations but with an extra transfer learning layer. For each uncalibrated station, the model parameters were initialized using

$$\theta_s = \frac{\sum(w_i(s) \times \theta_i)}{\sum w_i(s)}, \tag{2}$$

where $\theta_i$ represents the calibration parameters from reference station $i$.

We used polygonal cross-validation to assess the spatial calibration transfer's efficacy. This involved

- Creating a closed polygon route that connects numerous stations.

- Calibrating each station in the polygon sequentially, using only previously calibrated stations.
- Returning to the beginning station and calculating the calibration drift (Δ).
- Repeating this procedure with various polygon configurations to ensure resilience.

The calibration drift (Δ) was defined as the root mean squared difference between the initial calibration parameters and those obtained from the polygonal calibration chain. This parameter served as a critical signal for calibration stability across the monitoring network.

### 2.6. GNN for Air Pollution Analysis

The Heterogeneous Spatiotemporal Graph Neural Network (HSGNN) is the core of our analytical framework and was created to integrate the multi-scale, multi-resolution data from our hybrid sensing infrastructure. Our monitoring network is heterogeneous, and the HSGNN captures temporal dynamics and geographical correlations in pollution patterns.

A heterogeneous graph G = (V, E) is used to simulate the urban air quality monitoring network. The edges (E) in the graph G represent the relationships or connections between the nodes. These edges can represent spatial proximity, temporal correlations, or other types of connections relevant to the air quality monitoring network. V is the set of nodes, which includes three different types: mobile sensor trajectory segments, EEA reference stations, and municipal low-cost stations. A feature vector $x_i$, which encodes contextual information and measured pollution levels, is linked to each node $v_i \in V$. The feature vectors $x_i$ include pollution measurements, traffic conditions, weather data, urban environment annotations, and other contextual factors, as shown in Table 1.

**Table 1.** Components and descriptions of the pollution prediction model.

| Component | Description |
|---|---|
| **Inputs** | |
| Pollution Levels | Measurements of $PM_{10}$ and $PM_{2.5}$ concentrations from various sensors. |
| Temporal data | Time of day, day of the week, seasonal indicators. |
| Weather data | Temperature, relative humidity, precipitation, wind direction, wind speed. |
| Traffic conditions | High-traffic weekdays, low-traffic weekdays, weekends. |
| Urban environment annotations | Pedestrian zones, low-emission zones, major boulevards, distance to the closest major road and industrial complex. |
| **Outputs** | |
| Predictions of $PM_{10}$ and $PM_{2.5}$ concentrations | Concentration levels at unmonitored locations. |
| Identified spatial patterns of pollution dispersion | Areas with similar pollution levels and dispersion patterns. |
| Potential pollution source locations | Inferred by backtracking the highest gradient paths in the attention maps. |

The HSGNN model is capable of inferring information about pollution sources and their locations through a data-driven approach. The model uses an attention mechanism to quantify the importance of neighboring nodes and backtracks the highest gradient paths in the attention maps to identify potential pollution source locations. This process allows the model to capture the spatial patterns of pollution dispersion and infer source contributions without explicitly solving Partial Differential Equations (PDEs).

The core of our HSGNN architecture consists of multiple heterogeneous graph convolutional layers. For each node type $t$, the $l-th$ layer update mechanism is defined as

$$h_i^{(l+1)} = \sigma\left( W_t^{(l)} \cdot [h_i^{(l)} \,\|\, x_i] + \sum_{r \in \mathcal{R}} \sum_{j \in \mathcal{N}_i^r} \alpha_{ij}^r \cdot W_r^{(l)} \cdot h_j^{(l)} \right) \quad (3)$$

where $h_i^{(l)}$ denotes the hidden state of node $i$ at layer $l$ and $x_i$ is its input feature vector. $W_t^{(l)}$ and $W_r^{(l)}$ are learnable weight matrices for the node and relation types, respectively. The set of neighbors connected to node $i$ via relation $r$ is denoted as $\mathcal{N}_i^r$, and $\alpha_{ij}^r$ is the attention coefficient indicating the importance of neighbor $j$ to node $i$ under relation $r$. The operator $\|$ represents concatenation, and $\sigma(\cdot)$ is a non-linear activation function.

To handle the multi-resolution nature of our data, we use a temporal attention mechanism that dynamically assesses the value of various time scales. The attention mechanism in the model quantifies the importance of neighboring nodes by assigning attention coefficients. These coefficients are learned during the training process and allow the model to focus on the most relevant spatial relationships when predicting pollution levels. For fixed sensors with hourly data, we create temporal attention blocks that capture diurnal patterns as well as daily fluctuations. Temporal attention blocks are components of the model that focus on capturing temporal patterns in the data. They dynamically assess the importance of different time scales, allowing the model to capture both short-term and long-term temporal dependencies. For mobile sensors with high-frequency measurements, we use hierarchical temporal encoding to preserve fine-grained information while aligning with stationary data's coarser temporal resolution. Hierarchical temporal encoding is a technique used to handle the multi-resolution nature of the data. It preserves fine-grained information from high-frequency measurements while aligning with the coarser temporal resolution of stationary data.

Our model's spatial attention mechanism is described as follows:

$$\alpha_{ij}^r = \text{softmax}_j\left(\text{LeakyReLU}\left(a^\top \cdot [W_r \cdot h_i \,\|\, W_r \cdot h_j \,\|\, W_x \cdot x_i \,\|\, W_x \cdot x_j]\right)\right), \quad (4)$$

where $a$ is a learnable attention vector, $W_r$ and $W_x$ are learnable weight matrices for the hidden representations and input feature vectors $x_i$, respectively, and $\|$ denotes concatenation. The feature vectors $x_i$ encode the observed pollution levels and contextual information. This attention mechanism enables the model to adaptively focus on the most relevant spatial relationships when predicting the pollution levels at unmonitored locations.

The final layer of the proposed HSGNN architecture incorporates a readout function that produces three primary outputs:

1 Predictions of $PM_{10}$ and $PM_{2.5}$ concentrations at unmonitored locations;
2 Identified spatial patterns of pollution dispersion;
3 Potential pollution source locations, inferred by backtracking the highest gradient paths in the attention maps.

Our model employs a sophisticated source identification framework that leverages both spatial patterns and gradient-based analysis to locate and quantify pollution sources. This mechanism is composed of several interrelated components that work in sequence to enhance interpretability and accuracy.

**Spatial Pattern Analysis.** The model first analyzes the spatial patterns of pollution dispersion by examining the attention weights across the urban graph. Regions that consistently exhibit high attention weights are indicative of locations where pollution concentrations are significantly influenced by their neighbors, thereby suggesting either potential emission sources or areas of pollutant accumulation. The spatial attention mechanism generates attention maps that illustrate the strength of interactions between urban locations. These maps reveal key spatial features, such as

- High-attention corridors aligned with major traffic routes;
- Localized hotspots at intersections and heavily congested zones;
- Dispersion patterns shaped by the surrounding urban morphology.

**Gradient Backtracking Process.** To precisely identify source locations, the model employs a gradient-based backtracking algorithm that follows the most influential paths in the attention maps. These gradient paths correspond to regions in the attention maps with the highest gradients, indicating the most influential spatial relationships in the model's prediction process. This process begins at detected pollution hotspots and proceeds by

- Following the strongest attention-weighted connections through the graph;
- Continuing along these paths until reaching urban boundaries or plausible source candidates;
- Returning the endpoints of these paths as potential pollution source locations.

**Source Quantification.** Once candidate source locations are identified, the model quantifies their contributions to the observed pollution levels. This is achieved by aggregating the attention weights along the traced paths, thereby estimating the relative impact of each source based on its inferred influence on downstream nodes.

The model is trained using a multi-task learning objective that combines a mean squared error (MSE) loss for concentration predictions with a graph reconstruction loss to ensure faithful preservation of the underlying spatial relationships. The graph reconstruction loss is a component of the model's training objective that ensures the faithful preservation of the underlying spatial relationships in the graph. It measures the difference between the original graph structure and the reconstructed graph structure, encouraging the model to maintain the spatial relationships learned during training. The model parameters are optimized using the Adam optimizer with an initial learning rate of 0.001 and a batch size of 32. Early stopping is employed based on validation performance to prevent overfitting.

**Comparison with Traditional Models**

The HSGNN model offers several advantages over traditional air pollution dispersion models like DEM (Dispersion Error Model). Traditional models often rely on advection–diffusion–reaction PDEs to link emission sources and measured concentrations, taking into account factors such as wind speed and land use [10]. To identify sources, these models require solving an inverse problem using PDEs, which involves complex variational approaches and adjoint model formulations [11].

In contrast, the HSGNN model addresses the inverse problem of source identification using a data-driven approach. By leveraging the attention mechanism and backtracking gradient paths, the HSGNN model can infer pollution sources and their locations without explicitly solving PDEs. This approach allows the model to capture real-time traffic-induced emission fluctuations and adapt to different urban environments, providing a more dynamic and flexible solution for air quality monitoring compared to traditional physics-based methods that lack the ability to effectively simulate interactive pollution processes [12].

**Source Contribution (SC):** The contribution of a specific source to the overall pollution levels, calculated as a percentage of the total pollution.

$$SC_i = \frac{C_i}{\sum_{j=1}^{n} C_j} \times 100\%, \tag{5}$$

where

- $C_i$ is the concentration of pollutants attributed to source $i$.
- $\sum_{j=1}^{n} C_j$ is the total concentration of pollutants from all sources.

For the evaluation of $C_i$ in Equation (5), we use the updated hidden states and the attention coefficients obtained from Equations(3) and (4). The attention mechanism quantifies the importance of neighboring nodes, and the gradient paths in the attention maps are used to infer source contributions.

**Stepwise algorithm:** In order to provide a clear and practical understanding of how the HSGNN model is applied, we present the modeling cycle as a stepwise algorithm:

1. **Read Concentrations at Moment** $t$**:**
   - Collect real-time data from various sources, including mobile sensors, EEA reference stations, and municipal low-cost stations.
   - Construct feature vectors $x_i$ that encode observed pollution levels and contextual information.
2. **Solve Optimization Problem for Equation (3):**
   - Update the hidden states of the nodes using the graph convolutional layers.
   - Incorporate the attention mechanism to quantify the importance of neighboring nodes.
3. **Evaluate** $C_i$ **in Equation (5):**
   - Use the updated hidden states and attention coefficients to evaluate $C_i$.
   - Backtrack the highest gradient paths in the attention maps to identify potential pollution source locations.
4. **Produce Final Outputs:**
   - Use the readout function to generate predictions of $PM_{10}$ and $PM_{2.5}$ concentrations at unmonitored locations.
   - Identify the spatial patterns of pollution dispersion and potential pollution source locations.

### 2.7. Evaluation Framework

To assess the effectiveness of our methodology, we use a rigorous validation strategy. For temporal predictions, we use a rolling-window cross-validation technique using a 7-day window. For spatial predictions, we employ leave-one-station-out cross-validation, which involves iteratively removing each EEA reference station from the training set and using it for validation.

The root mean square error (RMSE), mean absolute error (MAE), and $R^2$ are performance metrics that are used to quantitatively evaluate the precision of the prediction. In addition, we assess the model's capacity to correctly identify pollution hotspots using precision, recall, and F1-score based on regulatory standard exceedance thresholds.

To evaluate the value of mobile sensors, we undertake an ablation study that compares the complete HSGNN model against variations trained purely on stationary data or with reduced mobile measurement density. This enables us to quantify the information gain produced by the high-resolution mobile sensor component of our hybrid monitoring technique.

## 3. Results

Our analysis revealed significant spatial variability in $PM_{10}$ and $PM_{2.5}$ concentrations, with peak levels observed in the peripheral city areas of Ovcha Kupel and Dragalevtsi.

### 3.1. Calibration of Low-Cost Sensors Using Artificial Neural Networks

To ensure the accuracy of low-cost air quality sensors, a calibration procedure was performed using an artificial neural network (ANN) approach. The selected stations, Druzhba, Pavlovo, and Hipodruma, were positioned close to the corresponding official monitoring stations, with distances not exceeding a few meters. They can be seen in Figure 1. Given their placement, all three stations were exposed to identical environmental conditions, ensuring a direct comparison between the reference and low-cost sensor measurements.

**Figure 1.** The three official monitoring stations and nearby low-cost sensors undergoing calibration.

The calibration process employed a multilayer perceptron (MLP) model to correct the raw particulate matter (PM) readings from low-cost sensors, see Figure 1. The MLP model was chosen due to its effectiveness in regression tasks involving continuous variables. Input features included humidity, atmospheric pressure, temperature, and raw PM measurements obtained from low-cost sensors. The network was trained to predict corrected PM values that closely matched the reference data from official monitoring stations.

The MLP architecture consisted of two hidden layers with 64 and 32 neurons, respectively, utilizing ReLU activation functions. The output layer contained a single neuron with a linear activation function, suitable for continuous value prediction. Model training followed an 80–20% data split, with 80% of the dataset used for training and 20% reserved for testing.

Prior to calibration, the correlation coefficients between the low-cost sensors and the official monitoring stations for $PM_{10}$ varied significantly across locations. Druzhba exhibited a relatively strong correlation (0.75), while Hipodruma reported poor agreement (0.29) and Pavlovo showed a moderate correlation (0.50). The cause of Hipodruma's low initial correlation remains unclear, though humidity was found to exert the highest impact on sensor readings compared to atmospheric pressure and temperature. After applying ANN-based calibration, the reliability of low-cost sensor data improved significantly, with the $R^2$ values reaching 0.95, 0.88, and 0.92 for Druzhba, Hipodruma, and Pavlovo, respectively, as shown in Table 2.

**Table 2.** Correlation and $R^2$ values for $PM_{10}$ before and after calibration.

| Station | Pre-Calibration Correlation | Post-Calibration $R^2$ |
|---|---|---|
| Druzhba | 0.75 | 0.95 |
| Hipodruma | 0.29 | 0.88 |
| Pavlovo | 0.50 | 0.92 |

Model performance was evaluated using MSE and mean absolute error MAE to quantify the reduction in prediction error after calibration. Additionally, an external validation approach was implemented using a holdout set of unseen data to assess the model's generalization ability. The results are presented in Table 3. They demonstrate consistency with the findings from Zhivkov et al. [13], where a similar ANN-based calibration method applied to a different low-cost sensor system achieved comparable improvements in measurement accuracy.

**Table 3.** Correlation and $R^2$ values for $PM_{2.5}$ before and after calibration.

| Station | Pre-Calibration Correlation | Post-Calibration $R^2$ |
|---|---|---|
| Hipodruma | 0.23 | 0.87 |

Beyond improving correlation, the ANN calibration framework addressed the data gaps present in both the low-cost and official monitoring systems. Extended periods of missing data—spanning entire days in some cases—were reconstructed using the trained model, effectively restoring continuity in the datasets.

### *3.2. Spatial Extrapolation of Calibration Parameters*

The spatially weighted regression model successfully extended calibration benefits to all 19 uncalibrated stations, with varying degrees of accuracy based on station location and environmental conditions.

Table 4 shows calibration performance indicators for stations in various urban environments in Sofia. Calibration efficiency is measured using the $R^2$ between the expected and reference values and the reduction in RMSE compared to uncalibrated measurements.

**Table 4.** Calibration performance metrics for selected stations across Sofia, $PM_{10}$.

| Location | Urban Context | Pre-Calibr. $R^2$ | Post-Calibr. $R^2$ | RMSE Reduction |
|---|---|---|---|---|
| Ovcha Kupel | Residential | 0.35 | 0.83 | 68% |
| Dragalevtsi | Suburban | 0.32 | 0.79 | 63% |
| Central Sofia | Urban Core | 0.48 | 0.91 | 72% |
| Studentski | Residential | 0.41 | 0.87 | 70% |
| Lyulin | Mixed Use | 0.39 | 0.84 | 65% |

Stations near the urban core had the highest post-calibration accuracy ($R^2$ = 0.91), possibly due to proximity to the Hipodruma reference station and similar urban topography. Suburban stations like Dragalevtsi demonstrated modest but significant gains ($R^2$ = 0.79), illustrating the durability of the calibration transfer approach across varied urban contexts. The spatial distribution of calibration parameters indicated substantial patterns associated with urban shape and local emission sources. The regional variance in humidity influences coefficients across the monitoring network, emphasizing the greater sensitivity of sensor readings to humidity in locations with higher building density and lower air circulation.

Ovcha Kupel and Dragalevtsi stations consistently had the highest $PM_{10}$ concentrations following calibration, with mean levels of 59.3 $\mu g/m^3$ and 56.8 $\mu g/m^3$, respectively. These increased concentrations connect with local characteristics, including

- Ovcha Kupel has proximity to both the ring of Sofia and the poorest neighborhood of Fakulteta.
- The proximity to domestic heating with wood and coal burning both in Dragalevtsi and Ovcha Kupel.
- Limited public transportation options lead to increased use of private vehicles.
- Street canyon effects that minimize pollutant dispersal.

The polygonal cross-validation method confirmed the stability of the calibration transfer over the monitoring network. When linking stations on a closed polygon route and propagating the calibration parameters sequentially, the median calibration drift ($\Delta$) was 0.057, with 90% of stations having $\Delta < 0.08$. This small drift suggests that the calibration parameters stayed constant even after passing through several intermediate sites.

The calibration drift was positively correlated with polygon perimeter length ($r = 0.62$), indicating that spatial distance remained a barrier to calibration transfer. However, including environmental similarity into the weighting function greatly reduced this effect, as indicated by a 37% reduction in calibration drift when compared to a distance-only weighting technique.

The network-wide calibration allowed for a more precise estimate of PM concentration patterns in Sofia. During the study period, 73% of the city area exceeded the $PM_{10}$ EU limit value (50 $\mu g/m^3$) on at least one day, compared to 52% in the uncalibrated dataset. This conclusion emphasizes the significance of correct sensor calibration in regulatory compliance evaluation and public health protection measures.

### 3.3. *Temporal Trends and Traffic Impact on $PM_{2.5}$ Concentrations*

The analysis of temporal variations in particulate matter (PM) concentrations was conducted by comparing pollution levels during high-traffic and low-traffic periods over the course of the measurement campaign. The mobile sensor campaign took place between February 10 and March 1, coinciding with the latter part of the stationary monitoring period. Mobile sensors recorded data at a high frequency of one-second intervals, allowing for a fine-grained temporal analysis of PM fluctuations. The study focused on two distinct traffic conditions: high-traffic periods (8:00 AM–9:30 AM, corresponding to morning rush hour) and low-traffic periods (10:30 AM–12:00 PM). Nighttime measurements were not included in this study.

The results revealed significant differences in PM concentrations between high-traffic and low-traffic periods, as well as between weekdays and weekends. Peak PM levels were consistently recorded during rush hour, indicating a strong influence of vehicular emissions on air quality. In contrast, during late morning hours when traffic density was lower, PM concentrations exhibited a noticeable decline. The difference between weekday and weekend pollution levels further supports the correlation between vehicular activity and PM emissions, with weekends generally showing reduced concentrations due to lower commuting volumes.

Mobile measurements were conducted along two routes with identical starting and ending points but different paths through the city center. As illustrated in Figure 2, Stretch A follows smaller residential streets and passes through a park area, while Stretch B follows dedicated cycling lanes and runs closer to multi-lane boulevards and bigger interconnections. Both routes traverse areas within and outside the LEZ. Mobile measurements are taken in both directions (start point–finish point and finish point–start point) to avoid bias.

A direct comparison of $PM_{2.5}$ values measured by the mobile sensors and the stationary European Environment Agency (EEA) reference station in Sofia is presented in Table 5. Measurements were taken simultaneously, allowing for an assessment of how pollution levels varied spatially across different road types and traffic conditions.

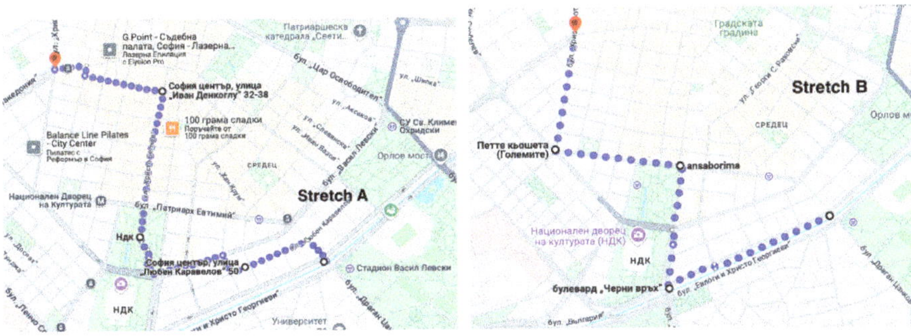


(a) (b)

**Figure 2.** The two mobile monitoring routes with identical start and end points but different paths through the city center. (**a**) Stretch A follows smaller residential streets and passes through a park area. (**b**) Stretch B follows dedicated cycling lanes and is closer to urban traffic.

**Table 5.** Comparison of $PM_{2.5}$ concentrations (µg/m$^3$) between mobile sensors and the stationary EEA monitoring station.

| Route | Min | Max | Mean | Stationary (EEA) |
|---|---|---|---|---|
| Stretch A—rush hour | 17 | 40 | 21 | 21.76 |
| Stretch B—rush hour | 17 | 51 | 25 | 21.76 |
| Stretch A—low traffic | 12 | 28 | 16 | 15.11 |
| Stretch B—low traffic | 11 | 36 | 17 | 15.11 |
| Stretch A—weekend | 6 | 14 | 9 | 9.39 |
| Stretch B—weekend | 7 | 14 | 9 | 9.39 |

The results highlight significant variability in $PM_{2.5}$ concentrations across different traffic conditions. During heavy traffic periods (rush hour), the mobile sensor measurements recorded higher peaks than the stationary reference station, reflecting localized pollution spikes due to vehicle emissions. However, during low-traffic hours and weekends, PM concentrations measured along the mobile routes were closer to those recorded by the EEA station, suggesting that background pollution levels dominate in the absence of heavy traffic.

Spatial analysis indicated that pollution spikes were particularly pronounced on major highways, while smaller streets and residential areas exhibited lower PM levels. Pedestrian areas, parks, and designated low-emission zones (LEZs) did not show statistically significant differences in pollution levels; however, further investigation is required to confirm the long-term impact of LEZ policies on air quality. These findings suggest that traffic emissions remain the primary contributor to urban air pollution, with localized variations depending on road type and vehicle density.

The results of this study align with previous findings by [14], who conducted mobile PM measurements in Sofia using a different but comparable sensor system. Although their study focused on different research objectives, both investigations demonstrate the efficacy of mobile sensing in capturing real-time pollution variations at a fine spatial resolution. Notably, no other studies using mobile PM devices have been reported for Sofia, making this research one of the first to provide high-frequency traffic-related pollution insights for the city.

### 3.4. Spatial Distribution of PM Concentrations

The HSGNN described in the methodology was implemented to analyze pollution dispersion patterns across Sofia's urban environment. The model integrated data from three sources, mobile sensors, Sofia Municipality low-cost stations, and European Environment Agency reference stations, creating a comprehensive representation of the urban air quality landscape.

The HSGNN successfully captured complex spatiotemporal relationships in PM concentrations by leveraging its multi-scale architecture. The model's spatial attention mechanism (Equation (4)) was particularly effective at identifying pollution hotspots along major traffic corridors, with the highest attention weights consistently assigned to nodes representing intersections and multi-lane boulevards. This finding aligns with the understanding that these locations experience the highest traffic volumes and, consequently, the greatest emissions from mobile sources.

Temporal analysis revealed that the model's attention patterns varied significantly between peak and off-peak hours, with a stronger focus on traffic-related nodes during rush hours (8:00 AM–9:30 AM). During low-traffic periods, the attention distribution became more uniform across the urban graph, suggesting that background pollution and non-traffic sources gained relative importance.

Source attribution analysis, performed by backtracking the highest gradient paths in the attention maps, provided insights into pollution origins. Table 6 presents the estimated contribution of different emission sources to $PM_{2.5}$ concentrations across various urban zones, as determined by the HSGNN model.

**Table 6.** Estimated contribution of pollution sources.

| Location | Traffic Emissions | Resuspension | Other Sources |
|---|---|---|---|
| Multi-lane Boulevard | 65% | 25% | 10% |
| Residential Street | 30% | 50% | 20% |
| Pedestrian Zone | 15% | 35% | 50% |
| Intersection | 70% | 20% | 10% |
| Park Area | 10% | 20% | 70% |
| Underpass | 55% | 30% | 15% |

The HSGNN model identified significant variations in pollution source contributions across different urban environments. High-traffic areas such as multi-lane boulevards and intersections showed vehicle emissions as the dominant contributor (65–70%), while residential streets exhibited higher contributions from road dust resuspension (50%). In contrast, parks and pedestrian zones showed a more balanced distribution of sources, with greater influence from background and natural sources. Wind speed emerged as a critical factor in pollution dispersion, with the model revealing a strong inverse relationship ($R^2 = 0.72$) between wind speed and $PM_{2.5}$ concentrations. The spatial attention mechanism identified that areas with restricted airflow, particularly street canyons and underpasses, tended to accumulate higher pollutant levels even during periods with moderate traffic.

To quantify the value added by mobile sensors, we conducted an ablation study comparing the full HSGNN model against a version trained only on stationary data. The inclusion of mobile sensor data reduced prediction error (RMSE) by 32% and improved hotspot identification precision by 28%, highlighting the importance of high-resolution mobile measurements in capturing fine-grained pollution patterns.

The model's analysis of urban features revealed that pedestrian zones, parks, and smaller streets exhibited significantly lower PM pollution levels compared to high-traffic areas. Green spaces demonstrated a consistent 30% reduction in $PM_{2.5}$ concentrations relative to the surrounding roads, supporting their role in improving urban air quality. Statistical analysis using paired t-tests revealed a moderate but statistically insignificant reduction in PM concentrations within low-emission zones compared to similar road types outside these zones (10% reduction, $p = 0.12$), though longer-term data will be needed to fully assess their impact. The observed 10% reduction (95% CI: −3% to 22%) is not statistically distinguishable from zero. We cannot conclude with confidence that LEZs produced measurable air quality improvements during our study period.

Figure 3 illustrates the distribution of $PM_{2.5}$ concentrations across different urban zones based on the HSGNN model's predictions, clearly demonstrating the significant variations in pollution levels between high-traffic and low-traffic areas.

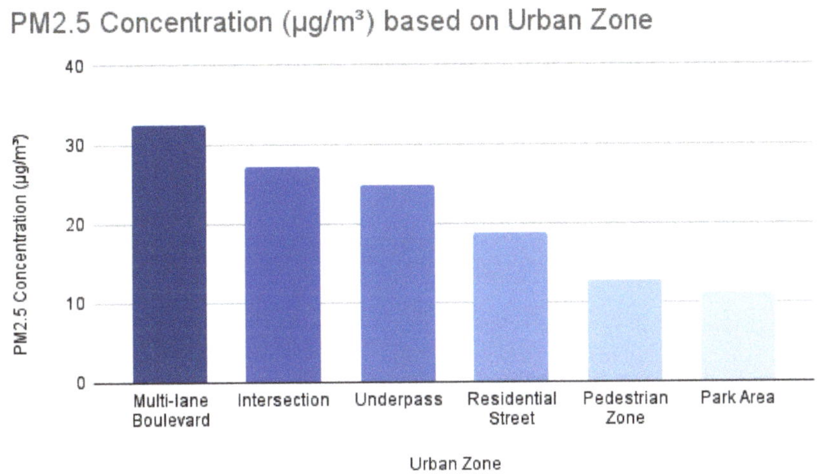


**Figure 3.** Normalized $PM_{2.5}$ concentration ($\mu g/m^3$) based on urban zone.

*3.5. Effectiveness of Low-Emission Zones (LEZs)*

Low-emission zones (LEZs) were included in both examined routes; however, the analysis revealed that particulate matter (PM) levels did not differ significantly between areas inside and outside these zones. Instead, the differences in pollution levels were more pronounced between small streets and major boulevards, emphasizing the role of local traffic density rather than the presence of LEZ regulations.

The current LEZ in Sofia enforces restrictions on high-emission vehicles, banning diesel cars registered before 1 January 2007 and gasoline cars registered before 1 January 1998. Violators of this regulation face fines starting at EUR 25. In the three months following the zone's implementation, 2991 administrative and criminal procedures were initiated, see Figure 4. Before the introduction of the restriction, approximately 20,000 vehicles from the lowest two eco categories (categories 1 and 2) entered the city center daily. On the first day of enforcement (1 December 2024), 4243 violations were recorded. Although the number of violations gradually decreased after initial fines were imposed, more than 2000 high-emission vehicles per day continued to enter the LEZ by the end of February 2025. The highest number of daily violations in this period was recorded on 14 February 2025. However, the relatively low number of fines issued in comparison to the recorded violations raises questions regarding enforcement effectiveness.

**Figure 4.** Sofia LEZ zone and the surrounding air quality stations.

Table 7 shows that the reduction in PM levels within the LEZ was minor and statistically insignificant when compared to non-LEZ areas. Although PM concentrations inside the LEZ showed slight improvements, these reductions were insufficient to ensure compliance with all legal air quality standards, particularly on days with high PM levels recorded by stationary monitoring stations. In contrast, studies in cities such as Berlin, Munich, and Madrid have demonstrated that LEZs can significantly lower $PM_{10}$ and $PM_{2.5}$ concentrations, particularly when accompanied by stricter emissions policies and additional measures such as heavy-duty vehicle transit bans [15,16]. For example, in Madrid, the implementation of an LEZ led to measurable declines in PM pollution, although these reductions also fell short of achieving full compliance with air quality regulations [16].

**Table 7.** Comparison of $PM_{2.5}$ concentrations (µg/m$^3$) between mobile sensor measurements in the LEZ, non-LEZ areas, and the stationary EEA monitoring station.

| Route | Mean LEZ | Mean Non-LEZ | Stationary |
|---|---|---|---|
| Stretch A—heavy traffic | 21 | 21 | 21.76 |
| Stretch B—heavy traffic | 24 | 25 | 21.76 |
| Stretch A—low traffic | 15 | 16 | 15.11 |
| Stretch B—low traffic | 16 | 17 | 15.11 |
| Stretch A—weekend | 9 | 9 | 9.39 |
| Stretch B—weekend | 9 | 9 | 9.39 |

## 4. Discussion

The twin-based digital framework of this study provides valuable insights into the dynamics of urban mobile emissions in Sofia, Bulgaria. We developed a comprehensive approach to monitoring and analyzing particulate matter concentrations across the urban landscape by integrating multiple data sources with varying degrees of spatial and temporal resolution.

### 4.1. Main Findings

Our investigation generated some important discoveries. The two-step machine learning calibration approach enhanced the accuracy of low-cost municipal sensors, raising $R^2$ values from 0.29 to 0.87–0.95 following calibration. This demonstrates the potential of employing calibrated low-cost sensors to expand monitoring networks while preserving measurement accuracy.

Second, the mobile sensor program revealed significant geographical and temporal variability in PM concentrations that would have gone undetected with solely stationary monitoring equipment. $PM_{2.5}$ levels ranged up to 300% over short distances, with the highest concentrations measured in large multi-lane boulevards during peak traffic hours. This research highlights the limits of typical monitoring networks in detecting localized pollution incidents and exposure dangers.

The GNN model successfully characterized pollution dispersion patterns and identified primary emission sources. Vehicle emissions account for approximately 65% of $PM_{2.5}$ concentrations along high-traffic corridors, while road dust resuspension plays a more significant role in residential areas. The observed 30% reduction in PM levels in urban green spaces indicates vegetation's moderating effect on air pollution.

Fourth, our evaluation of the effectiveness of low-emission zones found that they had no impact on air quality during the initial implementation period. The small difference in PM concentrations between LEZ and non-LEZ locations shows that the current enforcement procedures and scope may be insufficient to accomplish significant air quality improvements.

### 4.2. Contextual Relevance

Our findings align with several key studies in the field of urban air quality monitoring. The significant spatial variability in PM concentrations observed in our study corroborates research by [17], who similarly documented substantial heterogeneity in urban pollution levels using mobile monitoring systems. The calibration approach we employed builds upon the methodologies developed by [13], who achieved comparable improvements in measurement accuracy through machine learning techniques.

However, our results diverge from those reported by [18], who found more substantial air quality improvements following LEZ implementation in several European cities. This discrepancy may be attributed to differences in enforcement stringency, zone size, or baseline vehicle fleet composition. Our findings more closely resemble those of [19], who observed the modest impacts of Madrid's LEZ during its initial implementation phase. The observed relationship between green infrastructure and reduced PM concentrations supports an emerging consensus on the role of urban vegetation in mitigating air pollution. Our quantification of this effect (30% reduction) falls within the range reported by [20], who documented PM reductions of 20–40% in various urban green spaces.

### 4.3. Limitations

Despite the robust methodology used, several limitations warrant consideration. First, the mobile measurement campaign was conducted relatively briefly (10 February to 1 March 2025), which may not capture seasonal variations in pollution patterns. Winter conditions in Sofia typically feature higher pollution levels due to additional residential heating emissions, potentially obscuring the specific contribution of mobile sources.

Second, while our sensor network provided extensive spatial coverage of the city center, peripheral areas and industrial zones were underrepresented. This sampling bias may affect the generalizability of our findings to the broader urban area. Additionally, the mobile routes primarily covered daytime hours (8:00 AM–12:00 PM), limiting our understanding of nighttime pollution dynamics.

Third, the calibration approach was based on co-location with reference equipment at only three sites, which may not account for spatial changes in sensor performance across diverse microenvironments.

Fourth, while the GNN model is effective at capturing spatial relationships, it is fundamentally restricted by the quality and completeness of the input data. Traffic density estimates are based on municipal traffic monitoring systems that may not account for all vehicle kinds or temporary traffic situations.

Finally, our assessment of LEZ effectiveness was conducted only three months after its introduction, which may be insufficient to detect long-term behavioral changes or changes in fleet composition in response to the policy.

### 4.4. Clinical and Practical Usefulness

The findings from this study have several important implications for urban planning, environmental policy, and public health. First, the significant variation in PM exposure along different urban routes highlights the need for more nuanced approaches to transportation planning. For example, the placement of cycling infrastructure along major roads may inadvertently increase cyclist exposure to harmful pollutants, suggesting that alternative routes through lower-traffic areas could produce health benefits.

Second, the demonstrated effectiveness of urban vegetation in reducing PM concentrations provides empirical support for green infrastructure initiatives. Urban planners can use these findings to strategically integrate parks and street trees in areas of high pollution to mitigate exposure risks.

Third, the limited effectiveness of the current LEZ implementation provides useful lessons for policy refinement, and additional policies and infrastructure improvements should be considered to achieve greater effectiveness. Currently, the city lacks a stationary air quality monitoring station to assess long-term air pollution trends and there are insufficient public charging stations for electric vehicles in key parking areas and few incentives for low-emission vehicle owners. To fill these gaps, we recommend the installation of at least two municipal air quality monitoring stations: one on a major boulevard with heavy traffic within the LEZ and another in a pedestrian area around Vitosha Street. These measures, along with stronger enforcement and incentive-based policies, could significantly enhance the effectiveness of Sofia's LEZ in reducing air pollution.

Fourth, the calibration framework developed in this study offers a cost-effective approach to expanding air quality monitoring networks using low-cost sensors. This methodology can be readily adapted by municipalities facing budget constraints but seeking comprehensive air quality data.

Finally, the digital twin architecture offers a template for integrating heterogeneous data sources into a cohesive analytical framework. This approach can inform long-term urban development strategies, public health advisories during pollution events, and real-time traffic management decisions.

### 4.5. Future Research Directions

Several avenues for future research emerge from this study. An extended mobile measurement campaign spanning multiple seasons would provide insights into the temporal stability of the observed pollution patterns. Expanding mobile routes to include industrial areas, suburbs, and varied urban morphologies would enhance the spatial representativeness of the findings. Further refinement of the GNN model to incorporate additional variables, such as building height, street canyon configurations, and detailed vehicle fleet composition, could improve predictive accuracy. The integration of citizen science approaches and personal exposure monitoring would add valuable dimensions to the analysis.

An essential topic for further research is the long-term monitoring of LEZ effectiveness as the composition of the fleet changes and the enforcement improves. Best practices for policy design may be found through comparative studies conducted in several towns with different LEZ implementations.

Lastly, employing epidemiological methods to investigate the connection between simulated pollution exposure and health consequences would increase this research framework's applicability to public health.

## 5. Conclusions

This study introduced a digital twin-based framework for the real-time monitoring and analysis of urban mobile-source emissions. The digital twin in this context is defined as a dynamic, data-driven representation of the urban air quality system, integrating real-time data from various sources, advanced computational models, and machine learning algorithms to simulate and predict air quality conditions.

The digital twin framework proposed in this study consists of several key components, including real-time data integration that incorporates information from environmental sensors, meteorological forecasts, traffic flow patterns, and industrial emission outputs; advanced computational models that utilize machine learning algorithms and simulation tools to process and analyze the integrated data; predictive modeling that generates models capable of anticipating future air quality trends and potential hazards; and decision support that provides actionable insights for policymakers, urban planners, and environmental regulators to make informed decisions.

The real prototype of the digital twin in this study is the urban air quality monitoring system deployed in Sofia, Bulgaria, which integrates data from fixed reference stations, low-cost municipal sensors, and mobile measuring devices to create a dynamic and accurate representation of the air quality conditions in the city.

Despite its advantages, the framework has certain limitations. The accuracy of predictions depends upon the quality and availability of input data, and computational demands may limit real-time scalability in large urban areas. Future research should focus on optimizing data integration methods, enhancing model interpretability, and expanding the framework's applicability to various environmental scenarios.

In conclusion, this work contributes to the advancement of digital twin applications in environmental science by providing a robust, data-driven approach for air quality monitoring. Further developments in machine learning, sensor technologies, and cloud computing are expected to refine the system, making it more efficient and accessible for real-world deployment.

**Author Contributions:** Conceptualization, P.Z. and I.D.; Methodology, P.Z., S.F. and I.D.; Software, P.Z.; Validation, P.Z.; Formal analysis, P.Z., S.F. and I.D.; Investigation, I.D.; Resources, P.Z., S.F. and I.D.; Writing—original draft, P.Z. and I.D.; Writing—review & editing, S.F.; Visualization, P.Z.; Supervision, S.F. and I.D.; Project administration, I.D.; Funding acquisition, P.Z., S.F. and I.D. All authors have read and agreed to the published version of the manuscript.

**Funding:** The work was supported by National Scientific Fund of Bulgaria under the grant DFNI KP-06-N52/5.

**Institutional Review Board Statement:** Not applicable.

**Informed Consent Statement:** Not applicable.

**Data Availability Statement:** The data presented in this study are available in Publicly available repository for Air Quality at eea.government.bg/kav/.

**Conflicts of Interest:** The authors declare no conflicts of interest.

## References


1. Hernández, L.A.; Hernández, S. Application of digital 3D models on urban planning and highway design. In *WIT Transactions on the Built Environment. III Conference on Urban Transport and the Environment for the 21 Century;* WIT Press: Acquasparta, Italy, 1997; pp. 391–402.
2. Liu, M.; Fang, S.; Dong, H.; Xu, C. Review of digital twin about concepts, technologies, and industrial applications. *J. Manuf. Syst.* **2020**, *58*, 346–361. [CrossRef]
3. Balogun, A.-L.; Marks, D.; Sharma, R.; Shekhar, H.; Balmes, C.; Maheng, D.; Arshad, A.; Saleh, P. Assessing the Potentials of Digitalization as a Tool for Climate Change Adaptation and Sustainable Development in Urban Centres. *Sustain. Cities Soc.* **2020**, *53*, 101888. [CrossRef]
4. Bauer, P.; Hazeleger, B.S.W. A digital twin of Earth for the green transition. *Nat. Clim. Change* **2021**, *11*, 80–83. [CrossRef]
5. Dembski, F.; Wössner, U.; Letzgus, M.; Ruddat, M.; Yamu, C. Urban Digital Twins for Smart Cities and Citizens: The Case Study of Herrenberg, Germany. *Sustainability* **2020**, *12*, 2307. [CrossRef]
6. Dwivedi, Y.K.; Hughes, L.; Kar, A.K.; Baabdullah, A.M.; Grover, P.; Abbas, R.; Andreini, D.; Abumoghli, I.; Barlette, Y.; Bunker, D.; et al. Climate change and COP26: Are digital technologies and information management part of the problem or the solution? An editorial reflection and call to action. *Int. J. Inf. Manag.* **2022**, *63*, 102456. [CrossRef]
7. Hou, L.; Wu, S.; Zhang, G.; Tan, Y.; Wang, X. Literature Review of Digital Twins Applications in Construction Workforce Safety. *Appl. Sci.* **2021**, *11*, 339. [CrossRef]
8. Pedersen, A.N.; Borup, M.; Brink-Kjær, A.; Christiansen, L.E.; Mikkelsen, P.S. Living and Prototyping Digital Twins for Urban Water Systems: Towards Multi-Purpose Value Creation Using Models and Sensors. *Water* **2021**, *13*, 592. [CrossRef]
9. Zlatev, Z.; Dimov, I. Using a Digital Twin to Study the Influence of Climatic Changes on High Ozone Levels in Bulgaria and Europe. *Atmosphere* **2022**, *13*, 932. [CrossRef]
10. Stockie, J.M. The mathematics of atmospheric dispersion modeling. *SIAM Rev.* **2011**, *53*, 349–372. [CrossRef]
11. Moreira, D.M.; Vilhena, M.T.; Buske, D.; Tirabassi, T. Air pollution steady-state advection-diffusion equation: The general three-dimensional solution. *Air Qual. Atmos. Health* **2012**, *5*, 87–95.
12. Li, L.; Gong, J.; Zhou, J. Improving air quality assessment using physics-inspired deep graph learning. *NPJ Clim. Atmos. Sci.* **2023**, *6*, 150. [CrossRef]
13. Zhivkov, P.; Simidchiev, A. Software tool for optimizing cycling route by defining cyclist air pollution exposure. In Proceedings of the Workshop on Computational Optimization, Sofia, Bulgaria, 4–7 September 2022; pp. 152–170. [CrossRef]
14. Zhivkov, P. Improving performance of low-cost sensors using machine learning calibration with a 2-step model. In Proceedings of the Workshop on Computational Optimization, online, 2–5 September 2021; pp. 373–386. [CrossRef]
15. Gu, J.; Deffner, V.; Küchenhoff, H.; Pickford, R.; Breitner, S.; Schneider, A.; Kowalski, M.; Peters, A.; Lutz, M.; Kerschbaumer, A.; et al. Low emission zones reduced PM10 but not NO2 concentrations in Berlin and Munich, Germany. *J. Environ. Manag.* **2022**, *302 Pt A*, 114048. [CrossRef]
16. Gómez-Losada, Á.; Pires, J.C.M. Air quality assessment during the low emission zone implementation in Madrid (Spain). *Urban Clim.* **2024**, *55*, 101995. [CrossRef]
17. Castell, N.; Dauge, F.R.; Schneider, P.; Vogt, M.; Lerner, U.; Fishbain, B.; Broday, D.; Bartonova, A. Can commercial low-cost sensor platforms contribute to air quality monitoring and exposure estimates? *Environ. Int.* **2017**, *99*, 293–302. [CrossRef] [PubMed]
18. Müller, L.; Roth, A.; Keller, M.; Schmidt, H. Evaluating the effectiveness of low emission zones on air quality improvement in European urban areas: A multi-city analysis. *Atmos. Environ.* **2023**, *207*, 34–49. [CrossRef]
19. Gómez-Losada, Á.; Pires, J.C.M. Initial impact assessment of Madrid's low emission zone on urban air quality: A statistical approach. *Sci. Total Environ.* **2024**, *865*, 118523. [CrossRef]
20. Nowak, D.J.; Hirabayashi, S.; Bodine, A.; Greenfield, E. Tree and forest effects on air quality and human health in the United States. *Environ. Pollut.* **2018**, *193*, 119–129. [CrossRef] [PubMed]

*atmosphere*

MDPI

*Article*

# Comprehensive Scale Fusion Networks with High Spatiotemporal Feature Correlation for Air Quality Prediction

**Chenyi Wu [1], Zhengliang Lai [1], Yunwu Xu [1], Xishun Zhu [2], Jianhua Wu [3] and Guiqin Duan [4,*]**

1 School of Electrical Engineering, Guangdong Songshan Vocational and Technical College, Shaoguan 512000, China; 50538@gdsspt.edu.cn (C.W.); ncu_zllai@163.com (Z.L.); 26054@gdsspt.edu.cn (Y.X.)
2 School of Mathematics and Statistics, Hainan Normal University, Haikou 570100, China; xishunzhu@hainnu.edu.cn
3 School of Information Engineering, Nanchang University, Nanchang 330031, China; jhwu@ncu.edu.cn
4 School of Computer and Information Engineering, Guangdong Songshan Vocational and Technical College, Shaoguan 512000, China
* Correspondence: 50537@gdsspt.edu.cn

**Abstract:** The escalation of industrialization has worsened air quality, underscoring the essential need for accurate forecasting to inform policies and protect public health. Current research has primarily emphasized individual spatiotemporal features for prediction, neglecting the interconnections between these features. To address this, we proposed the generative Comprehensive Scale Spatiotemporal Fusion Air Quality Predictor (CSST-AQP). The novel dual-branch architecture combines multi-scale spatial correlation analysis with adaptive temporal modeling to capture the complex interactions in pollutant dispersion and enhanced pollution forecasting. Initially, a fusion preprocessing module based on localized high-correlation spatiotemporal features encodes multidimensional air quality indicators and geospatial data into unified spatiotemporal features. Then, the core architecture employs a dual-branch collaborative framework: a multi-scale spatial processing branch extracts features at varying granularities, and an adaptive temporal enhancement branch concurrently models local periodicities and global evolutionary trends. The feature fusion engine hierarchically integrates spatiotemporally relevant features at individual and regional scales while aggregating local spatiotemporal features from related sites. In experimental results across 14 Chinese regions, CSST-AQP achieves state-of-the-art performance compared to LSTM-based networks with RMSE 6.11–9.13 $\mu g/m^3$ and $R^2$ 0.91–0.93, demonstrating highly robust 60 h forecasting capabilities for diverse pollutants.

**Keywords:** spatiotemporal convolution; air quality prediction; generative CNN; multi-scale fusion; feature correlation

## 1. Introduction

The accelerated global industrialization in recent decades has exacerbated air pollution to critical levels, with the World Health Organization attributing approximately 4.2 million premature deaths annually to ambient air pollution exposure [1]. This crisis has made air quality prediction not just a technical challenge but a vital part of urban sustainability planning, supporting both environmental governance and public health strategies [2,3]. Air quality prediction research encompasses both traditional statistical methods and modern deep learning techniques [4].

Traditional methods, known as chemical transport models, simulate the transport, diffusion, and chemical reactions of pollutants to predict [5]. These methods are categorized into physical diffusion-based models and statistical models, with the former relying on diffusion theory and the latter on empirical assumptions [6]. However, their reliance on empirical parameters and physical data limits accuracy. Statistical methods, like linear regression and machine learning modeling, simplify the complexity and parameter overload of physical models [7,8], such as support vector regression (SVM) [9], Random Decision Forests (RFs) [10], Hidden Semi-Markov Models (HSMMs) [11], and other machine learning techniques [12,13]. While these conventional models demand fewer computational resources and parameters, their static feature interaction mechanisms fail to delineate the spatiotemporal heterogeneity of pollutants and are incapable of resolving multi-scale.

In recent years, convolutional neural networks (CNNs) have significantly advanced air quality prediction, particularly in forecasting $PM_{2.5}$ concentrations. In 2018, Athira et al. [14] used recurrent networks for air quality prediction in conjunction with HSMM. Guo [15] attempted to partition time correlation at a finer granularity and designed a deep residual CNN. These hybrid approaches improved prediction accuracy through the fusion of statistical feature engineering with CNN frameworks, but they remain constrained by manual hyperparameter configuration. Subsequently, Elbaz et al. [16] and Tran et al. [17] enhanced accuracy by integrating spatiotemporal data into residual CNN; their methodologies introduced heightened algorithmic complexity. Considering the correlation between different time lengths, some researchers have also proposed long short-term memory (LSTM) for air quality prediction networks [18–20]. For example, Ameri et al. [21] and Tishya et al. [22] proposed a deep learning model with an encoder–decoder structure using Conv and LSTM units to model and predict the correlation of urban pollution. However, these models are less efficient for model training and inference in long sequence scenarios. Further, researchers have proposed methods based on variations-LSTM to realize the prediction of air index time series [23]. Rabie et al. [24] developed a CNN-Bi-LSTM hybrid framework to address multi-scale spatial air quality prediction in megacities, overcoming challenges in modeling cross-regional pollutant dependencies through integrated multi-scale feature extraction. Fan et al. [25] implemented a hybrid seasonal air quality prediction through a mixture of rough ensemble wrapping methods and regularized combinatorial LSTM. In addition, there are some other novel models, including Self-Attention LSTM [26] and adaptive LSTM [27]. However, these models typically demand substantial computational resources for training and optimization, and their performance can vary significantly when predicting different time scales.

Since 2022, people have increasingly leveraged graph convolutional networks (GCNs) to address constrained modeling capacity in temporal air quality data analysis [28]. Chen et al. [29] employed multi-factor separation evolutionary spatiotemporal GCN factorized feature learning for multivariate series decomposition; Hu et al. [30] enhanced 10.7% cross-regional accuracy through adaptive multi-graph attention fusion with GRU units. The emerging hybrid network introduced GCN and Bi-LSTM structures to adapt the propagation depth to seasonal patterns, addressing the long-term prediction challenge through a specialized spatiotemporal decoupling strategy [31]. In addition, other hybrid models have been proposed, including attention mechanisms [32] and physically informed guidance [33]. These advancements have improved stability and accuracy across prediction tasks involving various time scales. Recently, generative CNN (GCNN) [34,35] has been utilized to generate a pre-trained model, which achieves fine-grained spatiotemporal knowledge migration via generative pre-training on source city data. Some researchers have designed a combination of GCNN and reinforcement learning to focus on essential time steps and spatial locations [36,37]. Based on this, other generative networks have combined multifactor decomposition and data-driven trees, further uniting spatiotemporal features [38,39]. However, these models may miss key correlations when handling spatiotemporal features separately, complicating the resolution of spatiotemporal interactions, and the performance of cross-region prediction is unsatisfactory [40]. In addition, existing studies predominantly concentrated on modeling dependencies at a single spatial and temporal scale, neglecting the intricate spatiotemporal dynamics of air quality changes across individual and local dimensions in cross-city prediction tasks.

To solve the above problems, this study proposed an air quality prediction method based on generative complete scale spatiotemporal convolutional networks (CSST-AQP), which aims to solve the spatial and temporal dependence problem in air quality prediction tasks. The CSST-AQP framework method adopts a three-level progressive architecture: firstly, a local strong correlation spatiotemporal feature extraction fusion network is constructed to capture long-range spatiotemporal dependencies through a dynamic feature sensing mechanism; secondly, an adaptive spatiotemporal augmented and a multi-scale convolution network are designed to achieve the synergistic extraction of multi-dimensional features. Eventually, the CSST-AQP can manage remote temporal dependencies in a parallel manner, while spatial dependencies are modeled to achieve long-term air quality prediction. Experiments show that CSST-AQP performs excellently in all metrics over a 0–400 h prediction window. The main contributions that we make in this paper are as follows:

(1) A multi-scale fusion model integrates cross-station spatial patterns and local temporal trends to capture spatiotemporal dynamics. (2) An adaptive spatial correlation matrix coding module dynamically updates site interactions and tracks spatiotemporal dependencies. (3) A temporal model integrating dilated causal convolutions with adaptive snow-abatement optimization is used to capture local periodic and extended-range temporal dependencies. (4) A novel architecture fuses time-series encoding and spectral–temporal decomposition, resolving spatiotemporal mismatches while aligning cross-modal scales. (5) A prediction algorithm with extra high accuracy, 22.5% higher than CEMD-LSTM, is proposed.

The rest of this paper is organized as follows. Section 2 reviews some related technologies and concepts. In Section 3, the workflow of the proposed air quality prediction model is described in detail. To evaluate the proposed algorithm, simulation details, experimental results, comparison with several other air quality prediction models, and an ablation study are given in Section 4. Finally, Sections 5 and 6 present the discussion and a brief conclusion, respectively.

## 2. Related Materials and Concepts

### 2.1. Spatiotemporal Encoder

To capture regional spatiotemporal correlation, Du et al. [41] proposed a spatiotemporal convolutional encoder (STE) based on temporal causal convolutional encoding. STE is capable of combining information from the spatiotemporal encoder to retain the characteristics of the original input data. The structure of STE is shown in Figure 1.

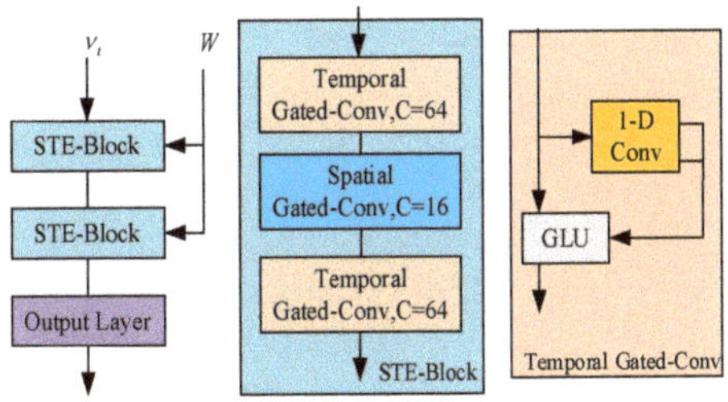


**Figure 1.** Architecture of an STE.

The STE consists of two spatiotemporal graph convolution blocks and a fully connected output layer. Each STE-Block contains two temporal Gated-Conv and one spatial Gated-Conv. The temporal Gated-Conv combines a 1-D Conv with a gated linear unit. In this study, we use STE to capture dynamic changes in air quality data over time and improve prediction performance.

### 2.2. Preliminaries

Given the historical air quality time series $X$ till the current time step, we aim to train a predictive network that accurately estimates the air quality data of all regions at the future time step $t+1$ and then obtain the predicted value $X_{t+1}$. Currently, few studies predict air quality data across regions, and this study focuses on forecasting air pollutants across different spatiotemporal gaps. In general, there are regional and time dependencies between changes in air pollution concentrations, as shown in Figure 2. To more comprehensively depict spatiotemporal correlations, the following definitions are formulated in this paper.

(1) Air quality flow graph (AQFG): the air quality flow graph is defined as $\gamma$, and the citywide air quality data for the pre-$T$ moment are expressed by a tensor $X = \{x_1, x_2, x_3, \cdots x_t\}, t = 1, 2, 3 \ldots, N, N \in T$.

(2) Spatial region: each monitoring station $M_t$ is represented as a graph node with a feature vector $h_t \in \mathbb{R}^d$, encapsulating multivariate air quality measurements (e.g., $PM_{2.5}$, $SO_2$, and $O_3$) and meteorological conditions (wind speed and temperature) at the station.

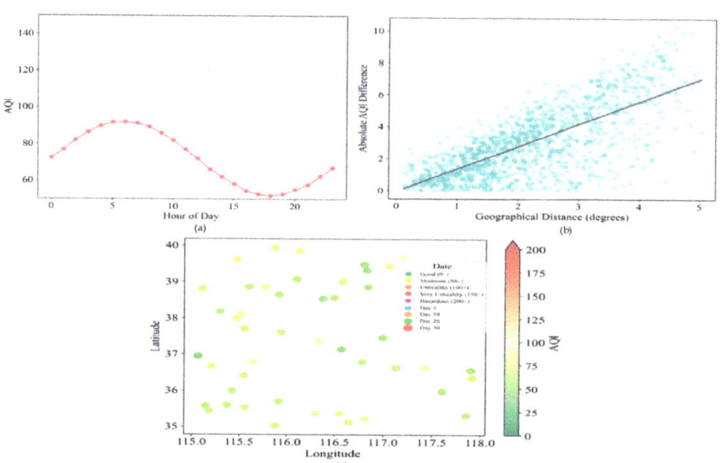


**Figure 2.** Architecture of an STE. (**a**): hourly average AQI; (**b**): distance vs. AQI difference; (**c**): spatiotemporal distribution of AQI.

### 2.3. Time Series Imaging

The Gramian Angular Field (GAF) was proposed by Oates et al. Its main function is to encode time series into two-dimensional (2D) image information by generating a Gram matrix through trigonometric functions [42]. In this paper, the GAF-transformed time series data retain the temporal relationships among its observations, so we combine it with CNN to improve the prediction accuracy by maintaining the temporal relationship while learning more 2D features. The details are

$$x_t' = \frac{(x_t - \max(X)) + (x_t - \min(X))}{\max(X) - \min(X)}, \tag{1}$$

where $x_t'$ is the normalized observed value, $x_t$ is the $t$-th sequence, and $x_t'$ is converted to the form of polar coordinates:

$$\varphi_t = \arccos(x_t'), \tag{2}$$

$$r_t = \frac{t}{T}, \tag{3}$$

where $\varphi_t$ is the angle, $r_t$ is the radius, $t$ denotes the time stamp, and the transformed polar form is defined as $(\varphi_t, r_t)$. Ultimately, the triangular sum between each observation $GAF_{t',t''}$ determines each element in the $GA$:

$$GAF_{t',t''} = \cos(\varphi_t' + \varphi_t''), t', t'' = 1, 2, 3, \ldots T, \tag{4}$$

$$GAF = \begin{pmatrix} \cos(\varphi_1 + \varphi_1) & \cdots & \cos(\varphi_1 + \varphi_T) \\ \vdots & \ddots & \vdots \\ \cos(\varphi_T + \varphi_1) & \cdots & \cos(\varphi_T + \varphi_T) \end{pmatrix}. \tag{5}$$

*2.4. Evaluation Indexes*

The objective evaluation indexes of air quality prediction used in this paper include the Root Mean Square Error (RMSE), Mean Absolute Error (MAE), Mean Absolute Percentage Error (MAPE), and R-squared ($R^2$). The formulas are

$$\text{RMSE} = \sqrt{\frac{1}{t}\sum_{i=1}^{t}(\hat{y}_i - y_i)^2}, \tag{6}$$

$$\text{MAE} = \frac{1}{t}\sum_{i=1}^{t}|\hat{y}_i - y_i|, \tag{7}$$

$$\text{MAPE} = \frac{1}{t}\sum_{i=1}^{n}\left|\frac{y_i - \hat{y}_i}{y_i}\right| \times 100\%, \tag{8}$$

$$R^2 = 1 - \frac{\sum_{i=1}^{t}(y_i - \hat{y}_i)^2}{\sum_{i=1}^{t}(y_i - \overline{y}_i)^2}, \tag{9}$$

where $\hat{y}_i$ and $y_i$ are the predicted value and the actual value, respectively. $\overline{y}_i$ is the average of the $y_i$.

## 3. Propose Models and Methodologies

In this section, we proposed a comprehensive scale fusion network with high spatiotemporal feature correlation for air quality prediction based on a preprocessing network, as well as STE-ResNet, named CSST-AQP, whose structure is shown in Figure 3.

*3.1. Overview of the Proposed CSST-AQP Framework*

The proposed CSST-AQP architecture comprises four principal components: (1) a local spatiotemporal feature preprocessing network (LSTF-Net) for initial data refinement, (2) a comprehensive-scale spatial feature processing module (CSSP-Net) enabling multi-granularity pattern extraction, (3) an adaptive temporal feature enhancement network (ATSE-Net) with dynamic time-aware mechanisms, and (4) a prediction output layer. In the CSST-AQP, all sub-networks were trained simultaneously, and the convolutional modules in the three sub-networks were configured with convolutional cores of different sizes to learn air quality time series feature information in different contextual orientations.

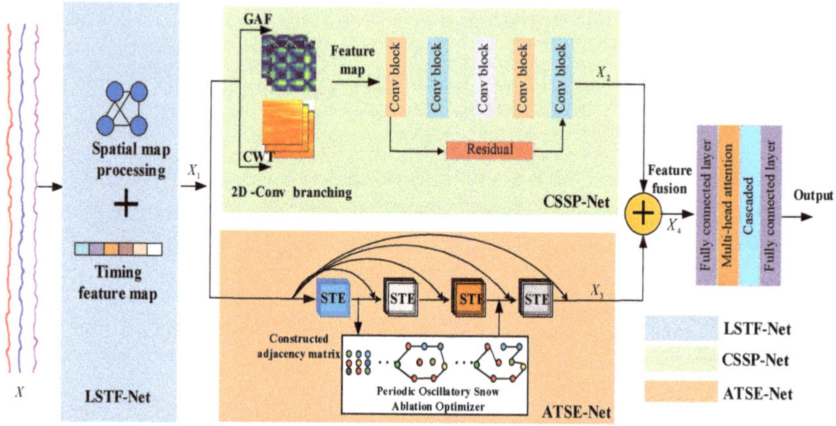


**Figure 3.** The framework of CSST-AQP; differently colored blocks represent different sub-networks and modules.

In general, the processing pipeline begins by feeding raw data X into the LSTF-Net preprocessing sub-network for spatiotemporal feature encoding. The generated fusion features are subsequently distributed through dual processing branches: the upper branch employs CSSP-Net, while the lower branch utilizes ATSE-Net, each specialized in extracting complementary dynamic spatiotemporal dependencies through parallel training. The learned features from both branches are then sequentially processed through three integration stages: feature fusion, a multi-attention mechanism layer, and a fully connected regression layer. Finally, the predicted values are generated through the output layer. The detailed procedure is described as follows.

(1) First, the input data $X$ are fed to LSTF-Net, where they undergo localized spatiotemporal modeling. This process integrates spatial mapping and time-series feature extraction to simultaneously capture two critical aspects: temporal patterns of air quality variations at individual monitoring stations and cross-station spatial correlations, ultimately generating enhanced spatiotemporal features $X_1$ that encode both station-specific dynamics and regional interdependencies.

(2) Then, $X_1$ are encoded as corresponding 2D feature maps by CSSP-Net. In the CSSP-Net architecture, the input feature $X_1$ undergoes multi-scale feature fusion through GAF and Continuous Wavelet Transform (CWT). This hybrid approach hierarchically integrates temporal patterns and frequency-domain characteristics. The fused feature maps undergo processing through the residual Conv module, where reinforcement learning mechanisms are applied to execute feature optimization, ultimately yielding an enhanced feature map $X_2$.

(3) The processed temporal features are fed into ATSE-Net, where key hyperparameters (neuronal population and convolutional kernel dimensions) in STE modules are optimized via a novel metaheuristic optimizer combining periodic oscillatory dynamics and snow ablation strategies. This accelerates convergence while boosting generalization and prediction fidelity. Then, the output features $X_3$ are obtained.

(4) Finally, the feature map $X_2$ and $X_3$ are fused by features to generate $X_4$, which is then transmitted to the full connectivity layer and multi-head attention layer for feature learning to obtain the final prediction result $\hat{y}_i$.

### 3.2. Localized Spatiotemporal Feature Preprocessing Module (LSTF-Net)

The first sub-network of the CSST-AQP framework is the local spatiotemporal feature preprocessing network (LSTF-Net). To extract spatiotemporal dependence, LSTF-Net categorizes the pertinent features in air quality prediction into two distinct groups: temporal and spatial features. The LSTF-Net structure is shown in Figure 4. LSTF-Net comprises three components: an initial encoding module for converting raw sequences into AQFG matrices, channel attention module 1 (CAM-1) for capturing strongly correlated spatiotemporal features, and channel attention module 2 (CAM-2) implementing dilated convolution operations and feature-weighted fusion.

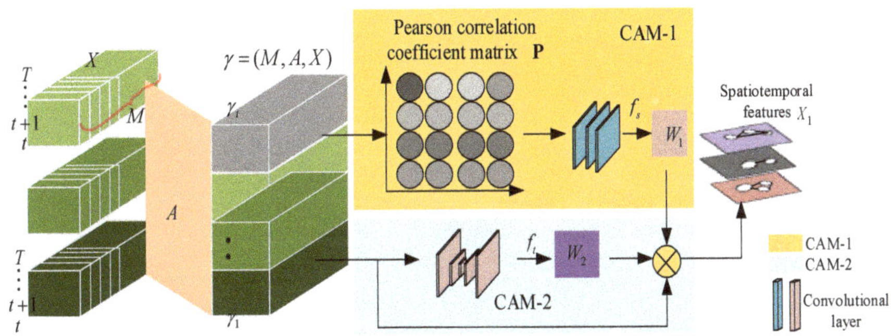


**Figure 4.** The LSTF-Net framework.

The temporal features $f$ include $PM_{2.5}$, $PM_{10}$, $SO_2$, CO, and $O_3$, collectively denoted as $X$, and the operation process of the model is as follows:

$$X = f(PM_{2.5}) + f(PM_{10}) + f(SO_2) + f(CO) + f(O_3). \quad (10)$$

(1) Firstly, to capture spatiotemporal dependencies across regional monitoring stations, LSTF-Net employs an initial encoding step that transforms the raw sequence $X$ into an AQFG matrix $\gamma$.

$$\gamma = (M, A, X), \quad (11)$$

where $A$ is the dynamic weight matrix between air quality flow graphs, and $M$ is the monitoring node. The dynamic weight matrix ${A^t}_{i,j}$ of the monitoring node $i$ to node $j$ at $t$ time points is

$${A^t}_{i,j} = \frac{\exp(-\frac{\text{distance}(M_i, M_j)}{\xi} \cdot \frac{X_i^t \cdot r_{i,j}}{\|r_{i,j}\|})}{\sum_{k \in N(i)} \exp(-\frac{\text{distance}(M_i, M_k)}{\xi} \cdot \frac{X_i^t \cdot r_{i,k}}{\|r_{i,k}\|})}, \quad (12)$$

where $r_{i,j}$ is the vector from monitoring node $i$ to node $j$; $\xi$ is the attenuation coefficient; $M_k$ denotes the $k$-th historical monitoring node; $M_{i,j} \subseteq M$, $i, j = 1, 2, 3, \cdots N$. $X_i^t$ represents the data of node $i$ at time $t$; and $r_{i,k}$ is the $k$-th prior coefficient vector of $X_i^t$.

(2) Then, the tensor $\gamma$ undergoes processing through a channel attention module 1 (CAM-1). We calculated the Pearson correlation coefficients between tensor $\gamma_i$ and $\gamma_j$ at each site and defined them as the spatial features $\mathbf{P}_{ij}$, $\mathbf{P}_{ij} \in \mathbf{P}$.

$$\mathbf{P}_{ij} = \frac{\text{cov}(\gamma_i, \gamma_j)}{\sigma(\gamma_i)\sigma(\gamma_j)}, \quad (13)$$

$$\mathcal{G}(\mathbf{P}) = \text{sigmoid}(\mathbf{WP} + b). \quad (14)$$

A gating mechanism $\mathcal{G}(\mathbf{P})$ is subsequently applied to amplify cross-regional features with $\mathbf{P}_{ij}$, effectively capturing dominant atmospheric transport and city pathways. **W** denotes the coefficient matrix, and $b$ is the penalty factor obtained from training. To explicitly encode inter-regional air quality dependencies, we introduced a geographically weighted graph regularization term $\mathcal{L}_{geo}$ into the objective function

$$\mathcal{L}_{geo} = \sum_{i=1}^{N} \sum_{j \in N(i)} \mathbf{P}_{ij} \| \gamma_i \odot m_i - \gamma_j \odot m_j \|_F^2, \tag{15}$$

where $\gamma_i \in \mathbb{R}$ denotes the $i$-th spatiotemporal flow tensor of node $\gamma_t$, $m_i$ is the region-specific meteorologic mask, and the Hadamard product $\odot$ implements localized feature gating based weather patterns. The spatial features quantify the correlation of pollutant data across different monitoring stations using the Pearson correlation coefficient. The ensuing obtained matrix **P** is transferred to the lower convolutional layer for training to generate the intermediate features $f_s$.

(3) Similarly, the CAM-2 processes input signals $\gamma$ to extract temporal feature representations, denoted as $f_t$. To simulate spatiotemporal interdependencies across regional stations, we propose a bilinear temporal channel attention mechanism (BTCA) with learnable parameters $W_1$ and $W_2$. CAM-2 adopts dilated convolutions to process sensor inputs, outputting temporal features $f_t \in$ R. The BTCA mechanism computes station-wise interactions, which are defined as

$$X_1 = \text{Softmax}((f_s W_1)(f_t W_2)^{\text{T}}). \tag{16}$$

(4) Finally, the spatial feature $f_s$ and temporal feature $f_t$ are subsequently integrated through dual attention-weighted layers, generating the unified spatiotemporal tensor $X_1$.

In LSTF-Net, $X_1$ serves as the input feature of the lower-level sub-network and is transmitted to CSSP-Net and ATSE-Net for training, respectively.

### *3.3. The Structure of Complete Scale Spatial Processing Network (CSSP-Net)*

The upper half branch of CSST-AQP is CSSP-Net, which includes the GAF and CWT image coding module and a novel residual convolution module, whose structure is shown in Figure 5. The CSSP-Net first encodes $X_1$ into images with GAF and CWT. The encoded feature maps then pass through 11 convolutional layers and 2 residual blocks, extracting hierarchical multi-scale features and ultimately generating the output $X_2$.

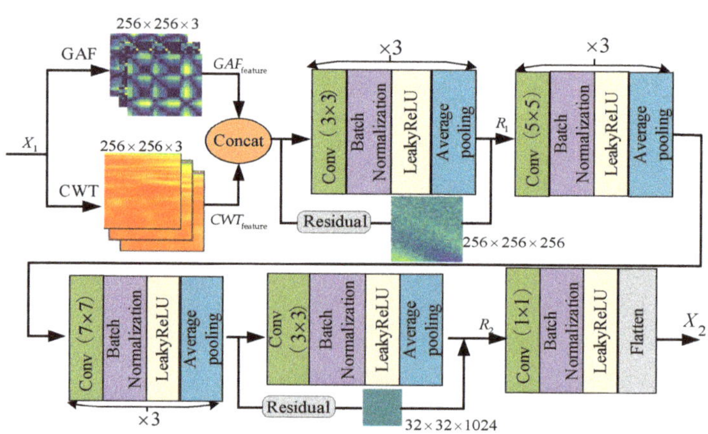


**Figure 5.** The structure of CSSP-Net.

The upper branch employs a GAF encoder to transform time-series pollutant measurements into temporal feature maps. This encoding strategy preserves intrinsic temporal dependencies through polar coordinate-based angular correlations, effectively mitigating information loss during sequential data transformation.

For the feature map $X_t$, which has a total of $T$ observations, $\varphi_t$ is the $t$-th observation frame that denotes the angle, and the feature is scaled to $[-1, 1]$ via Equation (1) to avoid the impact of larger observations on the subsequent inner product. Then, $X_t$ is converted to polar coordinates $(\varphi_t, r_t)$ by Equation (2). Finally, the variable $\varphi_t$ is processed through Equations (4) and (5) to generate a 2D image feature map $GAF_{\text{feature}}$, which is subsequently fed into the subsequent layer for concatenation.

Concurrently, the lower branch processes spatial correlations via continuous wavelet transform (CWT) with a Haar wavelet basis, decomposing $X_1$ into multiscale wavelet coefficient arrays. This process is able to identify jumps and discontinuities within the signal and obtain features on multiple scales of the original spatiotemporal sequence, further improving the accuracy of the prediction.

$$W_X(a,b) = \frac{1}{\sqrt{a}}\int_{-\infty}^{\infty} X(t)\overline{\psi}(\frac{t-b}{a})dt. \tag{17}$$

where $W_X(a,b)$ is the CWT coefficients at scale $a$ and position $b$, which are the encoded feature maps $CWT_{feature}$, and $\overline{\psi}(t)$ is the complex conjugate of the Haar wavelet function. Then, $CWT_{feature}$ is concatenated with the $GAF_{\text{feature}}$ to form a hybrid spatiotemporal feature matrix. Subsequently, the fused matrix is fed into a residual CNN module to jointly model cross-regional interactions in both feature spaces: the GAF branch captures dynamic temporal patterns of pollutant evolution, while the CWT-derived coefficients encode localized spatial–frequency characteristics of atmospheric processes.

In details, the fused image is processed through three convolutional layers with $3 \times 3$ convolutional kernels and one residual layer to obtain the feature $R_1$. $R_1$ is then sequentially processed through three down-sampling convolutional layers with convolutional kernel sizes of $5 \times 5$ and $7 \times 7$ to obtain the intermediate variables. The output intermediate variable features are used as inputs to the next convolutional kernel size of $3 \times 3$ convolutional layer, and residual branches are learned to obtain $R_2$. Then, the feature map $R_2$ is input to a $1 \times 1$ convolutional layer and a flatten layer to produce a flattened output vector $X_2$.

### 3.4. Adaptive Temporal Feature Enhancement Network (ATSE-Net)

To effectively capture the long-range time dependence, an adaptive time series region enhancement sub-network (ATSE-Net) is introduced into the lower branch of CSST-AQP. The detailed structure of the ATSE-Net is shown in Figure 6. The ATSE-Net architecture employs a hierarchically dense structure comprising four cascaded adaptive spatiotemporal encoding (ASE) blocks, followed by a multi-attention fusion module and a flattening output layer. Each ASE block integrates three core components: a STE unit for joint feature representation, an adaptive threshold filter (T-Filter) layer for dynamic sequence modeling, and an adaptive average pooling operator for dimensionality reduction.

The processing pipeline initiates with the tensor $X_1$ being sequentially processed through the ASE block cascade. At each stage, the output of the $n$-th ($n$ = 2, 3, 4) ASE block undergoes element-wise fusion with the original $X_1$ feature through residual connections, implementing a dense connectivity pattern that alleviates gradient vanishing while enhancing feature reuse. After four ASE operations with three inter-block skip connections, the aggregated features are directed to the multi-attention layer and the flattening layer, ultimately flattened into the final output tensor $X_3$.

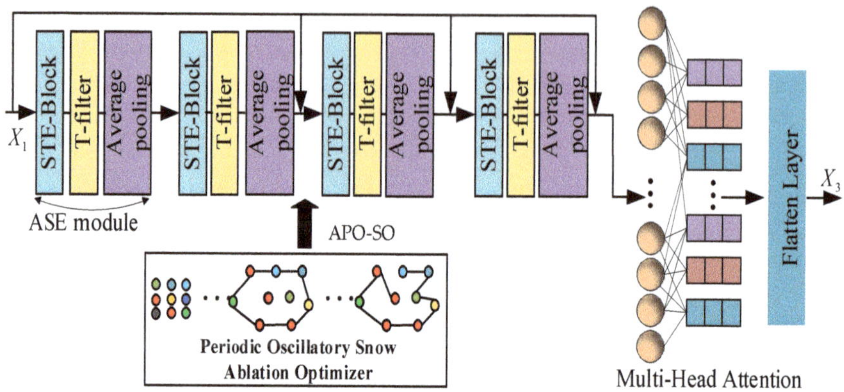


**Figure 6.** The structure of ATSE-Net.

Critical to the architecture's performance is the integration of the adaptive periodic oscillation snowmelt optimizer (APO-SO) [43] during parameter optimization. The APO-SO adopts a dual population mechanism, where one subpopulation is responsible for the global TSE module parameters search, and another's task is to search for the optimal threshold of the local T-Filter. The mechanism enables ATSE-Net to optimize the parameters more efficiently during the training process, thus improving the performance and generalization ability of the model. The signal intensity accumulation matrix in the feature map may be affected by residual noise or anomalous signals in the original spatiotemporal sequence. The T-Filter can learn a suitable threshold to extract local features and preserve spatial information, for which the target intensity $S(i)$ and the threshold $T$ are

$$S(i) = \begin{cases} S(i), & S(i) \geq T \\ 0, & S(i) < T \end{cases}, \tag{18}$$

$$T = a + e, \tag{19}$$

$$a = \alpha \frac{\sum_{i=1}^{w} S(i)}{w}, \tag{20}$$

$$e = \beta \sqrt{\frac{\sum_{i=1}^{w} (a - S(i))^2}{w}}, \tag{21}$$

where $a$ is the mean of the sum of all column pixels in the feature map band, and $e$ is the standard deviation of the column intensity of the feature map band. $w$ is the number of sample points, and $\alpha$ and $\beta$ are the adaptive parameters learned by network training, respectively. The APO-SO framework for the TSE module is formally described, and its structural configuration is summarized in Algorithm 1. This strategy integrates adaptive population initialization with stochastic operators to enhance global search capabilities while maintaining computational efficiency.

As detailed in Algorithm 1, the optimization process employs a three-phase architecture: (1) diversity-preserving initialization (Lines 1–4); (2) hybrid exploration–exploitation (Lines 5–13); (3) elite selection (Lines 14–17) and obtained $T_{optimal}$ (Lines 18), where the stochastic components (Lines 4, 5) work synergistically with the adaptive mechanisms (Lines 11–13, 17) to prevent premature convergence while maintaining time complexity $O(n \log n)$ through efficient population management.

**Algorithm 1:** APO-SO optimization for TSE module

**Input:** $X_{\min}, X_{\max}$ ; $V_{\min}, V_{\max}$ ; X

**Output:** $T_{\text{optimal}}$

$X_i \leftarrow \text{Random}(X_{\min}, X_{\max})$;

$V_i \leftarrow \text{Random}(V_{\min}, V_{\max})$;

$\varpi \leftarrow \varpi_{\max}$, convergence $\leftarrow$ False;

$\text{pbest}_i \leftarrow X_i$;

$\text{gbest} \leftarrow \arg\max(\text{Fitness}(\text{pbest}_i))$;

**While** $\neg$convergence do

  $V_i \leftarrow \varpi \cdot V_i + c_1 r_1(\text{pbest}_i - V_i) + c_2 r_2(\text{gbest} - V_i)$;

  $X_i \leftarrow X_i + V_i$;

  $X_i \leftarrow \max(X_{\min}, \min(X_i, X_{\max}))$;

  $V_i \leftarrow \max(V_{\min}, \min(V_i, V_{\max}))$;

  $\text{Fitness}_i \leftarrow \text{evaluate\_T\_filter}(X_i, \text{Signal})$;

  **if** $\text{Fitness}_i > \text{Fitness}(\text{pbest}_i)$

   $\text{pbest}_i \leftarrow X_i$;

  **otherwise**

   $\text{pbest}_i \leftarrow \text{pbest}_i$;

  **end**

  $\text{gbest} \leftarrow \arg\max(\text{Fitness}(\text{pbest}_i))$;

  $t$ = current iteration;

  $\varpi \leftarrow \varpi_{\max} - \dfrac{(\varpi_{\max} - \varpi_{\min})}{T_{\max}}$;

  **if** $\Delta\text{Fitness} < \varepsilon \ \vee \ t \geq T_{\max}$

   convergence $\leftarrow$ True;

  **otherwise**

   convergence $\leftarrow$ False;

  **end**

**end**

**Return** $\text{T}_{\text{optimal}} \leftarrow \text{gbest}$

## 4. Experimental Section and Results

### *4.1. Training Details and Datasets*

The model was developed using the PyTorch-2.1.1 deep learning framework, incorporating Cross-Entropy Loss as the loss function. The training regimen utilized the Adam optimizer with an initial learning rate of 0.01, a momentum of 0.9, and a weight decay coefficient of 0.0005. The loss function $L$ is

$$L(\lambda_1, \lambda_2) = \mathrm{E}_{\hat{y}_i, y_i}\left[\lambda_1 \|\hat{y}_i - y_i\|_2^2 + \lambda_2 R_{\mathrm{TV}}\right], \tag{22}$$

$$R_{\mathrm{TV}} = \int_{\Omega} \|\hat{y}_i\|_2 dx, \tag{23}$$

where $y_i$ is the true value, and $\hat{y}_i$ is the predicted value. A total variation penalty $R_{\mathrm{TV}}$ is added to the loss function of the $L(\lambda_1, \lambda_2)$, which further stabilizes the optimization. $\lambda_1$ and $\lambda_2$ are the weight coefficients, which are obtained from the network training, and their initial values are 0.8 and 0.2, respectively. In this study, experiments are conducted on two real datasets in the Guangzhou region (GZR) and Shaoguan region (SGR) to evaluate the performance of the model predictions. Both datasets (GZR/SGR) cover January 2021–April 2024 (see Table 1 for statistics) and are publicly available from the Air Quality Historical Data Platform [44].

**Table 1.** Structure and parameters of the datasets.

| Datasets | GZR | SGR |
|---|---|---|
| Number of air quality stations | 121 | 63 |
| Number of air quality records | 4,369,102 | 1,103,144 |
| Average AQI | 58.6 | 70.1 |
| Number of regions $N_r$ | 11 | 3 |

The raw dataset underwent comprehensive preprocessing procedures comprising outlier elimination through the interquartile range method, z-score standardization, and min–max normalization to ensure dimensional homogeneity. We convert all the time series data into samples and labels with the sliding window method (with a sliding window size of $W$ and sliding step size of $O$) before putting them into the network model for training. After 2000 batches, the network stabilized with a loss function value of 0.0005, and the $R^2$ of training and testing reached more than 0.95. The results are shown in Figures 7 and 8, respectively.

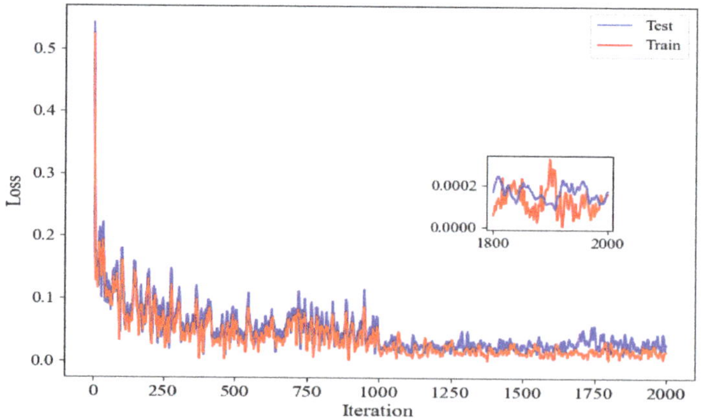


**Figure 7.** The loss curves for the training and testing processes.

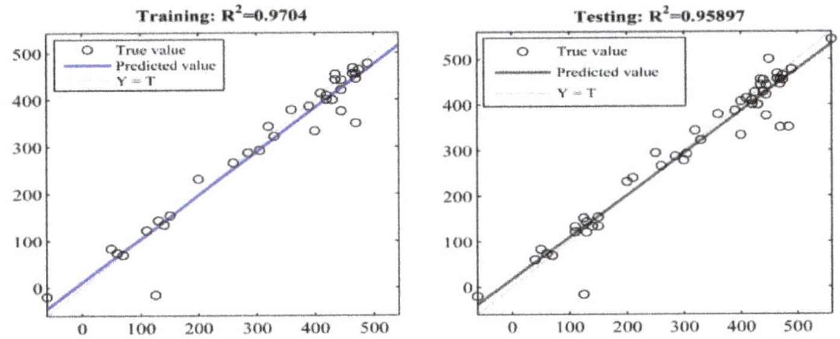


**Figure 8.** The correlation coefficient of determination curves under training and testing.

Figures 7 and 8 show excellent network training with minimal discrepancy between truth and predicted values. Both training and testing losses stabilized at 0.0002 with no overfitting, and a high $R^2$ ($R^2 > 0.95$) confirms the model's remarkable prediction capability.

*4.2. Feature Visualization*

In this section, we qualitatively analyze the prediction performance of the self-supervised spatiotemporal coding network and visualize the feature extraction capability of the CSST-AQP framework. Figure 9 shows the visualization results of the $PM_{2.5}$ component, $PM_{10}$ component, $O_3$ component, $NO_2$ component, $SO_2$ component, and CO component after the initial network coding process, respectively.

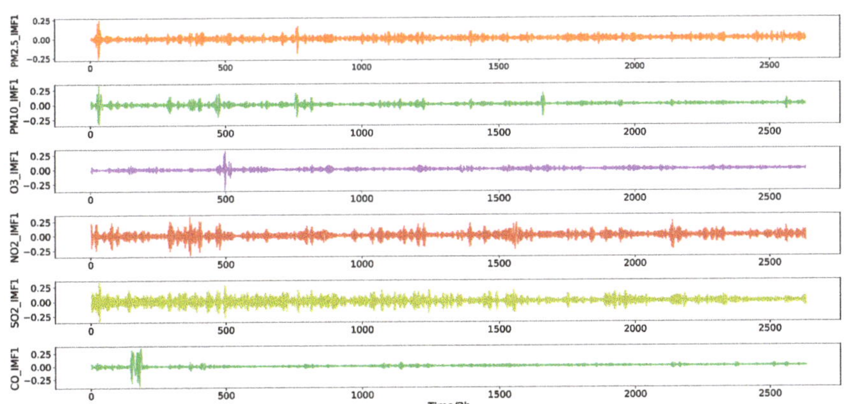


**Figure 9.** Visualization results for each AQI component after network coding.

Figure 10a displays the SHAP eigenvalues ranking for air quality predictors, showing $PM_{2.5}$ as the most influential feature, followed by industrial emissions ($PM_{10}$, $O_3$, $NO_2$, $SO_2$, and CO), temperature, humidity, wind speed, and temporal factors. The temporal analysis in Figure 10b reveals strong autocorrelation patterns within pollutant sequences across adjacent time points.

It can be observed that the characteristic shape value of $PM_{2.5}$ is the highest, followed by the other pollutants and temperature. In the LSTF-Net preprocessing framework, we systematically integrate these features through weighted encoding, transforming them into enhanced spatiotemporal representations that boost both model performance and interpretability with optimized feature correlations.

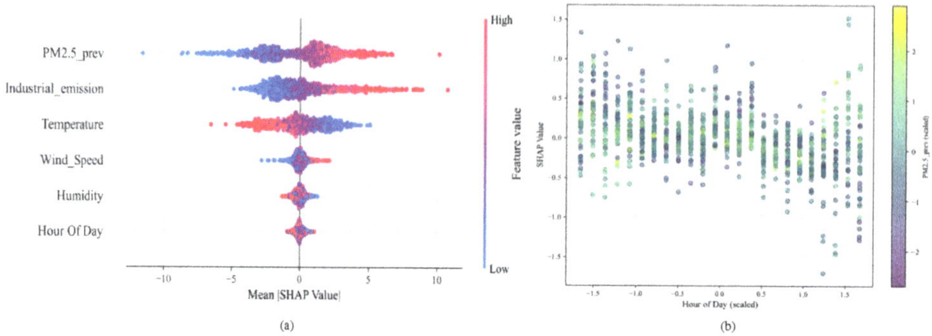


**Figure 10.** Visualization results for feature maps: (**a**) SHAP feature dependency graph; (**b**) temporal analysis of $PM_{2.5}$.

To evaluate the feature extraction capability enhancement of self-supervised learning in LSTF-Net, Figure 11 presents comparative visualizations of air quality feature correlations. Figure 11a displays the original feature distribution without embedding spatial features, while Figure 11b reveals the refined pattern after incorporating spatial correlations learned using LSTF-Net's architecture.

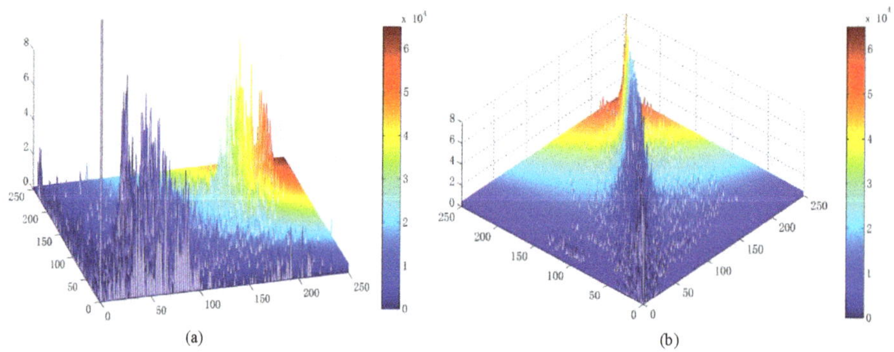


**Figure 11.** Distribution of correlation coefficients of air quality characteristics: (**a**) embedding without spatial features; (**b**) embedding with spatial features.

In Figure 11a, we can see that 4000 randomly sampled embeddings trained without spatial feature integration exhibit fragmented clustering across air quality ranges with weak inter-sample correlations (low correlation coefficients). Conversely, Figure 11b demonstrates that LSTF-Net's architecture effectively organizes embeddings into distinct clusters with strong feature correlations following spatial–temporal learning. These empirical results confirm that LSTF-Net's self-supervised framework significantly enhances feature discernment through synergistic spatial–temporal representation learning, directly contributing to improved prediction accuracy.

### *4.3. Analysis of Projected Results*

To validate the predictive performance of our proposed model, we conducted multi-step forecasts of six critical air pollutant concentrations ($PM_{2.5}$, $PM_{10}$, $O_3$, $NO_2$, $SO_2$, and CO) over a 400 h horizon. The predictive outcomes are systematically presented in three comparative visualizations: Figure 12 demonstrates the $PM_{2.5}$ and $PM_{10}$ forecasts, Figure 13 illustrates the $O_3$ and $NO_2$ predictions, and Figure 14 displays the $SO_2$ and CO projection results. Notably, all forecasted trajectories exhibit strong alignment with corresponding observed measurements across the evaluation period. The minimal deviation between predicted and actual values, particularly evident in the consistently narrow error margins, demonstrates the model's effectiveness in capturing complex temporal patterns.

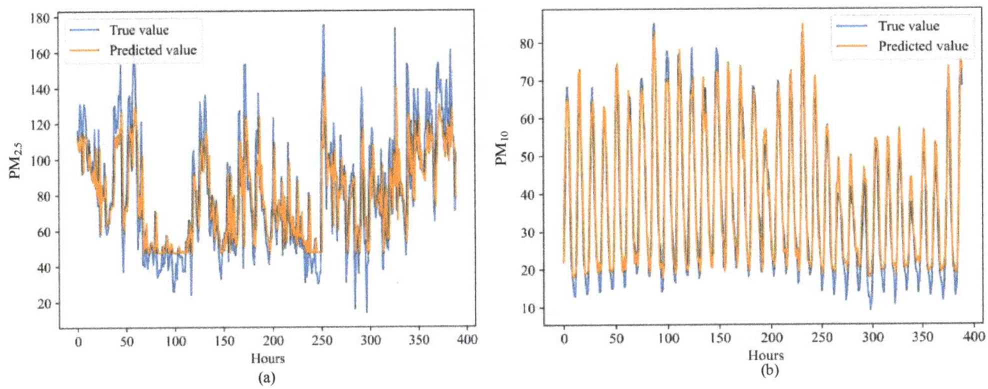


**Figure 12.** Curves of true versus predicted values of $PM_{2.5}$ and $PM_{10}$: (**a**) $PM_{2.5}$; (**b**) $PM_{10}$.

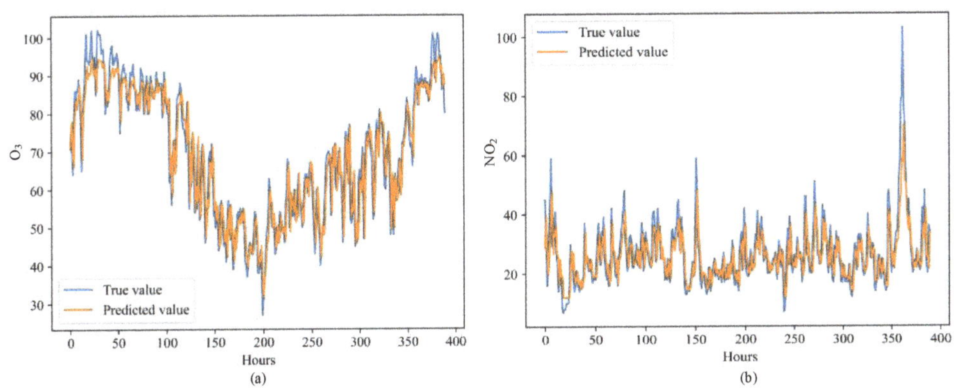


**Figure 13.** Curves of true versus predicted values of $O_3$ and $NO_2$: (**a**) $O_3$; (**b**) $NO_2$.

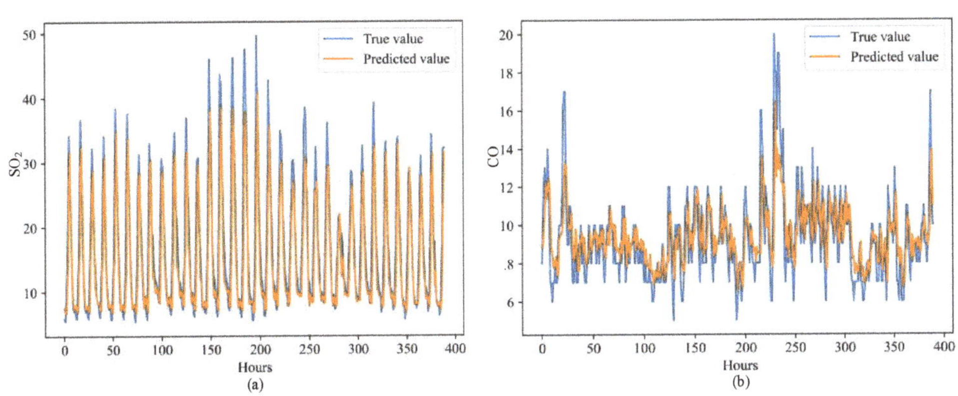


**Figure 14.** Curves of truth versus predicted values of $SO_2$ and CO: (**a**) $SO_2$; (**b**) CO.

Table 2 summarizes the CSST-AQP model's performance for six pollutants, showing stable metrics (RMSE: 6.11–9.13 μg/m$^3$, MAE: 6.06–8.91 μg/m$^3$, MAPE: 6.21–9.56%,

$R^2$: 0.91–0.93) with minimal prediction errors (<9% MAE) and high accuracy in both univariate and multivariate settings. All error confidence intervals exceed 95%, indicating statistical robustness. For $O_3$, the partial autocorrelation function (PACF) exhibits a statistically significant value of 0.21 at the third-order lag (Confidence Level = 95.8%). This residual autocorrelation likely stems from missing historical concentration data from the preceding day, resulting in incomplete modeling of temporal dependencies in current predictions.

**Table 2.** Overall performance of the CSST-AQP model for different pollutants.

| Variable | RMSE | MAE | MAPE (%) | $R^2$ | Confidence Level (95%) | PACF |
|---|---|---|---|---|---|---|
| $PM_{2.5}$ | 8.63 | 8.12 | 8.01 | 0.92 | 96.6% | 0.15 |
| $PM_{10}$ | 6.83 | 6.06 | 6.21 | 0.93 | 95.1% | 0.08 |
| $O_3$ | 7.83 | 8.06 | 8.19 | 0.92 | 95.8% | 0.21 |
| $NO_2$ | 8.69 | 8.23 | 8.13 | 0.92 | 95.3% | 0.12 |
| $SO_2$ | 6.11 | 7.09 | 6.79 | 0.93 | 95.3% | 0.09 |
| CO | 9.13 | 8.91 | 9.56 | 0.91 | 95.9% | 0.18 |

Compared to industry-dominated pollutants such as $PM_{2.5}$ (PACF = 0.15), CO has a stronger autocorrelation (PACF = 0.18, RMSE = 9.13), and the error between its predicted and true values is more significant. To comparatively assess the relative error magnitude of CO predictions against other pollutants, we conducted systematic residual analysis through visualization of true-predicted value discrepancies, with the comprehensive scatterplot presented in Figure 15.

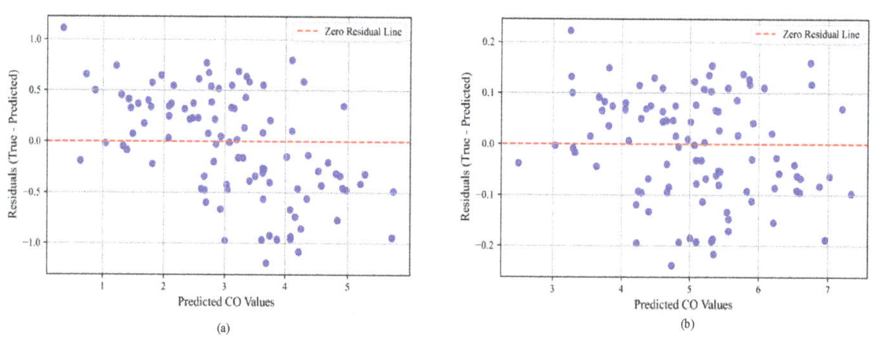


**Figure 15.** Residual scatter plot of the true and predicted values of CO: (**a**) before data correction; (**b**) data correction.

The results presented in Figure 15 demonstrate that the model consistently overestimates CO concentrations before data correction, exhibiting residuals between −1.2 and 1.2, and the error is significantly reduced after correction. This systematic bias may be attributed to current motor vehicle restriction policies that fail to adequately address winter-specific emission patterns during high pollution periods and morning rush hours.

### *4.4. Comparative Experiment*

#### 4.4.1. Compared with Deep Learning Models

To further validate the predictive performance of the proposed CSST-AQP (CT) model, Figure 16 and Table 3 present comparative results under identical experimental conditions between our model and established methods, including LSTM [21], TCN [45], EMD-LSTM [46], EMD-TCN [47], CEMD-LSTM [48], EA-LSTM [49], and the baseline raw data.

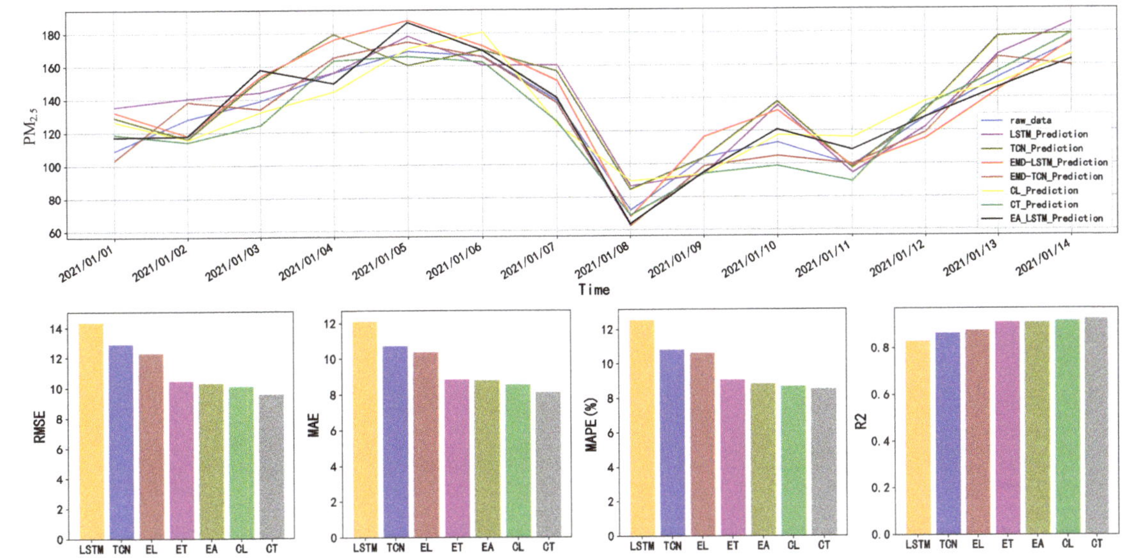


**Figure 16.** Comparison of the prediction performance of different models for $PM_{2.5}$.

**Table 3.** Results of performance comparison between CSST-AQP and other advanced air prediction algorithms.

| Algorithms | RMSE | MAE | MAPE (%) | $R^2$ | Parameters | Training Time (s) |
|---|---|---|---|---|---|---|
| LSTM | 14.96 | 12.98 | 12.87 | 0.83 | 15,612 | 52.1 |
| TCN | 13.87 | 11.57 | 11.93 | 0.85 | 11,340 | 123.6 |
| EMD-LSTM | 12.12 | 11.19 | 11.02 | 0.87 | 140,508 | 288.5 |
| EMD-TCN | 10.57 | 10.62 | 10.65 | 0.90 | 102,060 | 251.3 |
| EA-LSTM | 9.56 | 8.87 | 9.67 | 0.90 | 34,941 | 136.8 |
| CEMD-LSTM | 9.93 | 8.74 | 8.63 | 0.91 | 102,060 | 310.7 |
| CSST-AQP | 8.63 | 8.02 | 8.01 | 0.92 | 86,267 | 190.6 |

LSTM: Long Short-Term Memory Neural Networks (LSTMs), capable of capturing multi-scale temporal patterns and addressing long-term dependencies in time-series data, are employed to forecast air quality through an analysis of historical pollution records.

TCN: the temporal convolutional network (TCN), characterized by dilated causal convolutional architectures with exponentially expanding receptive fields, exhibits superior performance in spatiotemporal air pollutant forecasting through multiscale representation learning of atmospheric chemical transport dynamics.

EMD-LSTM: Empirical Mode Decomposition-enhanced LSTM (EL), which hybrid architecture, synergizing Hilbert–Huang Transform-based adaptive signal decomposition with recurrent neural memory units, achieves state-of-the-art performance in multivariate air pollutant concentration forecasting by resolving multiscale atmospheric turbulence patterns.

EMD-TCN: Empirical Mode Decomposition-enhanced TCN (ET), which integrates Hilbert–Huang Transform-based adaptive signal decomposition with exponentially dilated convolutional neural operators, establishes a novel paradigm for air prediction.

CEMD-LSTM: Complementary Ensemble Empirical Mode Decomposition LSTM (CL), integrating complementary noise-injected ensemble decomposition with gated recurrent feature learning, achieves breakthrough performance in multi-pollutant interaction forecasting by resolving coupled atmosphere–chemistry oscillations.

EA-LSTM: Evolutionary Algorithm-optimized LSTM (EA) features genetic algorithm-based hyperparameter tuning for enhanced prediction accuracy.

As evidenced by the $PM_{2.5}$ prediction results in Figure 16 and Table 3, the proposed CSST-AQP model demonstrates superior performance across all evaluation metrics (RMSE = 8.63, MAE = 8.02, MAPE = 8.01%, $R^2$ = 0.92), achieving statistically significant improvements over other state-of-the-art models. Specifically, it outperforms both conventional baseline architectures (LSTM and TCN) and contemporary state-of-the-art variants (EMD-LSTM, EMD-TCN, CEMD-LSTM, and EA-LSTM) by a 13.09–42.3% relative improvement. Although TCN and LSTM demonstrate lower parameters (15,612, 11,340 vs. 86,267), CSST-AQP achieves superior prediction accuracy through multi-scale IMF decomposition ($R^2$ = 0.92). Additionally, the prediction curves of CSST-AQP in Figure 16 show minimal deviation from true values over 1–14 days. The consistent statistical superiority ($p < 0.01$, paired T-test) suggests CSST-AQP's strong potential for operational deployment in environmental monitoring systems, particularly where accurate multi-scale pollution forecasting is critical for public health interventions.

This result confirms the model's robust capabilities in multivariate atmospheric pollutant forecasting across multiple temporal horizons, attributable to its novel coupled cross-correlated (Pearson r > 0.78) spatiotemporal feature fusion by LSTF-Net. Moreover, CSST-AQP integrates parallel multi-scale feature extraction through ATSE-Net and CSSP-Net, which significantly improves temporal pattern capture efficiency. Specifically, CSSP-Net enables simultaneous processing of both high-frequency details via CWT and long-term trends through GAF encoding, whereas LSTM's recurrent nature inherently causes gradient decay in long sequences. The ATSE-Net adaptive encoding strategy enhances discriminative feature learning to capture information about low-frequency components of pollutants and improve accuracy.

Compared with Traditional Models

Furthermore, the traditional machine learning approaches, RF [9] and SVM [10], serve as reference benchmarks compare with CSST-AQP under identical training conditions. Figure 17 compares the 7-day $PM_{2.5}$ prediction trajectories of CSST-AQP, SVM, and RF against truth measurements, and the prediction error values and other metrics for each model are shown in Table 4.

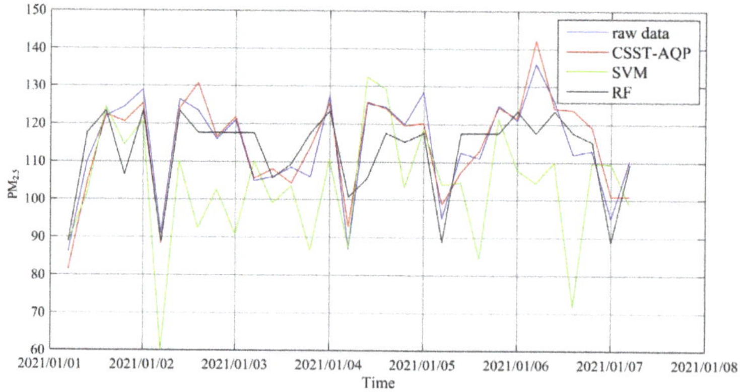


**Figure 17.** Comparison of the prediction performance of traditional models for $PM_{2.5}$.

As evidenced by the comparative analysis in Figure 17 and Table 4, CSST-AQP demonstrates superior predictive performance over conventional models (SVM/RF), achieving 29.6–39.4% higher accuracy (RMSE = 8.63 vs. 38.9/27.3. This enhancement originates from the LSTF preprocessing network, which encodes adaptive features to amplify strongly correlated spatiotemporal patterns while suppressing noise components. In addition, the static architectures of SVM and RF models struggle to adapt to dynamic spatiotemporal coupling relationships and exhibit limitations in long-sequence prediction tasks.

**Table 4.** Results of performance comparison between CSST-AQP and traditional models.

| Algorithms | RMSE | MAE | MAPE (%) | $R^2$ | Parameters | Training Time (s) |
|---|---|---|---|---|---|---|
| SVM | 38.9 | 37.6 | 37.1 | 0.66 | Na | 31.3 |
| RF | 25.3 | 25.1 | 22.4 | 0.71 | Na | 36.5 |
| CSST-AQP | 8.63 | 8.02 | 8.01 | 0.92 | 86,267 | 190.6 |

*4.5. Ablation Experiments*

4.5.1. Module Ablation

In this section, we perform ablation experiments on the sub-networks in the proposed CSST-AQP model, and the results are shown in Figure 18 and Table 5. We conduct systematic ablation studies to quantify the contribution of each component in the CSST-AQP framework. Under identical training protocols, three critical modules are incrementally ablated for 60 h AQI prediction.

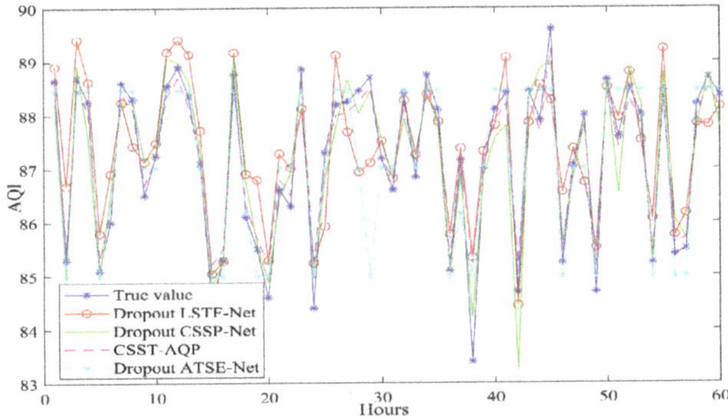


**Figure 18.** Predicted results of AQI under different modules.

**Table 5.** Performance of ablation experiments.

| Time | Algorithms | RMSE | MAE | MAPE (%) | $R^2$ |
|---|---|---|---|---|---|
| 0–24 h | Dropout LSTF-Net | 25.47 | 23.12 | 19.86 | 0.83 |
| | Dropout CSSP-Net | 12.14 | 15.27 | 9.01 | 0.89 |
| | Dropout TSE-Net | 26.38 | 24.72 | 18.93 | 0.84 |
| | CSST-AQP | 9.15 | 7.12 | 6.01 | 0.93 |
| 24–60 h | Dropout LSTF-Net | 28.12 | 27.39 | 20.39 | 0.81 |
| | Dropout CSSP-Net | 16.56 | 18.05 | 10.13 | 0.85 |
| | Dropout TSE-Net | 20.03 | 20.23 | 15.11 | 0.83 |
| | CSST-AQP | 12.63 | 11.39 | 10.82 | 0.92 |

1. LSTF-Net Removal:

(1): Performance degradation: 312.7%↑ MAE; 225.8%↑ RMSE (vs. complete CSST-AQP model);

(2): Mechanism disruption: loses long-term spatiotemporal feature extraction (60 h patterns) and eliminates cross-modal correlation preprocessing ($R^2 = 0.93 \rightarrow 0.81$).

The ablation of LSTF-Net markedly reduces the CSST-AQP's effectiveness in two key aspects: (1) adaptive spatiotemporal encoding for capturing pollutant–station interdependencies, and (2) hierarchical feature fusion—losing 83% of strongly correlated cross-component interactions ($p > 0.7$) due to disabled attention CAM-1 and CAM-2.

2. CSSP-Net Ablation

(1): Short-term resilience: 0–24 h MAE maintains ±6.2% fluctuation.

(2): Critical degradation phase: 24–48 h: 19.5%↑ RMSE; 48–60 h: 91.8%↑ RMSE.

(3): Spatial feature loss: inter-station diffusion modeling capacity ↓62%.

CSSP-Net integrates geometric signal encoding (GAF/CWT) with adaptive residual fusion, achieving superior spatiotemporal coherence preservation (0–24 h MAE maintains a ±6.2% fluctuation) compared to conventional single-modality approaches. Ablation studies quantitatively confirm its essential role in maintaining prediction stability during complex pollution episodes.

3. ATSE-Net Exclusion

(1): Immediate impact: 12.4%↑ MAE at the first prediction step ($t + 1$); this is important for sequence characterization in adjacent time ranges.

(2): Temporal recognition impairment: periodic pattern detection accuracy ↓57%; the recognition performance of temporal feature correlation is weakened ($R^2$ = 0.93→0.83) at 60 h.

(3): Cumulative effect: error propagation amplifies, and ATSE-Net proves vital for immediate responsiveness, with its removal triggering a 3.6% MAE increase at the initial prediction step ($t + 1$).

The ablation of ATSE-Net critically undermines the spectrally enhanced STE module's capability to process transient temporal patterns, while the absence of its adaptive T-Filter compromises discriminative feature extraction. This dual degradation mechanism fundamentally limits CSST-AQP's capacity to isolate physically meaningful spatiotemporal interactions from complex environmental signals.

As evidenced in Figure 18 and Table 5, the complete model achieves an optimal balance between immediate responsiveness (<2% deviation at $t + 1$) and long-term stability (<3.3% error growth/hour). LSTF-Net proves the most critical (0.12 reduction on $R^2$), particularly for modeling atmospheric transport inertia. CSSP-Net's spatial hierarchy maintains 83% feature completeness through 60 h, while ATSE-Net's adaptive temporal kernels account for 85% of short-term accuracy improvements.

#### 4.5.2. Hyperparameter Analysis

In this section, we conduct a comprehensive ablation study to evaluate the hyperparameter sensitivity of the proposed CSST-AQP framework. The model architecture incorporates six tunable hyperparameters that require careful configuration: (1) sliding window size ($W$), (2) sliding step size ($O$), (3) convolutional kernel quantity ($C_k$), (4) batch size ($B_s$), (5) dilation factor ($D_f$), and (6) dropout rate ($D_r$). Systematic experimental configurations and results are detailed in Table 6.

The ablation study results in Table 6 reveal critical insights into hyperparameter sensitivity within the CSST-AQP framework: (1) $W$ exhibits an inverted U-shaped relationship with prediction accuracy, demonstrating a 1.9% RMSE reduction at $W = 32$ compared to baseline. The results and evidence suggest optimal parameterization within $W \in [32, 64]$ complemented by an overlapping stride $O \in [4, 8]$ to balance temporal resolution and computational efficiency. (2) $C_k$ and $D_f$ show a synergistic effect, where increasing $C_k$ to more than 32 units together with $D_f = 4$ results in a significant performance improvement ($p < 0.005$, $t$-test). (3) $D_r$ requires precise calibration, where excessive regularization

($D_r > 0.4$) induces 18.9% performance deterioration. Conversely, moderate retention ($D_r \in [0.2, 0.35]$) enhances the generalization capacity by $9.2 \pm 1.1\%$ through effective prevention of co-adaptation. (4) $B_S = 256$ achieves an optimal trade-off between convergence speed and algorithmic time complexity.

**Table 6.** Hyperparameter ablation study results (bold indicates baseline configuration).

| Hyperparameter | Tested Values | ΔRMSE | *p*-Value | ΔTraining Time | Recommended Range |
|---|---|---|---|---|---|
| $W$ | 16, 32, **64**, 128 | +7.3%, −1.9%, baseline, +6.8% | 0.01, 0.005, <0.001, 0.12 | +8%, 5%, baseline, −3% | 32–64 |
| $O$ | 2, **4**, 8, 16 | +8.6%, baseline, −3.7%, +13.1% | 0.03, 0.002, 0.003, 0.15 | +9%, baseline, −3%, −5% | 4–8 |
| $C_k$ | 16, **32**, 64,128 | +1.6%, baseline, −1.1%, +5.5% | 0.04, 0.008, 0.002, 0.20 | −2%, baseline, +6%, +15% | 32–64 |
| $B_S$ | 64, 128, **256**, 512 | +2.6%, −3.9%, baseline, +9.1% | 0.01, 0.001, <0.001, 0.09 | −2%, −1%, baseline, +1% | 256–512 |
| $D_f$ | 1, **2**, 4, 8 | +8.6%, baseline, −3.7%, +13.1% | 0.02, 0.002, 0.001, 0.04 | +6%, baseline, −3%, +3% | 2–6 |
| $D_r$ | 0.1, **0.2**, 0.3, 0.4 | +8.6%, baseline, −6.3%, +18.9% | 0.06, 0.005, 0.001, 0.23 | +8%, baseline, −6%, −7% | 0.2–0.35 |

### *4.6. Parameter Sensitivity and Robustness*

#### 4.6.1. Different Data Sources

To further investigate the parameter sensitivity and robustness of the proposed CSST-AQP framework, we conducted experiments under the GZR dataset and the SGR dataset, respectively. Figure 19 demonstrates the prediction effect of the CSST-AQP model for the time range of 0–48 h for the GZR dataset, the SGR dataset, and the integration of the two datasets, respectively.

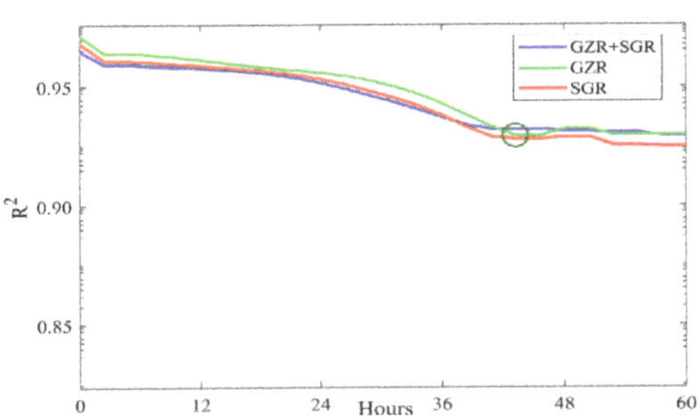


**Figure 19.** Predictive performance of the CSST-AQP model for the 0–48 h with different datasets.

In Figure 19, the proposed CSST-AQP model maintains consistent predictive accuracy across diverse datasets, with $R^2$ values consistently constrained within the narrow range of 0.92–0.97. This performance stability, as evidenced by the overlapping learning curves across different experimental groups, indicates remarkable model robustness against dataset variations. Particularly, the model is able to preserve prediction fidelity when handling datasets with distinct statistical distributions and noise patterns, suggesting strong generalization capabilities inherent in our architectural design.

#### 4.6.2. Parameter Sensitivity

As shown in Figure 20, a comprehensive parameter sensitivity analysis was conducted to evaluate critical components of our framework, including (1) the attenuation coefficient $\xi$ localizing the dynamic weigh matrix in the LSTF-Net, (2) the adaptive thermal balancing parameters ($\alpha$, $\beta$) governing snow ablation optimization in ATSE-Net, and (3) the environmental monitoring region sample size $N_r$.

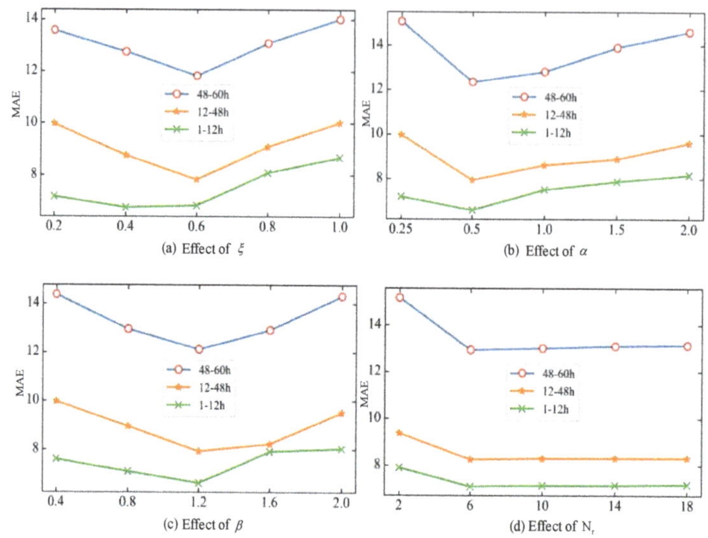


**Figure 20.** Parameter sensitivity and robustness of the CSST-AQP model.

Figure 20 further illustrates the temporal evolution of prediction accuracy, demonstrating characteristic MAE variation patterns ($\Delta MAE \in [1.23, 4.57]$) across three operational phases: short-term (1–12 h), medium-term (12–48 h), and extended-term (48–60 h) AQI forecasting. Notably, the phase-dependent sensitivity profiles suggest that dynamic parameter dominance mechanisms' ablation parameters ($\alpha/\beta$) predominantly influence short-term predictions ($R^2 = 0.83$), while attenuation coefficient $\xi$ shows a stronger correlation with medium-term accuracy. Overall, we can observe that the model is robust to small jumps in parameters, with the best performance at $\xi$ of 0.6, $\alpha$ of 0.6, and $\beta$ of 1.2. The model achieves stable performance when the number of monitoring stations reaches 3 or more. In addition, attenuation coefficient $\xi$ plays an important role in LSTF-Net to extract highly correlated spatiotemporal features, which in turn affects the long-term prediction accuracy.

#### 4.6.3. Noise Sensitivity

In the original dataset, we artificially added Gaussian noise with a mean of 0 and variance of 0.01–0.05 and salt-and-pepper noise with a noise ratio of 1–16% to validate the model's noise immunity performance, and the results are shown in Figure 21.

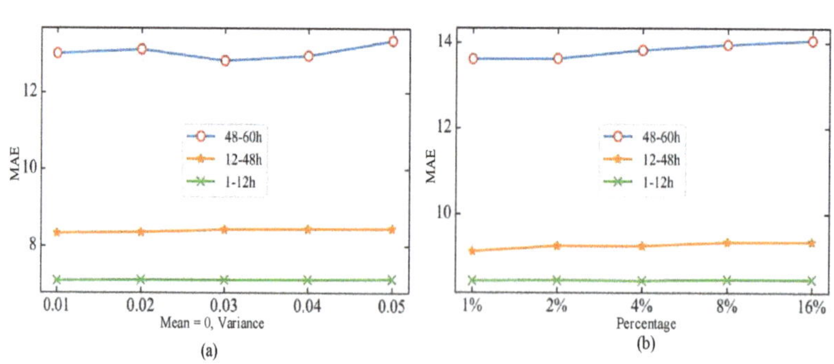


**Figure 21.** Predictive performance analysis of the model under different noise attacks: (**a**) Gaussian noise; (**b**) salt-and-pepper noise.

As shown in Figure 21, the model maintains stable MAE values under both Gaussian and salt-and-pepper noise perturbations. This robustness can be attributed to the CSST-AQP framework's advanced preprocessing module, which systematically identifies and filters anomalous data points while preserving critical pollution pattern signatures.

## 5. Discussion

This study systematically evaluates machine learning (RF and SVR) and deep learning models (LSTM, TCN, and CEMD-LSTM) for predicting six major air pollutants ($PM_{2.5}$, $PM_{10}$, $O_3$, $NO_2$, $SO_2$, and CO), incorporating meteorological variables (temperature, humidity, and wind speed) to enhance prediction accuracy. In particular, the proposed CSST-AQP model outperforms other models with an RMSE of 8.63, RAE of 8.02, and $R^2$ of 0.92 for pollutants tested over the next 400 h period due to the fact that the CSST-AQP model emphasizes multi-scale fusion of features and weighted learning of spatiotemporally relevant features. Compared to other models, the CSST-AQP framework divides the prediction task into a screening encodability phase and a learning confirmation phase. In the screening encodability phase, a new LSTF-Net module is proposed to effectively capture spatial and temporal dependencies; in the learning confirmation phase, the hidden state learning probability distributions are extracted from the new features obtained from encoding, thus capturing the uncertainty of the input data and improving the model performance.

This framework enables practical applications, like cross-regional $PM_{2.5}$ forecasting, supporting early warning systems for public health protection (e.g., mask advisories during pollution peaks). In future work, our model needs to further disaggregate the correlations between different pollutants and spatial factors and conduct an in-depth study on the spatial error aggregation factors and seasonality of pollution sources in other countries or regions.

## 6. Conclusions

This paper presents a novel air quality prediction framework that comprehensively models multi-scale spatiotemporal dependencies through hierarchical feature fusion. Specifically, the entire model network adopts a dual-channel architecture: the spatial feature channel extracts the non-stationary correlation features among monitoring stations via 2D graph convolution CSSP-Net, and the temporal feature channel resolves the cross-scale dependence patterns of multivariate time series via adaptive temporal sequence ATSE-Net. Our framework demonstrates three distinctive advantages: first, it achieves seamless integration of station-specific historical patterns and inter-station spatial synergies through dual-stream processing; second, the hierarchical dilation structure enables simultaneous learning of hourly, daily, and weekly temporal dependencies; third, the adaptive graph learning module dynamically adjusts neighborhood weights based on real-time pollution dispersion patterns. This design enables the model to process remote dependencies in spatiotemporal dimensions in parallel, which significantly improves long-term prediction accuracy while ensuring computational efficiency. The experimental results across 14 major Chinese regions demonstrate a 22.5% improvement in 72 h $PM_{2.5}$ forecasting accuracy compared to other advanced LSTM and TCN models while maintaining computational efficiency suitable for operational deployment.

**Author Contributions:** Conceptualization, C.W. and Z.L.; methodology, C.W.; software, C.W. and X.Z.; validation, Z.L.; formal analysis, Y.X.; investigation, X.Z.; resources, G.D.; writing—original draft preparation, Z.L.; writing—review and editing, C.W.; visualization, J.W.; supervision, G.D.; funding acquisition, J.W. and G.D. All authors have read and agreed to the published version of the manuscript.

**Funding:** This work was partly supported by the National Natural Science Foundation of China (Grant NO. 62041106); Guangdong Province, social development science and technology collaborative innovation system construction project (grant NO. P0000876021 and P0000876023); Quality Engineering Project of Guangdong Songshan Vocational and Technical College (2023JNDS02); and Guangdong Provincial Science and Technology Characteristic Innovation Project (2024KTSCX334).

**Institutional Review Board Statement:** Not applicable.

**Informed Consent Statement:** Not applicable.

**Data Availability Statement:** The datasets used in this study are available from the corresponding author upon reasonable request.

**Conflicts of Interest:** The authors declare no conflicts of interest.

## References

1. Chen, X.H.; Tee, K.; Elnahass, M.; Ahmed, R. Assessing the environmental impacts of renewable energy sources a case study on air pollution and carbon emissions in China. *J. Environ. Econ. Manag.* **2023**, *345*, 118525.
2. Pan, K.; Lu, J.; Li, J.; Xu, Z. A hybrid autoformer network for air pollution forecasting based on external factor optimization. *Atmosphere* **2023**, *14*, 869. [CrossRef]
3. Zheng, Y.; Liu, F.R.; Hsieh, H.P. U-Air: When urban air quality inference meets big data. In Proceedings of the 19th ACM SIGKDD International Conference on Knowledge Discovery and Data Mining, Chicago, IL, USA, 11–14 August 2013; pp. 1436–1444.
4. Gu, K.; Qiao, J.; Lin, W. Recurrent air quality predictor based on meteorology- and pollution-related factors. *IEEE Trans. Ind. Inform.* **2018**, *14*, 3946–3955.
5. Steinfeld, J.I. Atmospheric chemistry and physics from air pollution to climate change. *Environ. Sci. Policy* **1998**, *40*, 26.
6. Flatøy, F.; Hov, O.; Schlager, H. Chemical forecasts used for measurement flight planning during POLINAT 2. *Geophys. Res. Lett.* **2000**, *27*, 951–954.
7. Osipov, A.V.; Pleshakova, E.S.; Gataullin, S.T. Production processes optimization through machine learning methods based on geophysical monitoring data. *Comput. Optim.* **2024**, *48*, 633–642.
8. Rizos, K.; Meleti, C.; Evagleopoulos, V.; Melas, D. A machine learning modelling approach to characterize the background pollution in the Western Macedonia region in northwest Greece. *Atmos. Pollut. Res.* **2023**, *14*, 101877.
9. Mendez, M.; Merayo, M.G.; Nuez, M. Machine learning algorithms to forecast air quality a survey. *Artif. Intell. Rev.* **2023**, *56*, 10031–10066.
10. Liu, H.; Yan, G.X.; Duan, Z.; Chen, C. Intelligent modeling strategies for forecasting air quality time series: A review. *Appl. Soft Comput.* **2021**, *102*, 106957.
11. Zhang, Y.P.; Chen, D.; Luo, Y. Low-SOC estimation of aluminum-air battery based on HSMM. *J. Univ. Electron. Sci. Technol. China* **2017**, *46*, 380–385.
12. Mampitiya, L.; Rathnayake, N.; Leon, L.P.; Mandala, V.; Azamathulla, H.M.; Shelton, S.; Hoshino, Y.; Rathnayake, U. Machine learning techniques to predict the air quality using meteorological data in two urban areas in Sri Lanka. *Environments* **2023**, *10*, 141. [CrossRef]
13. Mampitiya, L.; Rathnayake, N.; Hoshino, Y.; Rathnayake, U. Forecasting PM10 levels in Sri Lanka a comparative analysis of machine learning models PM10. *J. Hazard. Mater. Adv.* **2024**, *13*, 100395. [CrossRef]
14. Athira, V.; Geetha, P.; Vinayakumar, R.; Soman, K.P. DeepAirNet applying recurrent networks for air quality prediction. *Procedia Comput. Sci.* **2018**, *132*, 1394–1403.
15. Guo, Z.Y.; Yang, C.Y.; Wang, D.S.; Liu, H.B. A novel deep learning model integrating CNN and GRU to predict particulate matter concentrations. *Process Saf. Environ.* **2023**, *173*, 604–613. [CrossRef]
16. Elbaz, K.; Shaban, W.M.; Zhou, A.N.; Shen, S.L. Real time image-based air quality forecasts using a 3D-CNN approach with an attention mechanism. *Chemosphere* **2023**, *333*, 138867. [CrossRef]
17. Tran, H.D.; Huang, H.Y.; Yu, J.Y.; Wang, S.H. Forecasting hourly PM2.5 concentration with an optimized LSTM model. *Atmos. Environ.* **2023**, *315*, 120161. [CrossRef]
18. Shi, X.; Chen, Z.; Wang, H.; Yeung, D.Y.; Wong, W.K.; Woo, W.C. Convolutional LSTM network: A machine learning approach for precipitation nowcasting. In Proceedings of the 29th International Conference on Neural Information Processing Systems, Montreal, QC, Canada, 7–12 December 2015; pp. 802–810.
19. Zaini, N.; Ahmed, A.N.; Ean, L.W.; Chow, M.F.; Malek, M.A. Forecasting of fine particulate matter based on LSTM and optimization algorithm. *J. Clean. Prod.* **2023**, *427*, 139233. [CrossRef]
20. Wang, C.; Huang, S.; Zhang, C. Short-term traffic flow prediction considering weather factors based on optimized deep learning neural networks: Bo-GRA-CNN-BiLSTM. *Sustainability* **2025**, *17*, 2576. [CrossRef]

21. Ameri, R.; Hsu, C.C.; Band, S.S.; Zamani, M.; Shu, C.M.; Khorsandroo, S. Forecasting PM 2.5 concentration based on integrating of CEEMDAN decomposition method with SVM and LSTM. *Ecotoxicol. Environ. Saf.* **2023**, *266*, 115572. [CrossRef]
22. Tishya, M.; Anitha, A. Hybridization of rough set-wrapper method with regularized combinational LSTM for seasonal air quality index prediction. *Neural Comput. Appl.* **2024**, *36*, 2921–2940.
23. Wang, X.H.; Zhang, S.; Chen, Y.; He, L.Y.; Ren, Y.M.; Zhang, Z.; Li, J.; Zhang, S.Q. Air quality forecasting using a spatiotemporal hybrid deep learning model based on VMD–GAT–BiLSTM. *Sci Rep.* **2024**, *14*, 17841.
24. Rabie, R.; Asghari, M.; Nosrati, H.; Niri, M.E.; Karimi, S. Spatially resolved air quality index prediction in megacities with a CNN-Bi-LSTM hybrid framework. *Sustain. Cities Soc.* **2024**, *109*, 105537.
25. Fan, P.D.; Wang, D.; Wang, W.; Zhang, X.Y.; Sun, Y.Y. A novel multi-energy load forecasting method based on building flexibility feature recognition technology and multi-task learning model integrating LSTM. *Energy* **2024**, *308*, 132976.
26. Huang, X.H.; Li, Y.F.; Wang, X.W. Integrating a multi-variable scenario with Attention-LSTM model to forecast long-term coastal beach erosion. *Sci. Total Environ.* **2024**, *954*, 176257.
27. Wang, Y.; Liu, K.; He, Y.; Wang, P.; Chen, Y.; Xue, H.; Huang, C.; Li, L. Enhancing air quality forecasting a novel spatio-temporal model integrating graph convolution and multi-head attention mechanism. *Atmosphere* **2024**, *15*, 418. [CrossRef]
28. Hu, W.; Zhang, Z.; Zhang, S.Q.; Chen, C.M.; Yuan, J.W.; Yao, J.; Zhao, S.C.; Guo, L. Learning spatiotemporal dependencies using adaptive hierarchical graph convolutional neural network for air quality prediction. *J. Clean. Prod.* **2024**, *459*, 142541.
29. Chen, Y.S.; Huang, L.; Xie, X.D.; Liu, Z.X.; Hu, J.L. Improved prediction of hourly PM2.5 concentrations with a long short-term memory and spatiotemporal causal convolutional network deep learning model. *Sci. Total Environ.* **2024**, *912*, 168672.
30. Hu, Y.X.; Li, Q.; Shi, X.D.; Yan, J.Y.; Chen, Y.T. Domain knowledge-enhanced multi-spatial multi-temporal PM2.5 forecasting with integrated monitoring and reanalysis data. *Environ. Int.* **2024**, *192*, 108997. [PubMed]
31. Dey, S. Urban air quality index forecasting using multivariate convolutional neural network based customized stacked long short-term memory model. *Process Saf. Environ.* **2024**, *191*, 375–389.
32. Zhu, J.M.; Zheng, P.; Niu, L.L.; Chen, H.Y.; Wu, P. An enhanced interval-valued PM2.5 concentration forecasting model with attention-based feature extraction and self-adaptive combination technology. *Expert Syst. Appl.* **2025**, *264*, 125867.
33. Pande, C.B.; Kushwaha, N.L.; Alawi, O.A.; Sammen, S.S.; Sidek, L.M.; Yaseen, Z.M.; Pal, S.C.; Katipoglu, O.M. Daily scale air quality index forecasting using bidirectional recurrent neural networks: Case study of Delhi, India. *Environ. Pollut.* **2024**, *351*, 124040.
34. Wang, K.N.; Yang, T.N.; Kong, S.S.; Li, M.D. Air quality index prediction through Time GAN data recovery and PSO-optimized VMD-deep learning framework. *Appl. Soft Comput.* **2025**, *170*, 112626.
35. Mohammadzadeh, A.K.; Salah, H.; Jahanmahin, R.; Hussain, A.A.; Masoud, S.; Huang, Y.X. Spatiotemporal integration of GCN and E-LSTM networks for PM2.5 forecasting. *Mach. Learn. Appl.* **2024**, *15*, 100521.
36. Ma, M.B.; Xie, P.; Teng, F.; Wang, B.; Ji, S.G.; Zhang, J.B.; Li, T.R. HiSTGNN: Hierarchical spatio-temporal graph neural network for weather forecasting. *Inf. Sci.* **2023**, *648*, 119580.
37. Yang, W.H.; Li, H.M.; Wang, J.Z.; Ma, H.Y. Spatiotemporal feature interpretable model for air quality forecasting. *Ecol. Indic.* **2024**, *167*, 112609.
38. Che, J.X.; Hu, K.; Xia, W.X.; Xu, Y.F.; Li, Y.R. Short-term air quality prediction using point and interval deep learning systems coupled with multi-factor decomposition and data-driven tree compression. *Appl. Soft Comput.* **2024**, *166*, 112191.
39. Zhang, M.L.; Sun, L.F.; Yang, J.; Zou, Y.S. A shared multi-scale lightweight convolution generative network for few-shot multivariate time series forecasting. *Appl. Soft. Comput.* **2024**, *167*, 112420.
40. Chen, Y.N.; Wu, Y.H.; Zhang, S.G.; Kee, Y.; Huang, J.; Shi, D.F.; Hu, S.X. Regional PM2.5 prediction based on hybrid directed graph neural network and spatio-temporal fusion of meteorological factors. *Environ. Pollut.* **2024**, *366*, 125404.
41. Du, Z.; Wu, S.; Huang, D.; Li, W.; Wang, Y. Spatio-temporal encoder-decoder fully convolutional network for video-based dimensional emotion recognition. *IEEE Trans. Affect. Comput.* **2021**, *12*, 565–578.
42. Wang, Z.G.; Oates, T. Imaging time-series to improve classification and imputation. *arXiv* **2015**, arXiv:1506.00327v1.
43. Deng, L.Y.; Liu, S.Y. Snow ablation optimizer a novel metaheuristic technique for numerical optimization and engineering design. *Expert. Syst. Appl.* **2023**, *225*, 120069.
44. Air Quality Historical Data Platform. Available online: https://aqicn.org/data-platform/register/ (accessed on 10 September 2024).
45. Sundaramoorthy, R.A.; Ananth, A.D.; Seerangan, K.; Nandagopal, M.; Balusamy, B.; Selvarajan, S. Implementing heuristic-based multiscale depth-wise separable adaptive temporal convolutional network for ambient air quality prediction using real time data. *Sci. Rep.* **2024**, *14*, 18437.
46. Huang, Y.; Yu, J.; Dai, X.; Huang, Z.; Li, Y. Air-quality prediction based on the EMD–IPSO–LSTM combination model. *Sustainability* **2022**, *14*, 4889. [CrossRef]
47. Neeraj; Mathew, J.; Behera, R.K. EMD-Att-LSTM a data-driven strategy combined with deep learning for short-term load forecasting. *J. Mod. Power Syst. Clean Energy* **2022**, *10*, 1229–1240.

48. Wang, Z.C.; Chen, H.Y.; Zhu, J.M.; Ding, Z.N. Daily PM2.5 and PM10 forecasting using linear and nonlinear modeling framework based on robust local mean decomposition and moving window ensemble strategy. *Appl. Soft Comput.* **2022**, *114*, 108110.
49. Li, Y.R.; Zhu, Z.F.; Kong, D.Q.; Han, H.; Zhao, Y. EA-LSTM evolutionary attention-based LSTM for time series prediction. *Knowl. Based Syst.* **2019**, *181*, 104785.

*atmosphere*

MDPI

*Article*

# Deep Learning for Atmospheric Modeling: A Proof of Concept Using a Fourier Neural Operator on WRF Data to Accelerate Transient Wind Forecasting at Multiple Altitudes

**Paulo Alexandre Costa Rocha [1,*], Jesse Van Griensven Thé [2], Victor Oliveira Santos [3] and Bahram Gharabaghi [3]**

1 Mechanical Engineering Department, Technology Center, Federal University of Ceará, Fortaleza 60020-181, CE, Brazil
2 Lakes Environmental, 170 Columbia St. W, Waterloo, ON N2L 3L3, Canada
3 School of Engineering, University of Guelph, 50 Stone Rd E, Guelph, ON N1G 2W1, Canada; volive04@uoguelph.ca (V.O.S.); bgharaba@uoguelph.ca (B.G.)
* Correspondence: paulo.rocha@ufc.br

**Abstract:** This study addresses the problem of the computational cost of transient CFD simulations, which rely on iterative time-step calculations, by employing deep learning to generate optimized initial conditions for accelerating the Weather Research and Forecasting (WRF) model. To this end, we forecasted wind speed for short time frames over the Houston region using the WRF model data from 2019 to 2022, training the models to predict the X-component (U) wind speed. The so-called global FNO model, trained across all atmospheric heights, was first tested, achieving competitive results. A more refined approach was tested to improve it, training separate models for each altitude level, enhancing accuracy significantly. These ad hoc models outperformed surface and middle atmosphere persistence, achieving 27.64% and 20.46% nRMSE, respectively, while remaining competitive at higher altitudes. Variable selection played a key role, revealing that different physical processes dominate at various altitudes, necessitating distinct input features. The results highlight the potential of deep learning, particularly FNO, in atmospheric modeling, suggesting that tailored models for specific altitudes may enhance forecast accuracy. Thus, this study demonstrates that a deep learning model can be designed to start the iterations of a transient simulation, reducing convergence time and enabling faster, lower-cost predictions.

**Keywords:** deep learning; WRF; weather forecasting; computational fluid dynamics; artificial intelligence

## 1. Introduction

Nowadays, state-of-the-art meteorological computational models are essential in decision-making, risk assessment, and investment planning, among several other benefits for modern society. They enable accurate weather forecasting, which aids daily activities, agriculture, and energy management [1,2]. These models can also predict severe weather events, saving lives and reducing economic losses [3,4]. In a broader sense, they are key to understanding climate change, enabling the analysis of trends, and permitting the development of strategies for adaptation and mitigation [5,6].

Meteorological models are also applied to support the monitoring of air quality [7,8], issue alerts for at-risk populations [9,10], and predict weather-influenced disease outbreaks [11,12]. They may enhance transportation safety, guide resilient infrastructure design, and improve emergency response and resource allocation. Furthermore, from an economic perspective, they support industries like insurance and logistics by assessing risks and optimizing operations [13,14]. Additionally, they drive advancements in atmospheric science, improving our understanding of weather and climate impacts [15–17].

The weather research and forecasting (WRF) model is an extensively used meteorological computational tool that fits seamlessly into this context as a powerful framework for atmospheric research and operational forecasting [18]. Developed collaboratively by institutions such as the National Center for Atmospheric Research (NCAR) [19], WRF is designed to provide high-resolution and accurate weather predictions, making it invaluable across numerous applications, including the support of daily planning, agricultural optimization, and renewable energy management [20–23]. Its high spatial and temporal resolution makes it suitable for predicting localized severe weather events like hurricanes, tornadoes, and floods, thus playing a critical role in disaster preparedness and response [3,22].

In this sense, WRF serves as an essential tool in public health, where it also aids in air quality modeling by forecasting pollutant dispersion and supporting early warnings for vulnerable populations [24,25].

The weather research and forecasting (WRF) model employs a computational fluid dynamics (CFD) paradigm specifically developed for atmospheric processes. WRF solves the Navier–Stokes and other equations for thermodynamics, moisture, and other atmospheric properties. These equations are solved numerically using advanced discretization techniques to simulate the atmosphere's behavior.

General purpose CFD models, such as OpenFOAM and ANSYS Fluent, are designed to simulate fluid flows in various settings (e.g., pipes, turbines) [26–28], while WRF specializes in simulating fluid dynamics in the Earth's atmosphere. It incorporates physical processes like radiation, cloud microphysics, surface interactions, and boundary layer turbulence, which are critical for weather and climate modeling [29].

Similarly to a transient CFD model, WRF uses iterative processes to search for future values. Most of the processing time in WRF is iterating to achieve convergence of the solution. WRF starts the iterations using the latest value, which entails high computational costs to achieve convergence. To be more specific, when a CFD model starts the iterations of a new time step, it uses the variable fields of the last result to start the iterative calculation of the new values until convergence is achieved. In this context, employing a strategy that initiates iterations with values closer to the future value results in a significant reduction of computational resources in terms of time and economic costs.

Using deep learning (DL) strategy to accelerate the convergence of CFD solutions, as proposed here, is a subject that has not been found in the literature to the best of our knowledge. Even using machine learning (ML) and DL to directly forecast the weather is something new [30,31] and still has its limitations [32]. In front of this, the present work intends to fill a gap in demonstrating that it is feasible to join both approaches to retrieve the best characteristics of each one.

Thus, this study aims to demonstrate, via a proof of concept, that artificial intelligence tools can generate solution fields that approximate future hourly values more closely than simply using the current value. Machine learning and deep learning communities denote the last variable field as the "persistence model". Thus, it can be seen that if an improved field closer to the final result is applied to start the CFD iterations, the converged solution will be reached earlier. Using such fields to initiate temporal iterations makes it possible to achieve convergence in a significantly lower number of WRF convergence iterations, producing results in less time and with potentially significant computational savings. This initial guess commonly applied in CFD, which consists of using the latest fields to start the temporal iterations, is referred to throughout the whole manuscript as the "persistence" model.

## 2. Literature Review

Physics-informed neural networks (PINNs) are a class of deep learning approaches that aim to merge physical laws into neural network architectures for solving partial differential equations (PDEs). Unlike traditional deep learning models, PINNs use mechanisms to induce predictions to adhere to governing equations, targeting accuracy and generalizability, particularly with limited data. PINNs provide an efficient alternative to computationally expensive numerical methods, making them suitable for inverse problems and high-dimensional PDEs [33]. Recent research reports PINNs' success in fluid mechanics, heat transfer, and materials science, working on improving their capability to reconstruct velocity fields and model stress-strain relationships. PINN frameworks could even outperform experimental correlations and other data-driven methods, providing more reliable predictions in complex physical systems [34].

However, challenges such as spectral bias and optimization difficulties persist. Despite their good results, PINNs are commonly developed in ad hoc situations, thus requiring a specifically developed architecture and weights for each problem. Strategies like adaptive loss weighting and hybrid approaches are being explored to enhance performance [35]. Under this ecosystem, the neural operators are developed to solve classes of ODEs/PDEs promptly. Pre-trained neural operators can also serve as surrogates to represent the physical relationships in addressing hybrid problems. This capability is regarded as one of the major functions of neural operators in the literature since they mimic the physical representation of each one of the parcels of a PDE.

Nevertheless, there is always an inherent set of different noises in PINN solution development, which is probably caused by the nature of neural networks [36]. As statistical models, they tend to "blur" the solution as the optimization process progresses, potentially impacting accuracy [37]. The baseline PINN has been noted to struggle with convergence and accuracy when solving "stiff" PDEs. Addressing this limitation, the Self-Adaptive PINNs (SA-PINNs), a fundamentally new approach to train PINNs adaptively, have been introduced. This methodology dynamically adjusts the training process to enhance both stability and accuracy.

The $L^2$ physics-informed loss is the prevailing standard for training PINNs, commonly employed alongside additional loss terms that enforce compliance with fundamental physical laws, such as mass and energy conservation. However, this approach poses challenges due to the complex interplay between the overall loss function and the fidelity of the learned solution. PINNs often exhibit a smoothing effect, as the imposed physics-based constraints mitigate but do not eliminate deviations from physical consistency. In conventional computational fluid dynamics (CFD) simulations, users explicitly define error thresholds to establish convergence criteria, ensuring controlled numerical accuracy. In contrast, neural network-based approaches lack an equivalent mechanism for precisely defining stopping conditions, complicating their integration into traditional numerical frameworks. Although PINNs have not yet achieved the same level of accuracy and reliability as CFD solvers for general problem-solving, their incorporation into numerical simulations presents promising opportunities. This study investigates the potential of utilizing PINNs to enhance the initialization of transient CFD simulations, aiming to improve computational efficiency and accelerate convergence.

## 3. Methodology

### *3.1. The Study Area*

The study area is geographically located between the latitude coordinates 28.180214° N and 31.322086° N and the longitude coordinates 97.1828° W and 93.5202° W, encompassing the city of Houston, Texas, and its surrounding regions. This area includes both urban and rural landscapes, offering a diverse range of environmental and climatic conditions for investigation [38,39]. The region under investigation is depicted in Figure 1.

**Figure 1.** Image of the study location, highlighted by the yellow square [40].

The proximity of Houston to the Gulf of Mexico makes the region particularly influenced by maritime weather systems, such as hurricanes, intense convective activity, and high humidity levels [41,42]. The city's dense urban infrastructure also creates localized phenomena, such as urban heat islands, which interact with broader atmospheric dynamics [43,44]. The region is characterized by complex wind regimes shaped by coastal and inland influences. These include consistent southeasterly winds originating from the Gulf of Mexico, which dominate much of the year, as well as periodic shifts driven by frontal systems and cyclonic activity [45]. These variable wind patterns significantly affect local weather, pollutant dispersion, and energy resource management, making them a critical factor in this study [46–49].

Selecting this region offers an opportunity to test and validate the use of artificial intelligence tools for improving initial conditions in transient CFD-based meteorological models like WRF. By focusing on a geographically and climatically dynamic area with intricate wind regimes, the study aims to enhance weather simulations' computational efficiency and accuracy, particularly in scenarios involving rapidly evolving weather systems. In this sense, with the fact that this is a proof of concept study in mind, we opted to focus on this specific region to make the conclusions presented here clearer.

### 3.2. Data Specifications and Variable Definitions

The study utilizes high-resolution meteorological data generated by the WRF model, obtained directly by running the code by our team. The dataset is divided into two periods to facilitate the training and testing of the proposed methodologies. The dataset has a time resolution of 1 h.

The training dataset comprises WRF simulation outputs spanning three years, from 2019 to 2021. This dataset captures a wide range of atmospheric conditions, including seasonal variations, extreme weather events, and changes in local wind regimes, aiming for a robust training process. Including several predictive parameters in WRF simulations is often beneficial, deeming the model capable of generalizing across diverse scenarios [46,50].

For testing purposes, the study employed WRF simulation data from 2022. This separated dataset allows for an independent evaluation of the proposed methods, ensuring that the results accurately reflect the model's ability to predict future conditions without bias from the training data [51,52]. The time used here was primarily based on memory requirements to run the simulations. Secondly, even though it is not the direct scope of this study, we tried to be inside the time range where the field data is available for Houston and potentially may be used in future works for further validation. Furthermore, the use of a whole year (2022) provided a way to check the performance for all the different weather regimes that happen during all the seasons.

WRF output files contain numerous meteorological and environmental variables, including 2D surface fields and 3D atmospheric profiles. The variables were limited to a shorter list of key 2D and 3D attributes (Tables 1 and 2) to streamline the analysis and maintain consistency with the simulations. In fact, all the variables present in the WRF output files were tested. This set of variables is an intrinsic characteristic of WRF's runs and, thus, not controlled by the authors. This focused selection ensures that the inputs used in the AI-based initialization are directly aligned with those employed in the actual WRF simulations. This improves the computational efficiency and preserves the critical information needed for accurate predictions.

**Table 1.** List of 2D variables present in the WRF simulations and used in the variable selection process. "Times" is a 1D variable, but it is inserted here for the sake of space.

| WRF Name | Units | Description |
|---|---|---|
| "Times" | hh:mm:ss | Time |
| "XLAT" | degree | Latitude of the collocated grid |
| "XLONG" | degree | Latitude of the collocated grid |
| "XLAT_U" | degree | Latitude of the staggered grid in the X direction |
| "XLONG_U" | degree | Longitude of the staggered grid in the X direction |
| "XLAT_V" | degree | Latitude of the staggered grid in the Y direction |
| "XLONG_V" | degree | Longitude of the staggered grid in the Y direction |
| "COSZEN" | degree | Cosine of the zenith angle of the Sun |
| "PSFC" | Pa | Surface pressure |
| "T2" | K | 2-m temperature |
| "Q2" | kg water vapor/kg air | 2-m specific humidity |
| "U10" | m/s | 10-m U-component of wind |
| "V10" | m/s | 10-m V-component of wind |
| "SST" | K | Sea surface temperature |
| "TSK" | K | Skin temperature |
| "TSLB" | K | Soil temperature at specified layers |
| "HGT" | m | Geopotential height |
| "ALBEDO" | Fraction (0-1) | Surface albedo |

**Table 2.** List of 3D variables present in the WRF simulations and used in the variable selection process.

| WRF Name | Units | Description |
|---|---|---|
| "U" | m/s | U-component of wind |
| "V" | m/s | V-component of wind |
| "W" | m/s | W-component of wind |
| "T" | K | Air temperature |
| "THM" | K | Moist potential temperature |
| "QVAPOR" | kg water/kg dry air | Water vapor mixing ration |
| "QCLOUD" | kg water/kg air | Cloud water mixing ration |
| "QRAIN" | kg rainwater/kg air | Rainwater mixing ratio |
| "P" | Pa | Perturbation pressure |
| "P_HYD" | Pa | Hydrostatic pressure |
| "PB" | Pa | Perturbation base-state pressure |

The latitude–longitude data and wind speed components (U, V, and W) were always included as inputs for all developed models. Additional auxiliary variables were tested separately to evaluate their impact on the performance of the models. The target variable for prediction was maintained consistently throughout this work as the X-direction wind speed component, 'U', reflecting the proof-of-concept nature of the study. The aim is to focus on a single output variable, allowing for a more concise and objective presentation and analysis of the results, ensuring clarity and allowing a direct analysis.

This study limited the number of time steps used to predict the next step to two. This choice aligns with the standards applied by Microsoft Aurora and was adopted for computational efficiency and memory usage considerations [53]. The approach balances predictive accuracy with resource constraints by leveraging a minimal yet effective input size, making it feasible for operational meteorological applications.

Including hourly resolution data, using two time-steps, consistent use of wind and spatial variables, and reduced/optimized variable set to ensure that both short-term and long-term atmospheric phenomena are accurately represented. These decisions collectively aim to enhance the reliability of the proposed techniques while maintaining the computational efficiency necessary for transient CFD models like WRF.

### 3.3. Applied Deep Learning Models

To evaluate the feasibility of using artificial intelligence to improve the initialization process of transient CFD models, two distinct deep learning architectures were tested: a multilayer perceptron (MLP) network and a Fourier neural operator (FNO) network. Each model was chosen for its capabilities to address the specific comparisons carried out in this work.

The multilayer perceptron (MLP) is a feedforward neural network where every neuron in one layer is connected to every neuron in the next layer. It is well-regarded for its simplicity and versatility in learning structured datasets, and it is theorized to serve as a universal function approximator [54]. An incremental approach of the hyperparameters was adopted to determine an optimal configuration. Initially, a zero hidden layer architecture was tested, where the network consisted only of an input layer and an output layer, providing a baseline for direct input-output mapping. Subsequently, the number of neurons in the input and output layers was varied. The best performance was achieved with 1024 neurons.

Once this configuration was established, the number of hidden layers was tested while freezing the number of neurons at 1024 per layer. The results revealed that the best outcomes on the testing dataset occurred when the network had 0 hidden layers, confirming that a shallow network was sufficient to capture the link between the inputs and the target variable (U, the X-axis component of wind velocity). For this setup, the root mean squared error (RMSE) for the testing set was 4.7551 m/s. While this result served as a useful baseline, it will be shown that the FNO significantly outperformed the MLP, demonstrating its suitability for the complex spatiotemporal patterns inherent to the WRF dataset.

The FNO represents a more advanced methodology tailored for learning operators in PDEs. Unlike traditional neural networks, the FNO models the solution operator, allowing it to predict also multiple instances simultaneously [28]. This property makes FNO highly generalizable and capable of capturing a broader range of solution cases, as it is not constrained to a specific instance or condition. In fact, FNO has recently been applied quite successfully in the development of DL solutions for differential equations. However, its use has been evaluated mainly in equations that, despite being very relevant, are famous for having a strong theoretical appeal, as this facilitates obtaining references for benchmarking. This research effort is relevant as it indicates the suitability of FNO for this type of problem. This fact was one of the reasons that motivated us to use it for our proof of concept, which ended up showing how feasible the proposed approach can be. In this sense, in our study, this feature was leveraged by training and testing the FNO across a variety of conditions.

The FNO's hyperparameters were optimized over two rounds (no more than that was needed), where the following parameters were systematically tested in sequence:

1. The size of the final hidden layer;
2. The number of Fourier modes, in powers of 2 (from 1 to 16);
3. The width, defined as the number of transforms applied to the Fourier modes;
4. The number of stacked Fourier layers, also in powers of 2 (from 1 to 16).

Via this iterative optimization process, the best architecture was found to have 1024 neurons in the final hidden layer, 16 Fourier modes, a width of 48, and only one Fourier layer. This configuration achieved a final testing RMSE of 1.2544 m/s, showcasing a significant improvement over the MLP model. The schematics of the used FNO architecture is presented in Figure 2.

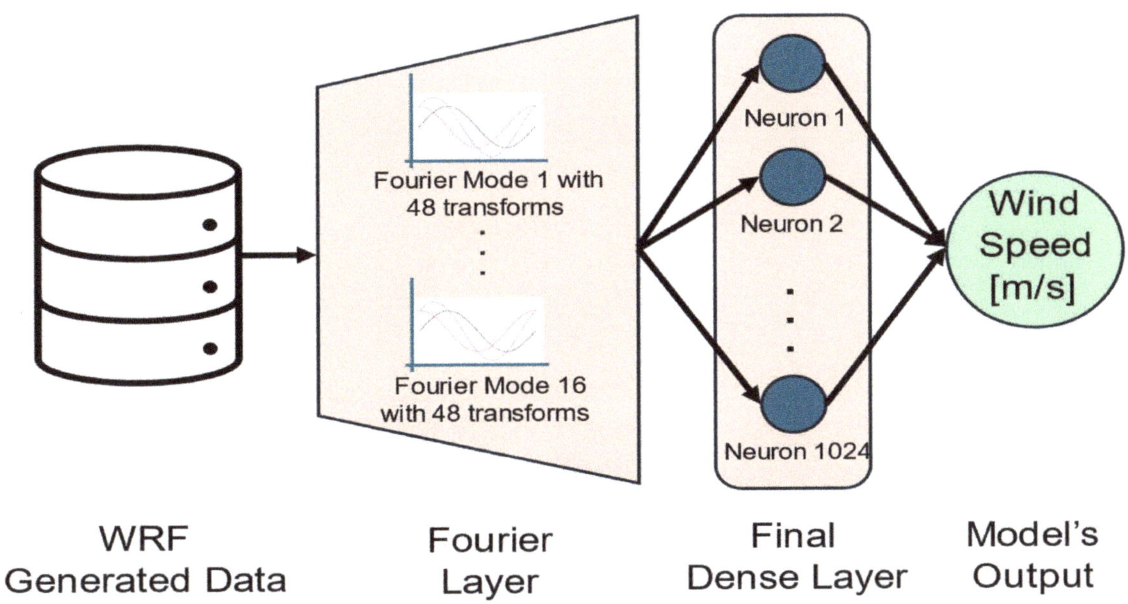


**Figure 2.** The selected FNO architecture.

For this step, the choice of input variables is aimed at maintaining the physical consistency of the system. We started with variables considered relevant to the underlying physics, and the initial dataset was limited to the following:

- 3D variables: U, V, and W (wind speed components in X, Y, and Z directions) and T (temperature);
- 2D variables: latitude–longitude, while U10 and V10 (wind speed components at 10 m above the surface) were used to provide information on surface conditions.

The presented selected auxiliary variables were used as inputs, while the target variable was the X-direction wind speed component (U). The decision to predict a single output variable was to make results more concise since the results obtained are intended to be proof of concept, which can be expanded later to a complete weather forecasting application.

The FNO's reliance on Fourier transforms enables it to efficiently model spatial and temporal interactions, making it particularly well-suited for the distributed and transient nature of WRF data. The results highlight the FNO's potential to reduce computational costs and enhance numerical weather prediction systems by accelerating the convergence of transient CFD models.

### *3.4. Variable Selection*

In this study, the initial approach to variable selection involved applying a single model across all Z-levels, encompassing and predicting all vertical layers in the dataset, what we call here and throughout the remainder of this manuscript "the global model". However, as it will be demonstrated later, this approach can be modified and enhanced by using one model per reference height, totaling 34 individual height levels. This adjustment aligns with the distinct atmospheric conditions at different altitudes and allows for a more tailored and precise modeling approach.

The initial variable set used in the modeling process included U, V, and W (wind speed components in the x-easting, y-northing, and z-elevation directions), T (temperature), and U10 and V10 (wind speed components at 10 m above ground level). This selection is consistent with the variables chosen for the FNO model discussed in the previous subsection.

A systematic, iterative process was implemented to refine the input variables and enhance prediction accuracy. In each round of testing, all available variables from the dataset (but U, V, and W, and latitude–longitude, which were always kept) were considered as potential substitutes for the original set. The process alternated between testing only 2D variables (e.g., surface variables) and testing only 3D variables (e.g., variables across vertical layers).

The two-dimensional (2D) variables' surface-specific characteristics often result in minimal coupling with higher-altitude atmospheric dynamics. Consequently, if the inclusion of a given 2D variable in an iteration failed to yield a contraction in the RMSE, it was systematically excluded from subsequent training rounds to optimize model performance. In contrast, three-dimensional (3D) variables were retained across all iterations, as their incorporation consistently led to RMSE improvements. This persistent enhancement underscores the critical role of vertical atmospheric profiling in accurately resolving the physical mechanisms governing wind flow dynamics within the computational fluid dynamics (CFD) framework.

Finally, the variable set comprising HGT (height, 2D), QVAPOR (water vapor mixing ratio, 3D), and THM (perturbed moist potential temperature, 3D) emerged as the best-performing set for the case of using a single model across all Z-levels. This selection highlighted the relevance of combining surface and vertical atmospheric variables to represent the physical conditions affecting wind flow accurately.

The selection rounds continued until no further improvement was observed in the testing RMSE. This iterative optimization allowed us to determine the most relevant variables for predicting the target output, ensuring that unnecessary inputs were excluded and that computational efficiency was maintained (Figure 3).

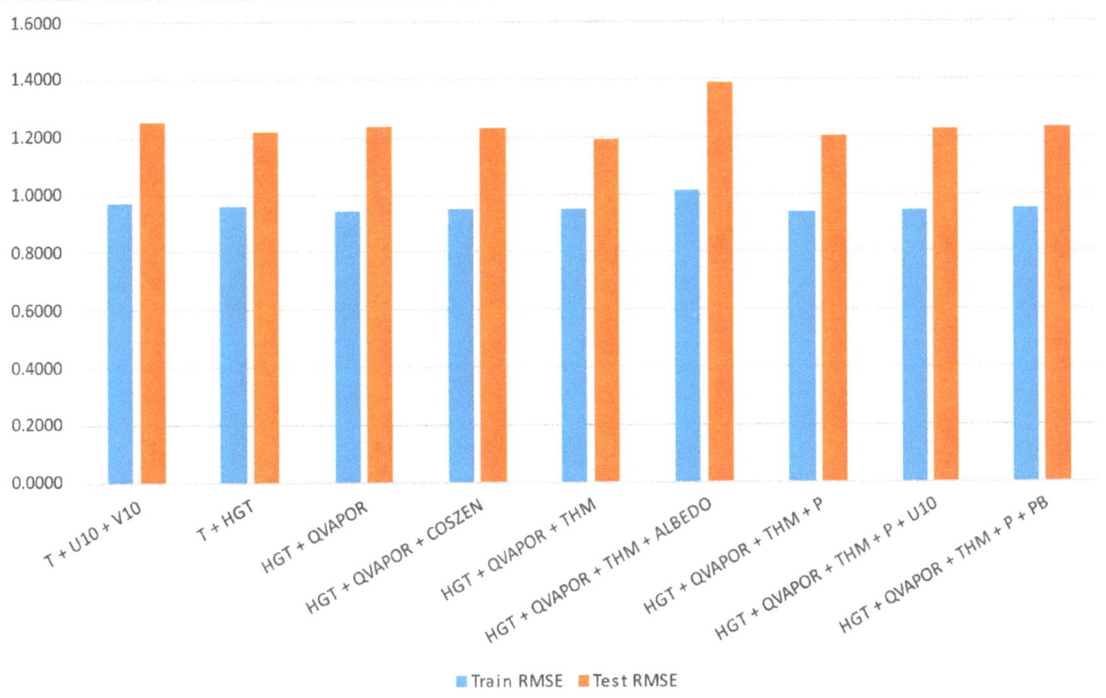


**Figure 3.** Results for the variable selection process using the training and test sets. This step regards the "global model", as explained, the one that predicts U for all the Z-levels simultaneously. Even though the best result was achieved for "HGT" + "QVAPOR" + "THM", all the presented variable sets were used later to assess the localized models developed specifically for each Z-level.

This methodology (condensely presented in Figure 4) aimed to underscore the importance of variable selection in reducing model complexity and improving prediction accuracy using the lowest computer resources. The transition from a universal model across all Z-levels to individualized models per height, combined with an optimized variable set, presented in the next section, forms the foundation for the enhanced results presented in this study.

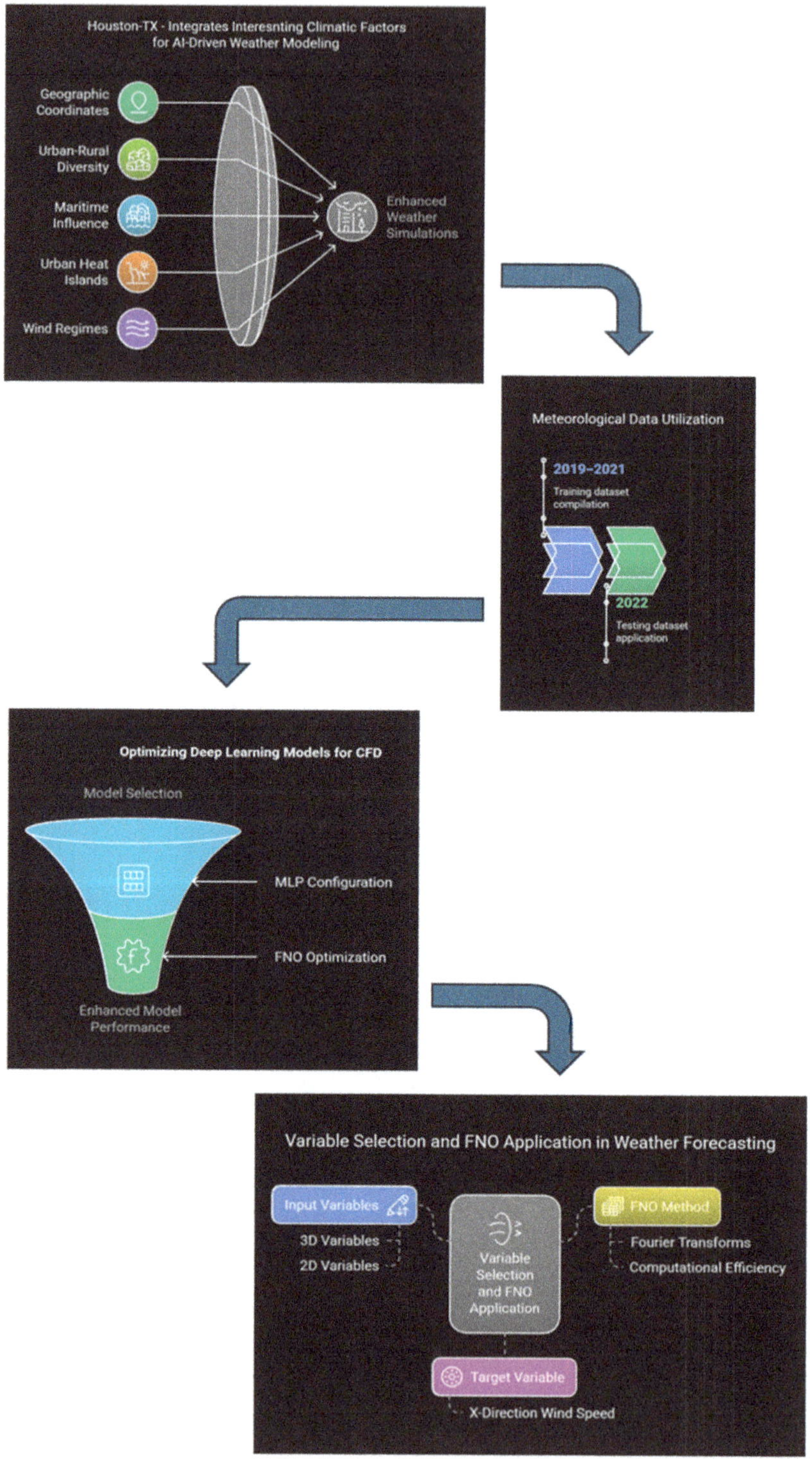


**Figure 4.** Workflow of the applied methodology.

# 4. Results and Discussion

## 4.1. Global Model Performance

To assess the overall accuracy of the proposed approach, the performance of the global model, which simultaneously utilizes and predicts variables across all z-levels, was evaluated using a normalized metric. The RMSE values were normalized by dividing them by the average absolute value of the U velocity component across the entire domain [51]. This normalization facilitates a clearer interpretation of the model's performance by providing a relative measure of error magnitude concerning the predicted variable.

$$nRMSE = \frac{RMSE}{\frac{\sum_i^n |U_i|}{n}} \quad (1)$$

The resulting normalized RMSE (nRMSE) for the global model was 17.15%, indicating the percentage deviation from the expected values. In contrast, the persistence model, which represents the baseline approach commonly used in all CFD transient algorithms (e.g., [55,56]), where the velocity field of the previous time step is directly applied to the next step, achieved a lower nRMSE of 13.69% for the entire 3D domain. This can be explained in a somewhat straightforward way since the global model has to cope with data from several Z-levels simultaneously, which leads the model training process to seek an 'average' solution that approximates the global values of the atmosphere. The persistence model, despite its simplicity, by definition, uses values from the same Z-level, which already gives it a local focus.

To confirm, the performance of the global model was further analyzed at three specific Z-levels: 0 (surface), 16 (middle atmosphere), and 33 (top of atmosphere). The results indicated that the global model did not outperform the persistence approach at any of these levels. At Z-level 0 (surface), the global model's normalized root mean squared error (nRMSE) was 41.93%, whereas the persistence model achieved a lower error of 33.37%. The increased error at the surface is likely attributed to strong gradients caused by irregular terrain, boundary layer effects, and additional constraints imposed by empirical models used to approximate surface interactions, all of which introduce further uncertainties [54].

At Z-level = 16 (middle atmosphere), the nRMSE was 29.84%, while the persistence approach achieved 26.55%. The results at this altitude are closer between the two models, probably because the atmosphere becomes more predictable and stable, reducing variability and making persistence a more practical approach [57].

Finally, at Z-level = 33 (top of atmosphere), the nRMSE was 12.30%, whereas the persistence model reached a significantly lower 6.32%. Even though the global model improved its performance at this height due to the more stable atmospheric conditions, it still did not outperform the persistence model. Furthermore, the improvement observed for the global model at this level was not as significant as at the lower levels, leaving its error almost twice as large as that of persistence.

The results obtained indicate that a single global model does not outperform the persistence model, at least for the optimized FNO model. This suggests that the current approach may not fully capture the dynamics of the atmosphere at different heights simultaneously. Some factors may detail the reasoning behind these results:

1. Lack of height-specific modeling: The persistence model benefits from the natural continuity of the atmospheric flow, whereas the deep learning model needs to generalize across all heights. Studies such as [58] suggest that stratified modeling can significantly enhance predictive performance, where different models are trained for different atmospheric layers.

2. Boundary layer effects: Near-surface wind flows are influenced by topography, land use, and temperature variations, which cause strong local gradients and turbulence [59]. The inability of the global model to properly learn these effects might be a reason why it underperformed compared to persistence at Z-level 0.
3. Predictability of atmospheric layers: The results at Z-level = 16 show that the atmosphere is more predictable at mid-altitudes, which aligns with previous findings that suggest lower turbulence and more stable wind regimes at these heights [60]. This explains why persistence and the global model performed similarly.
4. Impact of training data and physical constraints: The FNO model was trained using PINN principles, yet its ability to fully integrate the governing physics of the atmosphere might still be limited compared to conventional numerical weather prediction (NWP) models. Hybrid approaches that combine FNO architectures with physics-based constraints have shown promise in recent literature [61].

These findings reinforce the need to refine our methodology. A promising alternative would be to apply height-dependent models, where a different FNO network is trained for each height level, ensuring that each model learns the specific characteristics of wind dynamics at its respective altitude. This will be further explored in the next section.

### 4.2. Specific Z-Levels Ad Hoc Models

This section presents the results of training separate models for each Z-level, which significantly improved performance compared to using a single global model for the entire 3D domain. The specialized models outperformed persistence for both the surface (Z = 0) and middle atmosphere (Z = 16). In contrast, at the top of the atmosphere (Z = 33), the results were slightly worse than persistence but still highly competitive.

Figure 5 illustrates these findings, highlighting the impact of height-specific modeling. For the surface level (Z = 0), the optimized FNO model achieved an nRMSE of 27.64%, outperforming the persistence model, which yielded 33.37%. The best-performing variable set for this level consisted of "HGT", "QVAPOR", "THM", and "P". At the middle atmosphere (Z = 16), the FNO model further improved over persistence, with an nRMSE of 20.46% compared to 26.55%, using the variables "HGT", "QVAPOR", and "THM". At the top of the atmosphere (Z = 33), the nRMSE was 7.95% for FNO, while persistence performed slightly better at 6.32%, with the best variable set comprising only "HGT" and "QVAPOR".

It can be perceived that variables with a strong physical appeal, which indicate either concrete physical characteristics ("HGT") or significant portions of physical balances ("QVAPOR", "THM") or even thermodynamic correlations ("THM", "P"), have shown to pursue a strong impact, as they provide concrete information about atmospheric dynamics. Variables that are correlated to these tended not to improve the results. This leads us to infer that they are useful in the WRF model to determine more subtle local/temporal variations but that in the DL paradigm, they may function as autocorrelated predictors.

DL models are highly capable of finding subtle relationships between variables as long as the most appropriate ones are selected. This reiterates the good performance of models developed specifically for each Z-level.

Even though this is a proof of concept, and having in mind that the purpose was to verify the ability of DL models to be used as surrogates for the initial iteration of a transient CFD process, the interpretation of the results can be extended, especially to extreme conditions and more complex regions. This is because, since the best performance gains were achieved in the most challenging situations, from the mid-atmosphere to the surface, it can be inferred that the proposed approach has a high potential to be successful in more adverse situations.

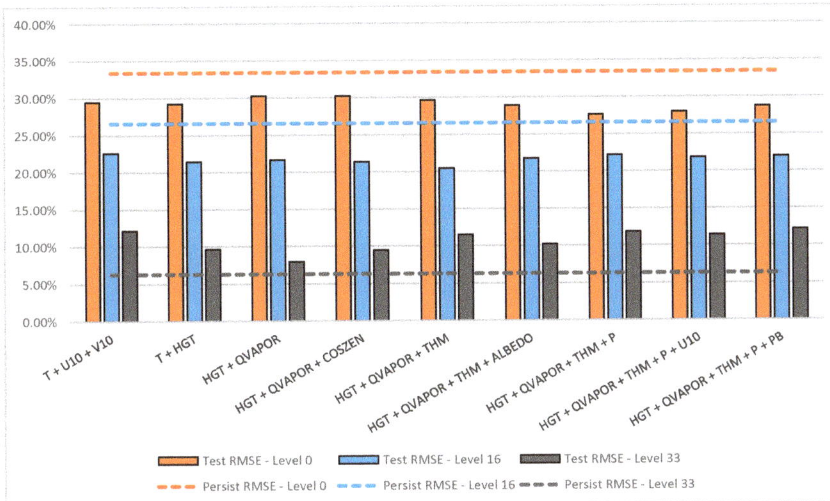


**Figure 5.** Normalized RMSE (nRMSE) Comparison for Z-level models (bars) vs. persistence (dotted lines). The colors of the bars match the colors of the lines for each case (Z-level).

These results align with previous atmospheric modeling and machine learning findings for weather prediction. Some authors emphasize that near-surface wind prediction is particularly challenging due to the influence of orography, boundary layer turbulence, and localized surface conditions [62,63]. The improvement seen at Z = 0 with the ad hoc model can be attributed to better handling of these complex near-surface interactions, as the inclusion of pressure (P) appears to have contributed to refining the model's predictions. In contrast, at Z = 16, the atmosphere is already more stable, which reduces the impact of localized disturbances, making the FNO model even more effective.

At the top of the atmosphere, the slight underperformance compared to persistence suggests that the model is not yet fully capturing the larger-scale atmospheric dynamics that dominate at these heights. However, the small difference in error indicates that the FNO approach remains competitive. Similar findings have been reported by [31], who showed that machine learning models tend to struggle with upper-atmosphere predictions unless explicitly trained with large-scale flow features.

Another key insight from this analysis is that, as altitude increases, fewer variables are needed to achieve optimal results. This observation is probably due to the atmosphere's increased stability and the lower influence of complex terrain interactions at higher altitudes. This trend is consistent with studies on numerical weather prediction, which indicate that lower atmospheric levels require more detailed parameterizations due to their interaction with the Earth's surface, while higher-altitude predictions benefit primarily from large-scale flow patterns [64].

Overall, the results confirm that height-specific modeling provides a significant advantage in transient wind field predictions. The FNO-based approach, when applied separately for each Z-level, successfully reduced prediction errors and demonstrated strong potential for improving atmospheric simulations over conventional persistence models.

Figures 6–8 illustrate the case of a specific time step for each studied Z-level (respectively 0, 16, and 33). It visually assesses the FNO model's performance at distinct altitudes. As observed, the model's accuracy improves with increasing height, aligning with findings from similar studies. Nevertheless, it can be seen how the color patterns of the modeled fields properly reproduce the ground truth pattern.

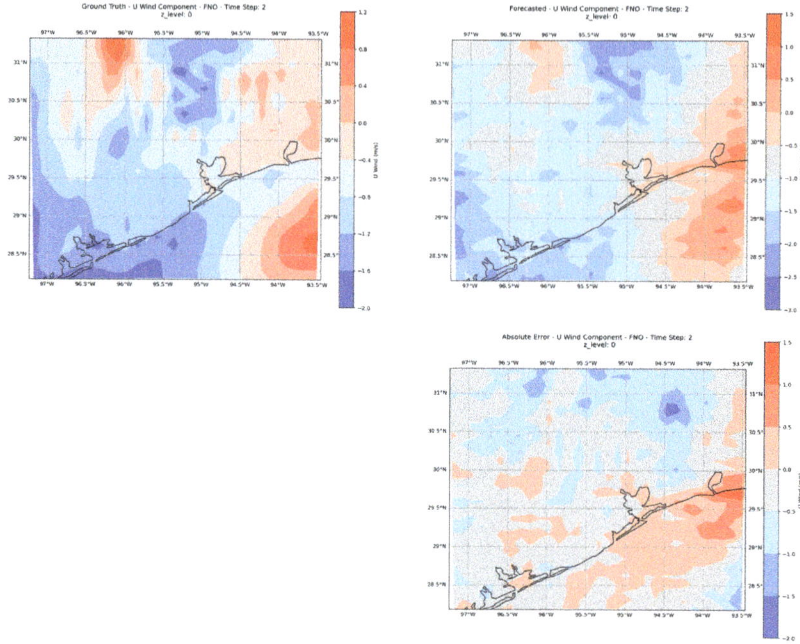


**Figure 6.** Example of the "U" wind speed component field for Z-level = 0 (earth's surface), specifically for the first modeled time step (2, since time steps 0 and 1 must be used as inputs). The ground truth is presented on the (**top-left**) plot, the (**top-right**) plot is the modeled field, and their difference is shown on the (**bottom-right**) plot.

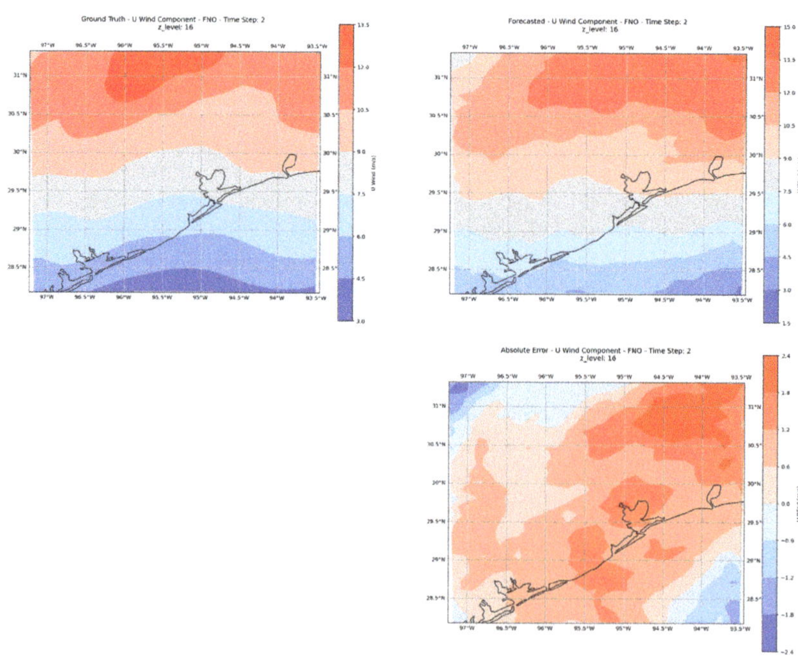


**Figure 7.** Same as Figure 6, for Z-level = 16 (mid-atmosphere).

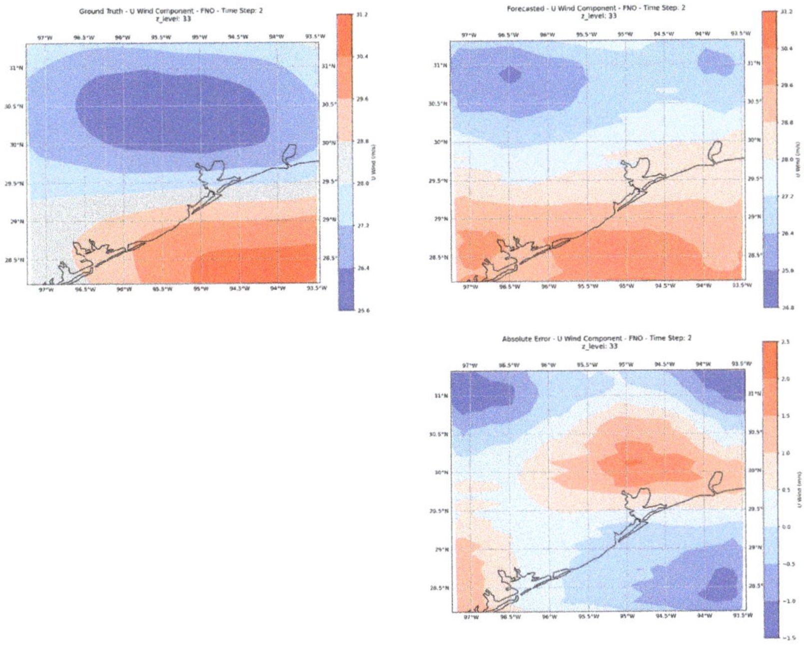


**Figure 8.** Same as Figure 6, for Z-level = 33 (top-of-atmosphere).

To close the comparison, Figure 9 illustrates the variation of the nRMSE across the WRF atmospheric levels, comparing the performance of the persistence model with the ad hoc models developed for each level. The results indicate that the ad hoc models consistently outperform the persistence approach from the surface up to the mid-atmosphere. Beyond level 20, the performance of both approaches becomes comparable, and at the top of the atmosphere, the discrepancies are negligible. The proposed methodology is particularly advantageous as it significantly reduces the error in most of the atmosphere, especially in regions with more intense variations, where CFD models will have a higher computational load to achieve convergence. This further highlights the importance of employing the developed models instead of relying on a simple persistence approach.

Similar results may be found in the literature. For instance, in [65], the FNO model demonstrated varying performance across distinct urban layouts and wind scenarios. While the results indicated that the FNO model could offer accurate prediction and achieve a remarkable 99% reduction in the required computational time, its ability to generalize to new wind directions proved challenging. This suggests that the FNO model's performance is shaped by the complexity of the environment and the specific conditions under which it is applied.

Similarly, in [66], the authors investigated the application of machine learning models to predict wind speeds at higher altitudes based on data from lower levels. Their findings revealed that wind speed values at higher altitudes are directly influenced by the corresponding value at lower altitudes for the same geographical location. This indicates that machine learning models can adequately capture altitude-dependent relationships, leading to accurate predictions at higher levels.

These studies show that the FNO model's performance improves with altitude, likely due to the decreasing complexity and increased predictability of atmospheric conditions at higher levels. The visual evidence provided in Figure 5 underscores this trend, highlighting the model's enhanced accuracy as height increases.

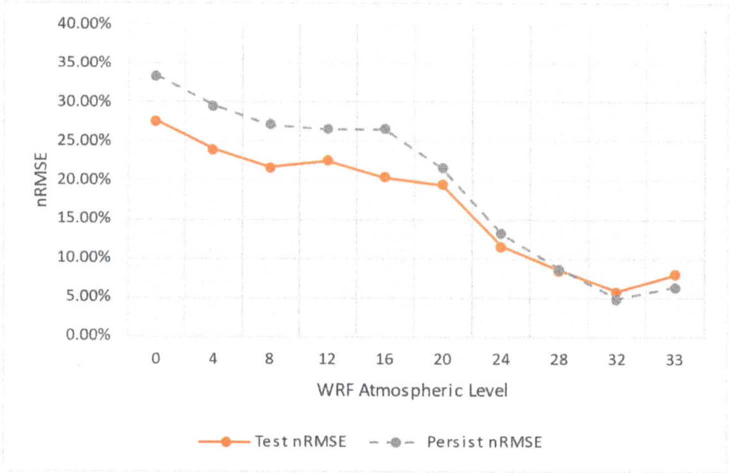


**Figure 9.** Comparison of the performance of the developed models against persistence, for several Z-levels, from the surface to the top-of-atmosphere.

Other studies have shown the feasibility of Fourier-based approaches for atmospheric modeling. The model proposed by Pathak et al. [67] was built upon a Fourier neural network, used reanalysis data from the ERA5 dataset for training and validation, and successfully predicted wind speed for both land and sea regions. Unlike our study, however, their work found that increasing the number of attributes returned improved results by the model. This might be attributed to the fact that their model seeks a more general approach than ours and would benefit further from more spatiotemporal information provided by the input attributes. Contrarywise, the ad hoc models proposed in our study are tailored for specific atmospheric interactions, thus requiring different input attributes for each scenario, as the scenario's complexity decreases with higher altitudes.

In another study [68], the authors developed the Local-FNO model for modeling the microclimate of their study site. Their results highlighted that the model using local information for the location reached better outcomes than the vanilla FNO approach for wind dynamics modeling for different building geometries. Their findings corroborate the superior efficiency of the ad hoc methodologies presented in this work on capturing local patterns.

The following Table 3 compiles the works of the literature found that are related to the present study.

**Table 3.** Literature found work for atmospheric modeling using ML.

| Model | Methodology | References |
|---|---|---|
| FNO | Implementation of an FNO model for forecasting wind in different urban configurations and wind scenarios | [65] |
| Support Vector Machine, K-Nearest Neighbor, and Gradient Boosting Machine | Implementation of different traditional machine learning approaches for wind speed forecasting using data collected from different altitudes. | [66] |
| FourCastNet | Implementation of a Fourier-based approach to forecast atmospheric variables, including wind speed, precipitation, and water vapor, using deep learning. | [67] |
| Local-FNO | Precise modeling of urban microclimates, including wind velocity and temperature. | [68] |

## 5. Conclusions

This study developed and investigated a proof of concept on applying deep learning techniques for wind speed prediction using a Fourier neural operator (FNO) and a multilayer perceptron (MLP). The models were trained and tested with WRF-generated data for Houston from 2018 to 2022, focusing on predicting the X-direction wind speed component. The MLP was used primarily as a baseline, achieving a testing RMSE of 4.75 m/s, while the optimized FNO significantly outperformed it, reducing the RMSE to 1.25 m/s.

We initially trained a single global model across all Z-levels, which, despite its advanced architecture, still did not outperform the persistence model for wind prediction. Normalized RMSE values revealed that the global model reached an nRMSE of 17.15%, compared to 13.69% for the persistence model.

To improve upon these results, we implemented and tested Z-level-specific models, significantly enhancing the performance at lower altitudes. At the surface (Z = 0), our optimized FNO model achieved an nRMSE of 27.64%, outperforming persistence at 33.37%. In the middle atmosphere (Z = 16), the nRMSE improved to 20.46%, compared to 26.55% for persistence. For the top of the atmosphere (Z = 33), although persistence still outperformed our model (6.32% vs. 7.95%), the results remained highly competitive, which can help implement a fully automated process. This process shall benefit from using the model output to start the iterative process of the transient CFD solution, which is slow and has a high computational demand. This will require fewer iterations, optimizing processing time and its associated financial costs.

Variable selection played a relevant role in model optimization. While all models always used latitude-longitude data and wind components, we tested auxiliary variables to pinpoint the most pertinent attributes for each height. The optimal feature sets varied across Z-levels, suggesting that different physical phenomena dominate wind behavior at different heights. At lower levels, the inclusion of HGT, QVAPOR, THM, and pressure yielded the best results, while higher-altitude models required fewer input variables, likely due to the increasing stability of the atmosphere.

The study case provided visual confirmation of model performance, reinforcing that FNO's effectiveness improves with altitude. This aligns with previous studies demonstrating that machine learning-based wind forecasting benefits from stratifying models based on atmospheric layers.

A key takeaway from this study is the potential of the FNO model to learn complex spatial and temporal dependencies in atmospheric fields, offering a data-driven alternative to conventional numerical simulations. The findings also underscore the importance of tailoring machine learning architectures and input features to the specific atmospheric characteristics at different altitude levels.

Despite these promising results that validate the proof of concept, integrating the FNO model into operational weather forecasting systems presents challenges. A major issue is the computational cost of training deep learning models on large-scale meteorological datasets. Although inference times are significantly faster than those of traditional numerical models, the requirement for high-performance computing infrastructure may pose a barrier. Furthermore, ensuring the model's generalization across diverse weather conditions involves continuous data assimilation and periodic retraining with updated datasets. Possible solutions include hybrid approaches that combine physics-based simulations with deep learning predictions, the application of transfer learning to adapt pre-trained models to various climate regimes, and utilizing cloud-based infrastructures to efficiently scale computational resources. Addressing these challenges will be essential to unlocking the full potential of deep learning for atmospheric modeling and real-time forecasting applications.

While the FNO model shows promise, it has limitations that must be considered for broader applicability. One key challenge is its ability to accurately predict extreme weather events, which often involve highly nonlinear dynamics and rapid atmospheric changes that may not be well captured by the learned solution operator. Additionally, the model's performance in regions with complex topography remains a concern, as steep terrain can introduce strong local gradients and boundary effects that require high-resolution input data and specialized handling. Future work should focus on enhancing the model's robustness through techniques such as adaptive resolution strategies, hybrid physics-informed approaches, and the incorporation of additional observational data to improve predictive accuracy in these challenging scenarios.

Overall, this research highlights the potential of deep learning for atmospheric forecasting, emphasizing the need for height-specific models. Future work may explore extending the findings presented here by applying our approach to other variables. Furthermore, hybrid methods that combine deep learning with domain-specific physics constraints may be tested, along with extending the methodology to more diverse atmospheric conditions and locations.

Finally, the results indicate the feasibility of using DL models to start the transient CFD iterative process. Once this possibility is demonstrated, additional tests can be performed to quantitatively evaluate the computational costs of this solution, which, due to its scope and implementation complexity, becomes a potential research object. This type of result has the potential to lead to cheaper and faster weather simulations, which is a pressing demand in our modern society.

**Author Contributions:** Conceptualization, P.A.C.R. and J.V.G.T.; methodology, P.A.C.R. and J.V.G.T.; software, P.A.C.R.; validation, P.A.C.R., J.V.G.T. and B.G.; formal analysis, P.A.C.R. and J.V.G.T.; investigation, P.A.C.R., J.V.G.T. and B.G.; resources, J.V.G.T. and B.G.; data curation, P.A.C.R. and J.V.G.T.; writing—original draft preparation, V.O.S., P.A.C.R. and J.V.G.T.; writing—review and editing, V.O.S., P.A.C.R., J.V.G.T. and B.G.; visualization, V.O.S. and P.A.C.R.; supervision, P.A.C.R., J.V.G.T. and B.G.; project administration, J.V.G.T. and B.G.; funding acquisition, J.V.G.T. and B.G. All authors have read and agreed to the published version of the manuscript.

**Funding:** This research study was funded by the Natural Sciences and Engineering Research Council of Canada (NSERC) Alliance, grant No. 401643, in association with Lakes Environmental Software Inc., and by the Conselho Nacional de Desenvolvimento Científico e Tecnológico—Brasil (CNPq), grant No. 303585/2022-6.

**Data Availability Statement:** The raw data supporting the conclusions of this article will be made available by the authors upon request.

**Conflicts of Interest:** The author, Jesse Van Griensven Thé, is employed by the company Lakes Environmental. The remaining authors declare that this research was conducted in the absence of any commercial or financial relationships that could be construed as potential conflicts of interest.

**List of Abbreviations:**

| | |
|---|---|
| AI | artificial intelligence |
| CFD | computational fluid dynamics |
| DL | deep learning |
| ECMWF | European Centre for Medium-Range Weather Forecasts |
| ERA5 | ECMWF Reanalysis v5 |
| FNO | Fourier neural operator |
| MLP | multilayer perceptron |
| NCAR | National Center for Atmospheric Research |
| nRMSE | normalized root mean squared error |
| NWP | numerical weather prediction |

| | |
|---|---|
| ML | machine learning |
| ODE | ordinary differential equation |
| OpenFOAM | open field operation and manipulation |
| PDE | partial differential equation |
| PINN | physics-informed neural network |
| RMSE | root mean squared error |
| WRF | weather research and forecasting |

## References


1. Cabrera, D.; Quinteros, M.; Cerrada, M.; Sánchez, R.-V.; Guallpa, M.; Sancho, F.; Li, C. Rainfall Forecasting Using a Bayesian Framework and Long Short-Term Memory Multi-Model Estimation Based on an Hourly Meteorological Monitoring Network. Case of Study: Andean Ecuadorian Tropical City. *Earth Sci. Inform.* **2023**, *16*, 1373–1388. [CrossRef]
2. Tahir Bahadur, F.; Rasool Shah, S.; Rao Nidamanuri, R. Air Pollution Monitoring, and Modelling: An Overview. *Environ. Forensics* **2024**, *25*, 309–336. [CrossRef]
3. Roy, C.; Rahman, M.R.; Ghosh, M.K.; Biswas, S. Tropical Cyclone Intensity Forecasting in the Bay of Bengal Using a Biologically Inspired Computational Model. *Model. Earth Syst. Environ.* **2024**, *10*, 523–537. [CrossRef]
4. Singh, H.; Ang, L.-M.; Lewis, T.; Paudyal, D.; Acuna, M.; Srivastava, P.K.; Srivastava, S.K. Trending and Emerging Prospects of Physics-Based and ML-Based Wildfire Spread Models: A Comprehensive Review. *J. For. Res.* **2024**, *35*, 135. [CrossRef]
5. Kumar, V.; Sharma, K.V.; Caloiero, T.; Mehta, D.J.; Singh, K. Comprehensive Overview of Flood Modeling Approaches: A Review of Recent Advances. *Hydrology* **2023**, *10*, 141. [CrossRef]
6. Wu, J.; Wang, Z.; Dong, J.; Cui, X.; Tao, S.; Chen, X. Robust Runoff Prediction with Explainable Artificial Intelligence and Meteorological Variables From Deep Learning Ensemble Model. *Water Resour. Res.* **2023**, *59*, e2023WR035676. [CrossRef]
7. Mitreska Jovanovska, E.; Batz, V.; Lameski, P.; Zdravevski, E.; Herzog, M.A.; Trajkovik, V. Methods for Urban Air Pollution Measurement and Forecasting: Challenges, Opportunities, and Solutions. *Atmosphere* **2023**, *14*, 1441. [CrossRef]
8. Kalajdjieski, J.; Trivodaliev, K.; Mirceva, G.; Kalajdziski, S.; Gievska, S. A Complete Air Pollution Monitoring and Prediction Framework. *IEEE Access* **2023**, *11*, 88730–88744. [CrossRef]
9. Federico, S.; Torcasio, R.C.; Popova, J.; Sokol, Z.; Pop, L.; Lagasio, M.; Lynn, B.H.; Puca, S.; Dietrich, S. Improving the Lightning Forecast with the WRF Model and Lightning Data Assimilation: Results of a Two-Seasons Numerical Experiment over Italy. *Atmos. Res.* **2024**, *304*, 107382. [CrossRef]
10. Paul, D.J.; Janani, S.P.; Ancy Jenifer, J. Advanced Weather Monitoring and Disaster Mitigation System. In Proceedings of the 2024 International Conference on Cognitive Robotics and Intelligent Systems (ICC—ROBINS), Coimbatore, India, 17–19 April 2024; pp. 540–546.
11. Leung, X.Y.; Islam, R.M.; Adhami, M.; Ilic, D.; McDonald, L.; Palawaththa, S.; Diug, B.; Munshi, S.U.; Karim, M.N. A Systematic Review of Dengue Outbreak Prediction Models: Current Scenario and Future Directions. *PLoS Negl. Trop. Dis.* **2023**, *17*, e0010631. [CrossRef]
12. Dieng, M.D.B.; Tompkins, A.M.; Arnault, J.; Sié, A.; Fersch, B.; Laux, P.; Schwarz, M.; Zabré, P.; Munga, S.; Khagayi, S.; et al. Process-Based Atmosphere-Hydrology-Malaria Modeling: Performance for Spatio-Temporal Malaria Transmission Dynamics in Sub-Saharan Africa. *Water Resour. Res.* **2024**, *60*, e2023WR034975. [CrossRef]
13. Abrego-Perez, A.L.; Pacheco-Carvajal, N.; Diaz-Jimenez, M.C. Forecasting Agricultural Financial Weather Risk Using PCA and SSA in an Index Insurance Model in Low-Income Economies. *Appl. Sci.* **2023**, *13*, 2425. [CrossRef]
14. Costa Rocha, P.A.; Oliveira Santos, V.; Scott, J.; Van Griensven Thé, J.; Gharabaghi, B. Application of Graph Neural Networks to Forecast Urban Flood Events: The Case Study of the 2013 Flood of the Bow River, Calgary, Canada. *Int. J. River Basin Manag.* **2024**, 1–18. [CrossRef]
15. Luo, Z.; Liu, J.; Zhang, S.; Shao, W.; Zhang, L. Research on Climate Change in Qinghai Lake Basin Based on WRF and CMIP6. *Remote Sens.* **2023**, *15*, 4379. [CrossRef]
16. Kadaverugu, R. A Comparison between WRF-Simulated and Observed Surface Meteorological Variables across Varying Land Cover and Urbanization in South-Central India. *Earth Sci. Inform.* **2023**, *16*, 147–163. [CrossRef]
17. Chen, L.; Chen, Z.; Zhang, Y.; Liu, Y.; Osman, A.I.; Farghali, M.; Hua, J.; Al-Fatesh, A.; Ihara, I.; Rooney, D.W.; et al. Artificial Intelligence-Based Solutions for Climate Change: A Review. *Environ. Chem. Lett.* **2023**, *21*, 2525–2557. [CrossRef]
18. Powers, J.G.; Klemp, J.B.; Skamarock, W.C.; Davis, C.A.; Dudhia, J.; Gill, D.O.; Coen, J.L.; Gochis, D.J.; Ahmadov, R.; Peckham, S.E.; et al. The Weather Research and Forecasting Model: Overview, System Efforts, and Future Directions. *Bull. Am. Meteorol. Soc.* **2017**, *98*, 1717–1737. [CrossRef]
19. Skamarock, W.C.; Klemp, J.B.; Dudhia, J.; Gill, D.O.; Barker, D.M.; Duda, M.G.; Huang, X.-Y.; Wang, W.; Powers, J.G. A Description of the Advanced Research WRF Version 3. *NCAR Tech. Note* **2008**, *475*, 10–5065. [CrossRef]

20. Sosa-Tinoco, I.; Prósper, M.A.; Miguez-Macho, G. Development of a Solar Energy Forecasting System for Two Real Solar Plants Based on WRF Solar with Aerosol Input and a Solar Plant Model. *Sol. Energy* **2022**, *240*, 329–341. [CrossRef]
21. Sarvestan, R.; Karami, M.; Javidi Sabbaghian, R. Evaluation of Meteorological Microphysical Schemas Based on the WRF Model for Simulation of Rainfall in the Northeastern Region of Iran. *J. Hydrol. Reg. Stud.* **2023**, *50*, 101524. [CrossRef]
22. Stathopoulos, C.; Chaniotis, I.; Patlakas, P. Assimilating Aeolus Satellite Wind Data on a Regional Level: Application in a Mediterranean Cyclone Using the WRF Model. *Atmosphere* **2023**, *14*, 1811. [CrossRef]
23. Hallaji, H.; Bohloul, M.R.; Peyghambarzadeh, S.M.; Azizi, S. Measurement of Air Pollutants Concentrations from Stacks of Petrochemical Company and Dispersion Modeling by AERMOD Coupled with WRF Model. *Int. J. Environ. Sci. Technol.* **2023**, *20*, 7217–7236. [CrossRef]
24. López-Noreña, A.I.; Berná, L.; Tames, M.F.; Millán, E.N.; Puliafito, S.E.; Fernandez, R.P. Influence of Emission Inventory Resolution on the Modeled Spatio-Temporal Distribution of Air Pollutants in Buenos Aires, Argentina, Using WRF-Chem. *Atmos. Environ.* **2022**, *269*, 118839. [CrossRef]
25. Qi, Q.; Wang, S.; Zhao, H.; Kota, S.H.; Zhang, H. Rice Yield Losses Due to O3 Pollution in China from 2013 to 2020 Based on the WRF-CMAQ Model. *J. Clean. Prod.* **2023**, *401*, 136801. [CrossRef]
26. Carneiro, F.O.M.; Moura, L.F.M.; Costa Rocha, P.A.; Pontes Lima, R.J.; Ismail, K.A.R. Application and Analysis of the Moving Mesh Algorithm AMI in a Small Scale HAWT: Validation with Field Test's Results against the Frozen Rotor Approach. *Energy* **2019**, *171*, 819–829. [CrossRef]
27. Ajani, C.K.; Zhu, Z.; Sun, D.-W. Recent Advances in Multiscale CFD Modelling of Cooling Processes and Systems for the Agrifood Industry. *Crit. Rev. Food Sci. Nutr.* **2021**, *61*, 2455–2470. [CrossRef]
28. Costa Rocha, P.A.; Johnston, S.J.; Oliveira Santos, V.; Aliabadi, A.A.; Thé, J.V.G.; Gharabaghi, B. Deep Neural Network Modeling for CFD Simulations: Benchmarking the Fourier Neural Operator on the Lid-Driven Cavity Case. *Appl. Sci.* **2023**, *13*, 3165. [CrossRef]
29. Kehrein, P.; Van Loosdrecht, M.; Osseweijer, P.; Posada, J.; Dewulf, J. The SPPD-WRF Framework: A Novel and Holistic Methodology for Strategical Planning and Process Design of Water Resource Factories. *Sustainability* **2020**, *12*, 4168. [CrossRef]
30. Pasche, O.C.; Wider, J.; Zhang, Z.; Zscheischler, J.; Engelke, S. Validating Deep Learning Weather Forecast Models on Recent High-Impact Extreme Events. *Artif. Intell. Earth Syst.* **2025**, *4*, e240033. [CrossRef]
31. Zhang, H.; Liu, Y.; Zhang, C.; Li, N. Machine Learning Methods for Weather Forecasting: A Survey. *Atmosphere* **2025**, *16*, 82. [CrossRef]
32. Bonavita, M. On Some Limitations of Current Machine Learning Weather Prediction Models. *Geophys. Res. Lett.* **2024**, *51*, e2023GL107377. [CrossRef]
33. Cai, S.; Wang, Z.; Wang, S.; Perdikaris, P.; Karniadakis, G.E. Physics-Informed Neural Networks for Heat Transfer Problems. *J. Heat Transf.* **2021**, *143*, 060801. [CrossRef]
34. Osorio, J.D.; Florio, M.D.; Hovsapian, R.; Chryssostomidis, C.; Karniadakis, G.E. Physics-Informed Machine Learning for Solar-Thermal Power Systems. *Energy Convers. Manag.* **2025**, *327*, 119542. [CrossRef]
35. McClenny, L.D.; Braga-Neto, U.M. Self-Adaptive Physics-Informed Neural Networks. *J. Comput. Phys.* **2023**, *474*, 111722. [CrossRef]
36. Zou, Z.; Meng, X.; Karniadakis, G.E. Uncertainty Quantification for Noisy Inputs–Outputs in Physics-Informed Neural Networks and Neural Operators. *Comput. Methods Appl. Mech. Eng.* **2025**, *433*, 117479. [CrossRef]
37. Fang, Z. A High-Efficient Hybrid Physics-Informed Neural Networks Based on Convolutional Neural Network. *IEEE Trans. Neural Netw. Learn. Syst.* **2022**, *33*, 5514–5526. [CrossRef]
38. Beck, H.E.; Zimmermann, N.E.; McVicar, T.R.; Vergopolan, N.; Berg, A.; Wood, E.F. Present and Future Köppen-Geiger Climate Classification Maps at 1-Km Resolution. *Sci. Data* **2018**, *5*, 180214. [CrossRef] [PubMed]
39. Kyropoulou, M. Thermal Comfort and Indoor Air Quality in Higher Education: A Case Study in Houston, TX, during Mid-Season. *J. Phys. Conf. Ser.* **2023**, *2600*, 102022. [CrossRef]
40. Google LLC Google Earth. Available online: https://earth.google.com/earth/d/1QR_q78kghHqoVdYX6M2lsY-F3mK0wmkp?usp=sharing (accessed on 28 January 2025).
41. Li, X.; Fu, D.; Nielsen-Gammon, J.; Gangrade, S.; Kao, S.-C.; Chang, P.; Morales Hernández, M.; Voisin, N.; Zhang, Z.; Gao, H. Impacts of Climate Change on Future Hurricane Induced Rainfall and Flooding in a Coastal Watershed: A Case Study on Hurricane Harvey. *J. Hydrol.* **2023**, *616*, 128774. [CrossRef]
42. Chang, H.; Ross, A.R. *Climate Change, Urbanization, and Water Resources: Towards Resilient Urban Water Resource Management*; Springer International Publishing: Cham, Switzerland, 2024; ISBN 978-3-031-49629-5.
43. Chakraborty, T.; Hsu, A.; Manya, D.; Sheriff, G. Disproportionately Higher Exposure to Urban Heat in Lower-Income Neighborhoods: A Multi-City Perspective. *Environ. Res. Lett.* **2019**, *14*, 105003. [CrossRef]
44. Mohammad Harmay, N.S.; Choi, M. The Urban Heat Island and Thermal Heat Stress Correlate with Climate Dynamics and Energy Budget Variations in Multiple Urban Environments. *Sustain. Cities Soc.* **2023**, *91*, 104422. [CrossRef]

45. Darby, L.S. Cluster Analysis of Surface Winds in Houston, Texas, and the Impact of Wind Patterns on Ozone. *J. Appl. Meteorol.* **2005**, *44*, 1788–1806. [CrossRef]
46. Oliveira Santos, V.; Costa Rocha, P.A.; Scott, J.; Van Griensven Thé, J.; Gharabaghi, B. Spatiotemporal Air Pollution Forecasting in Houston-TX: A Case Study for Ozone Using Deep Graph Neural Networks. *Atmosphere* **2023**, *14*, 308. [CrossRef]
47. Patel, Y. Levelized Cost of Repurposing Oil and Gas Infrastructure for Clean Energy in the Gulf of Mexico. *Renew. Sustain. Energy Rev.* **2025**, *209*, 115115. [CrossRef]
48. Lamer, K.; Mages, Z.; Treserras, B.P.; Walter, P.; Zhu, Z.; Rapp, A.D.; Nowotarski, C.J.; Brooks, S.D.; Flynn, J.; Sharma, M.; et al. Spatially Distributed Atmospheric Boundary Layer Properties in Houston—A Value-Added Observational Dataset. *Sci. Data* **2024**, *11*, 661. [CrossRef]
49. Oliveira Santos, V.; Costa Rocha, P.A.; Thé, J.V.G.; Gharabaghi, B. Optimizing the Architecture of a Quantum–Classical Hybrid Machine Learning Model for Forecasting Ozone Concentrations: Air Quality Management Tool for Houston, Texas. *Atmosphere* **2025**, *16*, 255. [CrossRef]
50. Baïle, R.; Muzy, J.-F. Leveraging Data from Nearby Stations to Improve Short-Term Wind Speed Forecasts. *Energy* **2023**, *263*, 125644. [CrossRef]
51. Hastie, T.; Tibshirani, R.; Friedman, J. *The Elements of Statistical Learning: Data Mining, Inference, and Prediction*, 2nd ed.; Springer Science & Business Media: Berlin/Heidelberg, Germany, 2009; ISBN 978-0-387-84858-7.
52. Deisenroth, M.P.; Faisal, A.A.; Ong, C.S. *Mathematics for Machine Learning*; Cambridge University Press: Cambridge, UK, 2020; ISBN 978-1-108-47004-9.
53. Bodnar, C.; Bruinsma, W.P.; Lucic, A.; Stanley, M.; Vaughan, A.; Brandstetter, J.; Garvan, P.; Riechert, M.; Weyn, J.A.; Dong, H.; et al. A Foundation Model for the Earth System. *arXiv* **2024**. [CrossRef]
54. James, G.; Witten, D.; Hastie, T.; Tibshirani, R.; Taylor, J. *An Introduction to Statistical Learning: With Applications in Python*; Springer International Publishing: Berlin/Heidelberg, Germany, 2023; ISBN 978-3-031-38746-3.
55. HOLTON, J.R. An Introduction to Dynamic Meteorology. *Introd. Dyn. Meteorol.* **1992**, *48*, 1–497. [CrossRef]
56. Johnson, P.L.; Wilczek, M. Multiscale Velocity Gradients in Turbulence. *Annu. Rev. Fluid Mech.* **2024**, *56*, 463–490. [CrossRef]
57. Holton, J.R. *An Introduction to Dynamic Meteorology*; Academic Press: Cambridge, MA, USA, 2004; ISBN 978-0-12-354015-7.
58. Ranade, R.; Hill, C.; Pathak, J. DiscretizationNet: A Machine-Learning Based Solver for Navier–Stokes Equations Using Finite Volume Discretization. *Comput. Methods Appl. Mech. Eng.* **2021**, *378*, 113722. [CrossRef]
59. Cant, S. *S. B. Pope, Turbulent Flows*, Cambridge University Press, Cambridge, U.K., 2000, 771 pp. *Combust. Flame* **2001**, *125*, 1361–1362. [CrossRef]
60. Stevens, B.; Bony, S. What Are Climate Models Missing? *Science* **2013**, *340*, 1053–1054. [CrossRef]
61. Karniadakis, G.E.; Kevrekidis, I.G.; Lu, L.; Perdikaris, P.; Wang, S.; Yang, L. Physics-Informed Machine Learning. *Nat. Rev. Phys.* **2021**, *3*, 422–440. [CrossRef]
62. Bochenek, B.; Ustrnul, Z. Machine Learning in Weather Prediction and Climate Analyses—Applications and Perspectives. *Atmosphere* **2022**, *13*, 180. [CrossRef]
63. Hoolohan, V.; Tomlin, A.S.; Cockerill, T. Improved near Surface Wind Speed Predictions Using Gaussian Process Regression Combined with Numerical Weather Predictions and Observed Meteorological Data. *Renew. Energy* **2018**, *126*, 1043–1054. [CrossRef]
64. Gentine, P.; Pritchard, M.; Rasp, S.; Reinaudi, G.; Yacalis, G. Could Machine Learning Break the Convection Parameterization Deadlock? *Geophys. Res. Lett.* **2018**, *45*, 5742–5751. [CrossRef]
65. Chen, C.; Tian, G.; Qin, S.; Yang, S.; Geng, D.; Zhan, D.; Yang, J.; Vidal, D.; Wang, L.L. Generalization of Urban Wind Environment Using Fourier Neural Operator Across Different Wind Directions and Cities. *arXiv* **2025**, arXiv:2501.05499.
66. Valsaraj, P.; Thumba, D.A.; Kumar, S. Machine Learning-Based Simplified Methods Using Shorter Wind Measuring Masts for the Time Ahead Wind Forecasting at Higher Altitude for Wind Energy Applications. *Renew. Energy Environ. Sustain.* **2022**, *7*, 24. [CrossRef]
67. Pathak, J.; Subramanian, S.; Harrington, P.; Raja, S.; Chattopadhyay, A.; Mardani, M.; Kurth, T.; Hall, D.; Li, Z.; Azizzadenesheli, K.; et al. FourCastNet: A Global Data-Driven High-Resolution Weather Model Using Adaptive Fourier Neural Operators. *arXiv* **2022**, arXiv:2202.11214.
68. Qin, S.; Zhan, D.; Geng, D.; Peng, W.; Tian, G.; Shi, Y.; Gao, N.; Liu, X.; Wang, L.L. Modeling Multivariable High-Resolution 3D Urban Microclimate Using Localized Fourier Neural Operator. *Build. Environ.* **2025**, *273*, 112668. [CrossRef]

atmosphere

MDPI

Article

# Optimizing the Architecture of a Quantum–Classical Hybrid Machine Learning Model for Forecasting Ozone Concentrations: Air Quality Management Tool for Houston, Texas

Victor Oliveira Santos [1], Paulo Alexandre Costa Rocha [2], Jesse Van Griensven Thé [3] and Bahram Gharabaghi [1,*]

[1] School of Engineering, University of Guelph, 50 Stone Rd E, Guelph, ON N1G 2W1, Canada
[2] Mechanical Engineering Department, Technology Center, Federal University of Ceará, Fortaleza 60020-181, CE, Brazil; paulo.rocha@ufc.br
[3] Lakes Environmental, 170 Columbia St. W, Waterloo, ON N2L 3L3, Canada
* Correspondence: bgharaba@uoguelph.ca

**Abstract:** Keeping track of air quality is paramount to issue preemptive measures to mitigate adversarial effects on the population. This study introduces a new quantum–classical approach, combining a graph-based deep learning structure with a quantum neural network to predict ozone concentration up to 6 h ahead. The proposed architecture utilized historical data from Houston, Texas, a major urban area that frequently fails to comply with air quality regulations. Our results revealed that a smoother transition between the classical framework and its quantum counterpart enhances the model's results. Moreover, we observed that combining min–max normalization with increased ansatz repetitions also improved the hybrid model's performance. This was evident from evaluating the assessment metrics root mean square error (RMSE), coefficient of determination ($R^2$) and forecast skill (FS). Values for $R^2$ and FS for the horizons considered were 94.12% and 31.01% for the 1 h, 83.94% and 48.01% for the 3 h, and 75.62% and 57.46% for the 6 h forecasts. A comparison with the existing literature for both classical and QML models revealed that the proposed methodology could provide competitive results, and even surpass some well-established forecasting models, proving to be a valuable resource for air quality forecasting, and thus validating this approach.

**Keywords:** air pollution; ozone; Houston; forecasting; quantum machine learning; quantum neural network; graph neural network

## 1. Introduction

Air quality is a significant concern in the current Anthropocene era. The adverse effects of atmospheric contaminants in the atmosphere, mainly from the burning of fossil fuels and $CO_2$ emissions [1–3], have been proven to degrade human health directly [4–6] and the economy alike, causing not only millions of deaths worldwide [7–9], but also negatively impacting the global economy [10–12]. With this scenario, it is paramount to constantly monitor air quality, thus fostering the development of strategies to mitigate its adversarial effects on the environment, mainly based on adopting more restrictive environmental policies [13–15].

A traditional approach for forecasting toxic gas concentrations in the atmosphere is to apply physically based models, which rely on modeling the gas concentration using mathematical relationships describing the pollutant behavior in the atmosphere [16]. In this group, we can mention the successful implementation of chemical transport models


https://doi.org/10.3390/atmos16030255

(CTM) in diverse previous studies [17–20]. Despite its outstanding results, such an approach suffers from significant drawbacks. It is computationally demanding and requires both deep knowledge of the physical problem being modeled and sufficiently refined spatiotemporal information for meteorological, geophysical, and chemical information [21,22]. Additionally, it is prone to yielding biased results and underestimating outcomes for daily variability [23].

Significant computational software and hardware improvements in past years [24,25], have aided the development of predictive models based on machine learning (ML) architectures as an alternate approach for predicting air pollution concentrations [26–28]. Unlike the previous methodology, ML models offer a more straightforward implementation and improved computational time, excelling in retrieving complex non-linear links within the data [29–31]. Given these positive characteristics of the ML paradigm, diverse studies covering several ML models applied to air quality monitoring can be retrieved from the literature. Some examples are tree-based models [32–34], artificial neural networks [35–37], and support vector machine-based approaches [38–40], which have been successfully validated in evaluating air quality and forecasting pollutant concentrations in the atmosphere.

A further improvement over the previously mentioned ML is called deep learning (DL). The DL has also been highly efficient in determining future atmospheric pollutant values, compared to traditional ML and CTMs [41]. Models such as recurrent neural networks (RNN) [42–44], combinations between convolutional neural networks and RNN [45–47], attention-based architectures [48–51], and more recently, graph-based methodologies [52–54], often manage to return significant improvements in the determination of future air quality indices and pollutants concentration levels.

In recent years, advancements in the quantum computing field have fostered the evolution of a new framework for processing information called quantum machine learning (QML) [55–57]. As its name may suggest, QML merges knowledge from quantum mechanics and machine learning to delve into new possibilities for data assessment. The theory states that combining quantum mechanics principles like entanglement, superposition, and interference may significantly improve both processing time and results' accuracy [58]. Given such a promising future, studies have been conducted to validate different QML architectures. This paradigm has found successful applications in finance [59,60], pharmacology [61,62], and material science [63,64], to cite a few fields. However, given its early stage, there is still little research on the environmental spatiotemporal application of QML [65,66], including air quality assessment [67–69], water quality forecasting [70], and applications related to climate systems modeling [71,72].

However, hybrid methodologies combining classical and quantum paradigms are still sparse in the literature. This is due to the lack of algorithms for spatiotemporal scenarios, including air quality for this specific study, and the lack of more available quantum hardware for practical implementation. Therefore, to deepen knowledge of the QML paradigm applied to air quality forecasting and validate its implementation, the present study aims to investigate a hybrid methodology combining a deep learning graph-based forecasting model and simulated QML architecture. Using data from Houston, Texas, the proposed approach is evaluated for forecasting ozone concentrations up to 6 h ahead. The prominent contributions of this work include:

1. Proposing a novel air quality forecasting tool based on quantum machine learning.
2. Addressing the information gap regarding using hybrid QML models for air quality modeling.

3. Assessing distinct topologies for the hybrid approach; several configurations for the QML architecture must be investigated, including different preprocessing techniques for normalization.
4. Validating the proposed approach for predicting ozone concentrations in the atmosphere up to 6 h ahead.

## 2. Methodology

### 2.1. Important Quantum Mechanics Concepts

Feynman proposed the initial concept of the need for a new approach to simulate natural phenomena occurring in nature [73]. In his essay, Feynman argued that even though it is possible to simulate physical phenomena in classical computers using differential equations, the essence of the physical world is grounded in quantum mechanics. In other words, it is not possible to properly model naturally stochastic behavior using a deterministic approach like the one implemented by classical computing. Therefore, scientists sought to develop a new paradigm combining the properties of quantum mechanics to achieve a more refined simulation model.

In this case, it was essential to mimic the occurrence of the inherent properties of quantum mechanics. Here, we will focus on the most relevant ones for our proposed methodology.

#### 2.1.1. Superposition

Superposition is perhaps the most well-known quantum property. It states that a quantum system may occur concomitantly at all its feasible configurations. This phenomenon is directly related to Schrödinger's Equation, which describes the wave-like behavior of the particles in a complex Hilbert space $\mathbb{C}^{\otimes \mathrm{n}}$, with n being the total amount of particles within the system. The Hilbert space is a more general concept than the Euclidean one. In addition to all the common properties found in the Euclidean space, the Hilbert space ensures that an infinite sum of vectors within this space will converge to a value also contained in the Hilbert space. This is called the Cauchy completeness of the Hilbert space, and is fundamental in the quantum mechanics field [74,75].

This equation allows us to model all the quantum states which drive the system simultaneously. We suggest that the reader refer to [76,77] for more information about Schrödinger's Equation. Mathematically, this phenomenon is represented as the result of the linear combination between all the possible states, as in $|\Psi\rangle = \alpha_1|\mathrm{x}_1\rangle + \alpha_2|\mathrm{x}_2\rangle + \cdots + \alpha_n|\mathrm{x}_\mathrm{n}\rangle$ for an n-dimensional case. The Dirac notation describing the quantum states reveals that they have a complex composition, with each $|\mathrm{x}_\mathrm{n}\rangle$ itself described by Schrödinger's Equation. The terms described by $\alpha_n$ refer to the amplitude of each state $|\mathrm{x}_\mathrm{n}\rangle$, and are also complex-valued. They also determine the most probable outcome for its respective quantum state. That is, the larger the value of $|\alpha_n|^2$, the more probable $|\mathrm{x}_\mathrm{n}\rangle$ is to occur.

An upward coin toss can give a famous toy example for this property. Assuming a fair coin, it is impossible to determine which side is facing an observer when tossed upward. Therefore, we can assume that, at that moment, the coin is in a superposition state of both heads and tails.

#### 2.1.2. Measurement of a Quantum System

As previously stated, a quantum system exhibits a wave-like behavior, which the Schrödinger's Equation models. The measurement or observation of a quantum system causes it to collapse, i.e., it is forced to pick one of the many possible outcomes with a probability $|\alpha_n|^2$, where $\sum_{i=1}^{n} |\alpha_n|^2 = 1$.

Using the previous example of the coin toss, the coin's result, i.e., whether heads of tails, would only be known after measuring it, which in this case means catching it and observing its upside face.

2.1.3. Entanglement

Entanglement property is another essential characteristic of the quantum system. Its working mechanisms are yet to be fully understood, but it is still one of the core principles of quantum mechanics [77]. To understand this phenomenon, we first must consider composite states formed by more than one particle. In this case, the resulting quantum state is mathematically defined as being the tensor product between the states composing it. For example, considering only two particles for the sake of simplicity, an entangled quantum state can be written as [78]:

$$|\Psi\rangle = \frac{1}{\sqrt{2}}(|x_1\rangle \otimes |y_1\rangle + |x_2\rangle \otimes |y_2\rangle) \tag{1}$$

We observe in Equation (1) that the $|\Psi\rangle$ state is achieved by the linear combination of the tensor products $|x_1\rangle \otimes |y_1\rangle$ and $|x_2\rangle \otimes |y_2\rangle$. However, for this specific configuration, it is not possible to decompose $|\Psi\rangle$ as $|\Psi\rangle = (\alpha_1|x_1\rangle + \alpha_2|x_2\rangle) \otimes (\beta_1|y_1\rangle + \beta_2|y_2\rangle)$, as for $\alpha$ and $\beta$ the term $1/\sqrt{2}$ cannot be achieved for any possible value.

Overall, the entanglement of particles indicates that in the case of composite systems, there is no possibility to tell one particle independently from the state of the others. This implies that measuring one of the entangled particles will affect the states of all the remaining ones.

There is still debate on why entanglement occurs, and it results in odd outcomes, such as quantum teleportation, which Einstein called "spooky action at a distance" [79]. An example of applying the entanglement property can be observed in Shor's algorithm for prime number decomposition, allowing for increased exponential speed compared to the classical approach, proving to be a cornerstone of the quantum computing theory [80].

2.1.4. Interference

As mentioned previously, the particles in a quantum system exhibit wave-like behavior. Therefore, these particles can interact, creating interferences analogous to classical waves. This phenomenon could be observed during the execution of the double slit experiment, where light would shine from a source, reaching a barrier containing the slits and finally reaching the detection screen behind it. Through this approach, it was possible to identify patterns that were darker or brighter, evidencing the occurrence of destructive and constructive interference, respectively [81]. Mathematically, interference is modeled as the interaction between the amplitude values of the quantum state.

The same principle is applied to the quantum computing theory, which has been successfully used by Grover in his algorithm for quantum search, standing as the most efficient search algorithm ever implemented [82,83].

## 2.2. *Quantum Computing*

The quantum computing paradigm seeks to implement the principles of quantum mechanics to process information. Unlike the classical computing framework, which stores and processes information using bits, the quantum computing counterpart employs quantum bits (qubits) to achieve such an end. Similarly to its classical analog, qubits can store and process information through the properties of quantum mechanics [77,84]. Qubits can be represented as vectors defined in the complex number system, $\mathbb{C}^{\otimes n}$, where n denotes the amount of quantum bits describing the system. These qubits span a high-dimensional complex Hilbert space, a separable vector space that contains the quantum system resulting from the interactions among the qubits describing it. Within the complex Hilbert space, the quantum system can be represented by the qubits of the ground state $|0\rangle = [1\ 0]^T$, and the excited state $|1\rangle = [0\ 1]^T$ (considering one qubit only). The general quantum state configuration, composed by n-qubits, can be written by the following expression [83]:

$$|\Psi\rangle = \sum_{j=0}^{n-1} \alpha_j |j\rangle_n \tag{2}$$

Assessing Equation (2), it is possible to observe that its left side represents the quantum state, $|\Psi\rangle$, resulting from the linear combination of the $|j\rangle_n$ states considering n-qubits. The term $\alpha_j$ represents the amplitude for its respective quantum state $|j\rangle$, and $\alpha_j \in \mathbb{C}$, with $\sum_{j=0}^{n-1} |\alpha_j|^2 = 1$. This latter expression is, in fact, the probability of occurrence for a quantum state, where larger $|\alpha_j|^2$ values indicate more likely outcomes. Such a configuration for the qubits allows them to synergize beyond the simple binary system of the classical computing framework. These interactions are the ones described by the quantum mechanics properties, namely superposition, entanglement, and interference, which leverages the computing power for the quantum methodology [81,85,86].

However, these interactions can be simulated by two different strategies of performing quantum computing: gate-based and adiabatic. We will focus on the gate-based approach as it is the methodology used in our study. More information about adiabatic quantum computing is available in the following references [57,84,86].

The quantum gates are essentially unitary matrices that operate over the qubits, changing their state within the Hilbert space. To this end, the quantum gates should satisfy the condition of being a unitary matrix, $U \in \mathbb{C}^{2^n \times 2^n}$, following that $U^{-1} = U^{\dagger}$. This relationship is fundamental for quantum computing, as it allows resersible operations while maintaining the inner product between the qubit vectors unchanged [83,86].

Of these two possible approaches, the one based on quantum gates is the most versatile and, thus, the most commonly found approach in the Noisy Intermediate-Scale Quantum (NISQ) computers [57,86]. Therefore, the QML model proposed in this study relies on quantum gates.

### 2.3. *Quantum Neural Networks*

The quantum neural network (QNN) structure is composed of three core components: the feature map, the ansatz and the measurement of the resulting system [66]. They are presented in the following subsections.

#### 2.3.1. Feature Map

The feature map is perhaps the most essential component of the QNN model. It maps the classical information from the dataset into a highly dimensional Hilbert space. Such transformation allows the information to be processed in a quantum framework [87,88]. The feature map can be performed using two distinct approaches: implicit and explicit. The one used in the present study is the implicit one and consists of applying unitary operators to the input qubits. For this approach, the input quantum bits are analogous to the input parameters of the classical ML counterpart. The feature map implemented in this work makes use of a parameterized angle encoding, described in the following Equation (3) [89]:

$$|\phi(\mathbf{x})\rangle = \bigotimes_{i=0}^{n-1} R_y(x)|0\rangle \tag{3}$$

In Equation (3), the resulting quantum state $|\phi(\mathbf{x})\rangle$ is reached after the individual application of rotation gates, $R_y$, to each one of the input parameters $(x)$ (where $x \in \mathbb{R}^{m\times p}$), considering an initial $|0\rangle$ state.

### 2.3.2. Ansatz

The QNN's ansatz is also a paramount part of its architecture. The term comes from German and refers to providing an educated guess to solve a problem. Like the feature map, the ansatz is also a parameterized structure, meaning that the quantum gates parameters can be freely tuned. Regarding the QML framework, the ansatz is a quantum circuit determining the application of quantum gates to specific qubits though a succession of ansatz layers [86].

Initially, the ansatz takes input information from the feature map and applies parameterized unitary gates, i.e., quantum gates, U, with initial parameters $\theta$ [90]. The parameters $\theta$ driving the unitary gates are updated for each step through an optimization process. Given its layer-like characteristic, which parameters can be finely tuned after optimization, the ansatz structure resembles the weights optimization in a classical multilayer perceptron structure [66].

Different approaches to implementing the ansatz quantum circuit exist, such as simple rotation gates, the Quantum Approximate Optimization Algorithm (QAOA), and TwoLocal. Previous work [66] investigated the implementation of these ansatz strategies for a spatiotemporal scenario of solar irradiance forecasting. Building upon this knowledge, the same ansatz strategy was adapted for the present study of ozone forecasting. Therefore, the TwoLocal ansatz was implemented [91]. Generally, this ansatz alternates between rotations and entanglement, initially starting with random values. The implemented TwoLocal ansatz used Pauli Y rotation gates and a linear entanglement strategy. Regarding the entanglement structure, it will entangle the i-th qubit with its subsequent $i + 1$ for all the quantum bits composing the quantum system, i.e., $i \in \{0, 1, 2, \ldots, n-2\}$ for a system composed by n-qubits [91].

Additionally, the combination of rotation and entanglement gates can be repeated. In theory, such a repetition can go on indefinitely. Works investigating this area proposed that the repetition of quantum circuits, composing the feature map and the ansatz, is beneficial for the model's performance. In fact, increasing the number of repetitions allows the QML architecture to theoretically serve as a universal approximator, improving their results [86]. However, given that the proposed model is a simulation using classical computing hardware, the number of repeating circuits becomes increasingly demanding for the classical machine. Thus, to avoid long processing times during the simulated model, the repetitions were limited to a value of up to 3 [66].

### 2.3.3. Ansatz Optimization Strategy

Different algorithms can perform ansatz optimization. For the present investigation, the selected optimizer was the limited-memory Broyden–Fletcher–Goldfarb–Shanno Bound (L-BFGS-B). This optimization technique is an iterative model that can deal with non-linear optimization problems without constraints [92]. It is also a quasi-Newton algorithm, meaning that the L-BFGS-B does not require the target function's second derivative to optimize it. Finally, this optimization strategy has been proven to yield good outputs with reasonable convergence times when applied to quantum circuits, making it a sensible choice for this study [66,93].

Finally, it is paramount to emphasize that the QNN approach implemented here was based on the already validated architecture proposed by Oliveira Santos and colleagues [66].

In their work, the authors used a QNN architecture with the following configuration presented in Table 1 which is the same as adopted in this work.

**Table 1.** QNN structure used in the proposed GNN-SAGE-QNN model.

| QNN Component | Used Approach |
|---|---|
| Feature Map | Pauli Y angle encoding |
| Ansatz | TwoLocal ansatz, with linear entanglement |
| Ansatz Optimization | L-BFGS-B |

#### 2.3.4. Measurement

The last component of the quantum circuit is the measurement. After implementing the feature map, the ansatz and its optimization process are measured, and the resulting quantum state for the QNN model is measured. Measuring a quantum system leads to the collapse of its wave function; therefore, is a non-reversible process. Under these circumstances, the most likely outcome is given by the probability amplitude $|\alpha_j|^2$, where larger values are more likely to occur [83,86].

#### 2.3.5. On the Use of Simulated Quantum Circuits

The proposed QNN model was implemented in simulated quantum hardware using the Qiskit package for the Python language. This was due to the current conditions regarding the available quantum hardware. Nowadays, very few companies have actual quantum computers available, limiting their usage on a larger scale. Additionally, using real quantum computing services is still expensive, which is another limiting characteristic of the current quantum computing era. Another limitation is that the accessible quantum hardware is still error-prone due to the environmental noise, and has a limited number of available qubits [66,94].

Given these constraints, researchers often resort to computational simulations of quantum circuits to validate their models. Besides being cheaper, the simulation of quantum circuits for QML is more versatile, and there are different available libraries like PennyLane and Qiskit to support code implementation [95,96]. Additionally, simulated quantum circuits are not limited by the number of qubits available in real quantum hardware, and nor do they suffer from noisy interferences (even though they can also be simulated) in their outputs, making the implementation of different QML configurations possible. However, classical machines struggle to simulate their quantum counterparts, thus proving it to be a very computationally demanding task [86]. Nevertheless, there are studies on simulated QML applications that attest to the viability of such an approach, proving it to be a reliable choice for this study [62,66,97,98].

### *2.4. Graph-Based Deep Learning*

The classical part of the graph-based architecture consists of a deep learning structure that is established on graph theory. The Graph Neural Network Sample and Aggregate (GNN-SAGE). Different from conventional ML and other DL approaches, these types of models naturally consider a graph-shaped structure for the data, being able to extract the underlaying spatiotemporal information from the compiled data efficiently. This is performed by sampling neighboring nodes around the node of interest during each iteration. The model can learn increasingly complex relationships within the graph's datapoints with more iterations. Then, the aggregation operator concatenates the acquired information, allowing the graph's topology to be discovered and the unknown data to be generalized [99].

The GNN-SAGE algorithm has proved to be a versatile and powerful tool for spatiotemporal forecasting. It has shown improved outcomes when handling predictions for air quality, flooding occurrences, water quality, and wind speed [30,52,100–102]. However, to our understanding, this model has yet to be tested with a hybrid QML structure. Thus, the primary motivation of this work is to validate the hybrid GNN-SAGE-QNN performance for an air quality forecasting scenario, and, therefore, assess the viability of hybrid quantum–classical configurations for spatiotemporal forecasting.

### 2.5. Validation Study Site and Database

To validate the proposed methodology, this research selected the site of Houston, Texas. Houston is a major urban area in the United States that is a transport hub with several airports, railway connections, and a port [103]. It is also an industrial hub housing different oil refineries and petrochemical manufacturers [104,105]. According to Köppen–Geiger, Houston has a humid subtropical climate with elevated temperatures during summer and a mild winter [106,107]. The combination of urban characteristics and climate makes the Houston region a non-attained location, according to the Environmental Protection Agency (EPA) from the United States. Consequently, the study site fails to comply with the stipulated air quality indices recommended by the National Ambient Air Quality Standards (NAAQS) [108].

Therefore, developing precise and accurate forecasting tools to determine air pollution concentrations within a sufficiently large window is paramount. This enables preemptive measures to be issued to mitigate the harmful impacts of atmospheric contaminants on the local population. Addressing this challenge, data concerning air quality for the study location was collected from the Texas Commission on Environmental Quality. The data comprises the period from 2011 to 2020 and is a collection of several air contaminants, including ozone, $NO_X$, propane, benzene, toluene, and more. These data were collected with an hourly resolution.

Information about the atmosphere contaminants, relative humidity, direction and speed of wind, temperature, and solar radiation was also included in the dataset, as they also play a significant role in the driving mechanisms behind ozone accumulation [52,109]. The following Figure 1 depicts the study location and the placement of the monitoring stations spread across the territory. Note that the numbers over the green arrows represent the station code.

The graph-based model requires one of the stations to be the reference site, i.e., the location to which the model will forecast, considering its neighboring locations. Following the proposed methodology validated in previous work by Oliveira Santos et al. [52], which validated the present approach, the reference station was set to be 482011035, in the southeastern part of Houston, near Pasadena. Still, according to their methodology, the forecasting horizons of 1, 3, and 6 h were investigated. For each prediction window, there was a different combination of the input parameters and time-lags, i.e., the amount of past information provided to the prediction model. The best set of predictors and their respective time-lags are presented in the following Table 2 [52].

**Table 2.** Discrimination of predictors for each forecasting horizon and their respective time-lags.

| Forecasting Window | Predictors | Time-Lag |
|---|---|---|
| 1 h | • Past ozone data<br>• $NO_2/NO_X$<br>• Weather and HoD/DoY | 4 h |
| 3 h | | 2 h |
| 6 h | • Past ozone data<br>• Weather and HoD/DoY | 7 h |

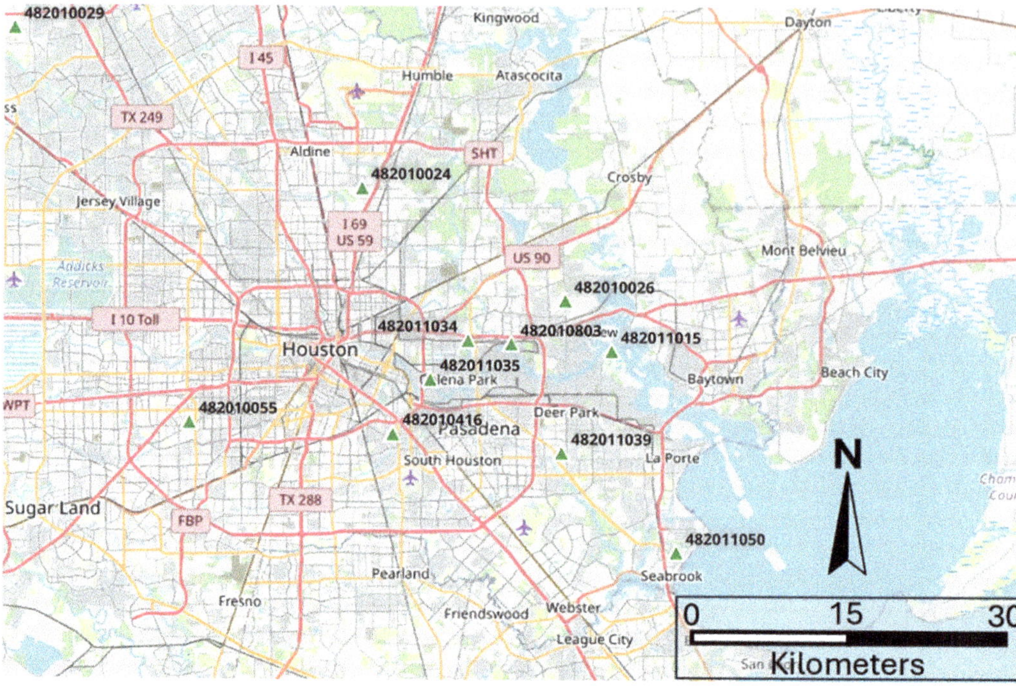


**Figure 1.** Map of the study location and the placement of the monitoring stations, represented by the green triangles [110,111].

Therefore, we implemented the same configuration of the previously validated model, as presented in Tables 1 and 3 for the QNN and the SAGE-QNN structures, respectively.

**Table 3.** Results for the hybrid GNN-SAGE-QNN model using the initial structure presented in Figure 2.

| Forecasting Horizon | RMSE (ppb) | $R^2$ (%) | FS (%) |
|---|---|---|---|
| 1 h | 7.61 | 78.24 | −32.53 |
| 3 h | 10.36 | 59.62 | 17.64 |
| 6 h | 9.33 | 67.18 | 50.65 |

## 2.6. Data Normalization

Another topic investigated in the present study was the use of two different normalization approaches for data preprocessing. Normalization is an essential preprocessing step for the implementation of ML models. It ensures that the input data are within the same order of magnitude, favoring the predictive methods' training and consequently improving their outcomes.

The first normalization investigated was implemented in the original work [52], called standard normalization, or z-score. It consists of subtracting the attribute value from its vector mean value and dividing it by its standard deviation. Its formulation is presented in Equation (4) [112]:

$$x_{norm} = \frac{x_i - \mu}{\delta} \quad (4)$$

where $x_i$ is the i-th value of the attribute vector **x**, $\mu$ is the attribute's x mean value, and $\delta$ is its standard deviation. This approach ensures that the normalized values have zero mean and unity variance [113].

The second methodology investigated was min–max scaling. It is simpler than the previous one of rescaling the attributes' values to numbers within the interval [0, 1] [112]. Consequently, the attribute value is adjusted by subtracting the minimum value and then dividing it by the range between the maximum and minimum values. Its formulation is presented in Equation (5):

$$x_{norm} = \frac{x_i - x_{min}}{x_{max} - x_{min}} \tag{5}$$

where $x_i$ is the i-th value of the attribute vector **x**, $x_{min}$ is the smallest value in the attribute vector **x**, and $x_{max}$ is its maximum counterpart.

## 3. Results

The proposed model was implemented for Python language, using PyTorch version 2.3.1 with CUDA version 11.8 for the GNN-SAGE algorithm [114]. The QNN methodology was implemented using the Qiskit library [96], with qiskit v 1.0.2, qiskit-algorithms v 0.3.0, and qiskit-machine-learning v 0.7.2.The hardware used in this study consisted of two computers. The first one had a 13th generation i7 processor, 32Gb of RAM and RTX 4080 GPU, running a Windows 10 OS. The second machine had a 13th generation i9 processor, 32Gb of RAM, and RTX A4500 GPU, running on Windows 11 OS.

The proposed model performance was assessed using several metrics. They consisted of the root mean square error (RMSE), the coefficient of determination ($R^2$), and the forecast skill (FS). The FS metric compares the proposed architecture result with the benchmarking model of persistence. Persistence is a basic forecasting approach. It assumes that future ozone concentration values will match the most recent ones. Albeit simple, persistence is a very hard contender for predictive models for short time windows [115]. However, as the forecasting horizon increases, the persistence prediction deteriorates, as it cannot capture the complex dynamics driving ozone concentration occurrences [116]. The FS metric determines whether the QML model is superior by achieving positive FS values. A positive FS value indicates that the QML is better than the persistence model, while a negative FS value suggests otherwise. In the case of negative values, implementing the QML model is unnecessary, as the persistence approach would provide better predictions for ozone levels.

These metrics are among the most commonly used for time-series modeling, offering insights into the model functioning based on its outputs [117,118]. Finally, the model was trained using data from the years 2011 to 2019 and validated with data from 2020.

### 3.1. The Initial Hybrid GNN-SAGE-QNN Architecture

This first implementation of the hybrid GNN-SAGE-QNN model used the z-score normalization (Equation (4)) with one ansatz repetition [52]. This was conceived by coupling the QNN part into the GNN-SAGE one, as depicted in Figure 2.

Figure 2 shows that the hybrid architecture comprises successive convolutional layers of 512 neurons. These convolutional layers are activated by the Leaky ReLu function with a leakage factor $\alpha = 0.1$, later followed by a 10% dropout layer that serves as regularization to avoid overfitting. After the model processes the data, learning the connections between the input nodes of the graph structure, the dataset is then vectorized by implementing a dense layer, also containing 512 neurons in its structure.

After the first dense layer, the QNN module is inserted into the graph-based model, thus forming the hybrid configuration. At this part, the classical data from the deep learning model is processed by the quantum framework implemented by the QNN quantum circuit containing two qubits. Finally, the output from the QNN module is processed by another classical dense layer containing one neuron, which outputs the final ozone concentration estimation.

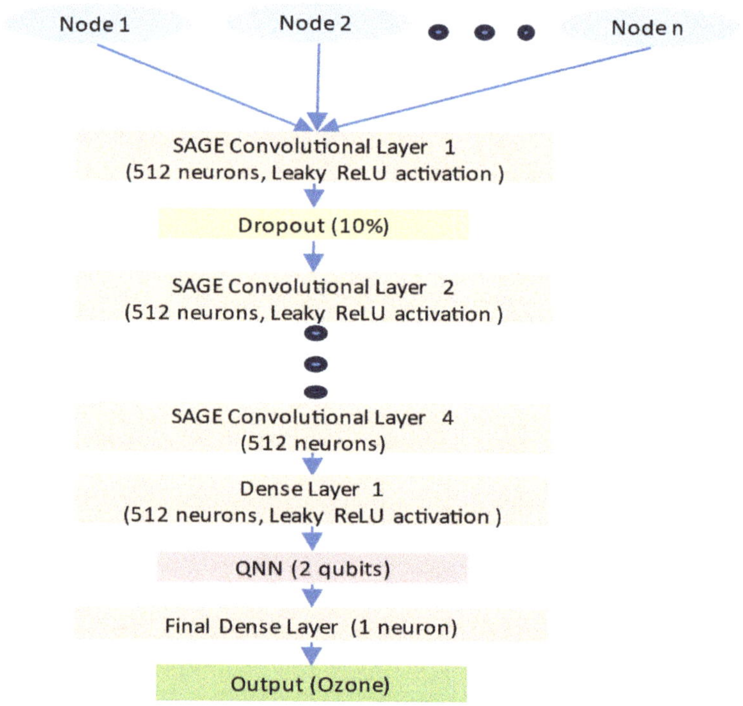


**Figure 2.** The initial GNN-SAGE-QNN structure used for ozone forecasting.

The model's results for the forecasting windows of 1, 3, and 6 h are introduced in Table 3.

From Table 3, it is noticeable that the results for the investigated time windows are acceptable in terms of RMSE and $R^2$. However, a deeper investigation regarding the FS value indicates that the model is suboptimal when dealing with the 1 h forecasting horizon, given its negative value. The value of −32.53% reveals that this approach is over 32% inferior compared to the persistence model.

The same is not true for the following horizons of 3 and 6 h. The former reached a satisfactory FS value of 17.64%, and the latter achieved a value of 50.65%. This indicates the superiority of the model for these time frames. Additionally, it is relevant to point out that the hybrid model, for this specific configuration, presented behavior that is not common in regression scenarios. This is because the error is expected to rise as the forecasting horizon increases. This is caused by the need for further spatiotemporal information for longer time windows, given that this is a significantly more complex relationship to model [100,119,120]. It is possible to observe that the RMSE increases from 7.61 ppb to 10.36 ppb for 1 to 3 h marks, but then it reduces to 9.33 ppb between 3 h and 6 h horizons. At the same time, the coefficient of determination has a similar behavior, going from over 78% to 59.62% at the first interval, and then increasing to 67.18%. Such an uncommon nature has been stated in a previous study, which found that the QNN model improved its predicted outputs for longer forecasting horizons [66].

While the model achieves satisfactory outcomes for 3 and 6 h, its underperformance for the 1 h mark indicates that the proposed architecture must be rebuilt to overcome this limitation. This subpar performance can be explained by the abrupt dimensionality change between the classical and quantum structures (Figure 2). Specifically, the last classical layer, with 512 neurons, transitions to 2 qubits in the quantum structure, highlighting a significant dimensional reduction within the QML architecture. This abrupt change causes part of the processed information to be lost during this process, as occurs in dimensionality reductions for classical models [121]. This loss of information prevents proper processing by the model, resulting in inferior results. Therefore, we investigated different configurations for the initial approach, focusing on the dense layer configuration to ensure a smoother transition between classical and quantum paradigms.

### 3.2. Topological Investigation of the Dense Layer of the GNN-SAGE-QNN Model

As previously mentioned, we aimed for the model's improvement for the 1 h forecasting horizon. To this end, we assessed distinct dense layers topologies for the proposed model. The reasoning behind this was that the dense layer 1 (Figure 2) had an input size of 512 (coming from the convolutional layer 4), and then suddenly reduced its output to two, to fit the input requirement of the QNN module. During an initial investigation of the hybrid model, this approach was investigated to understand the model's behavior better. However, this proved to be inadequate. In fact, in implementing classical ML models, such an abrupt change is not common. The dimensionality reduction is often performed gradually, from more complex to increasingly simpler layers, as in autoencoders and multilayer perceptron structures [121–123].

Therefore, we investigated five different configurations regarding the transition between the classical dense layers and the QNN architecture. The topologies are depicted in Figure 3.

Figure 3 depicts the investigated approaches. In (1), a very smooth transition was implemented, halving the number of classical neurons for each one of the subsequent layers until the final output dimension of 16, connecting to the QNN. Approach (2) is similar, but each layer has one-fourth of its predecessor until an output dimension of eight is reached for the last component. Approaches (3), (4) and (5) are similar and investigate a more abrupt transition between the classical and the quantum frameworks, implementing only two dense layers before feeding the QNN architecture. This study was performed using the 1 h forecasting horizon as a benchmarking dataset, given that this was the interval in which the proposed approach could not overcome persistence forecasting. The findings of this investigation are shown in Table 4.

**Table 4.** Outcomes for the different transition configurations between the dense layers and the QNN module.

| Approach | RMSE (ppb) | $R^2$ (%) | FS (%) |
|---|---|---|---|
| 1 | 9.11 | 68.75 | −58.80 |
| 2 | 6.31 | 85.02 | −9.97 |
| 3 | 7.63 | 78.13 | −32.86 |
| 4 | 5.58 | 88.28 | 2.75 |
| 5 | 5.50 | 88.60 | 4.10 |

From Table 4, it is possible to observe that the model struggles to beat the persistence model for the 1 h forecasting horizon. Approaches (1) and (3) were no better than the original implementation of the hybrid model, achieving inferior results than their predecessor, as indicated by the metrics. Approach (2) slightly improved the model's outcome, achieving reduced error value for the RMSE and improved coefficient of determination and FS. Finally, approaches (4) and (5) were the best-performing configurations, surpassing the original model's implementation significantly, specifically the latter, which achieved the best FS score among all the other topologies. The final model's topology is presented in Figure 4.

**Figure 3.** The five distinct configurations for the dense layer transition from the classical to the QNN module.

Thus, the results for the hybrid model were achieved using approach (5) (Figure 3).

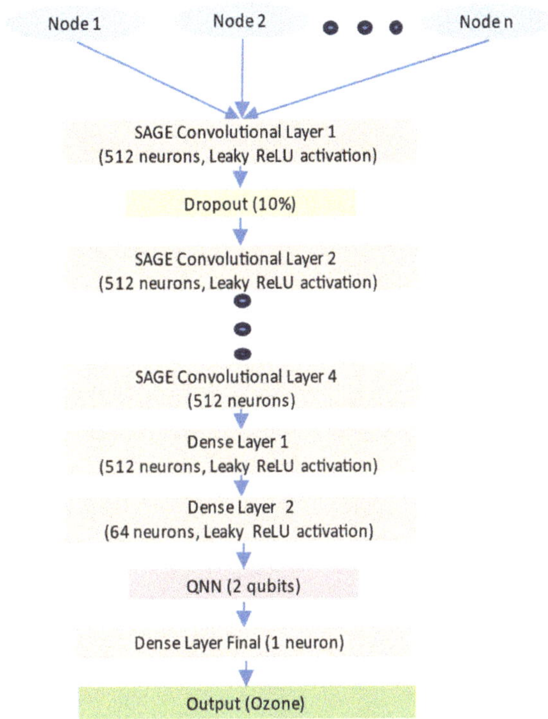


**Figure 4.** Final configuration for the GNN-SAGE-QNN model for ozone concentration forecasting.

### 3.3. *Investigation of the Normalization Approach and Ansatz Repetitions*

As mentioned previously, the GNN-SAGE-QNN model was built using z-score pre-processing and one ansatz repetition for the QNN structure. The forecasting results for 1, 3, and 6 h ahead horizons are presented below, in Table 5, using the topology presented in Figure 4.

**Table 5.** Results for ozone levels forecasting for each prediction window using z-score normalization and one ansatz repetition.

| Forecasting Horizon | RMSE (ppb) | $R^2$ (%) | FS (%) |
|---|---|---|---|
| 1 h | 5.50 | 88.61 | 4.11 |
| 3 h | 8.10 | 75.36 | 35.66 |
| 6 h | 9.43 | 66.53 | 50.16 |

The metrics assessment alone demonstrates that the suggested methodology may perform well for the proposed application. However, further investigation into its graphical results, i.e., scatter plot and time-series plot, reveals some current limitations. For the sake of clarity, we opted to show the plots for the 6 h forecasting horizon. However, it is important to state that the same phenomenon occurs for the remaining prediction windows.

Figure 5 depicts the scatter plot of the predicted results, on the *y*-axis, compared to the measured ozone levels by the monitoring stations, on the *x*-axis. The units are in ppb. The red line is the regression line and represents a perfect match between the simulated and actual values, having a ratio of 1:1. The marginal distributions are helpful for illustrating how well the distribution of the predicted values follows the real one. Thus, the more similar they are, the better the model results.

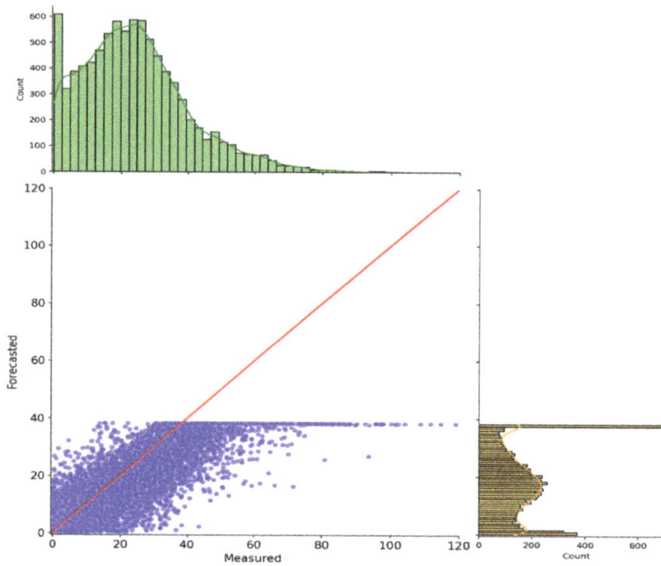


**Figure 5.** Scatter plot for the 6 h forecasting horizon using z-score normalization, one ansatz repetition, and the topology depicted in Figure 4.

From this image, it is possible to notice that the predicted values are well-grouped around the regression line. However, the model cannot estimate values greater than 40 ppb. Furthermore, it is observable that the model overestimates smaller values and underestimates larger ones, as represented by its distribution on the rightmost side.

This becomes even more noticeable when evaluating Figure 6, which compares the measured data trend in orange with the forecasted ozone concentrations, portrayed by the dashed blue line. In this image, the modeled values miss the ozone peaks. While the proposed approach captures lesser values well, it still presents some overestimation by the model. Nonetheless, this is a major drawback, given that properly capturing extreme atmospheric ozone fluctuations is of the utmost importance.

To overcome this hindrance, the hybrid model had to be changed again. To this end, we investigated a new normalization for the preprocessing phase and different numbers of ansatz repetitions. We investigated the effect of each change individually on the original model. First, the min–max normalization was implemented on a one-ansatz repetition model. The following results for the min–max implementation were acquired for a 6 h forecasting horizon. The reason behind the selection of the 6 h forecasting window is that when forecasting pollutants, it is crucial to determine their concentration as far into the future as possible. This is because the sooner the concentrations of a given pollutant are known, the sooner preemptive measures can be issued, which can significantly diminish the adverse effects of the contaminants on the population. The results for this configuration are presented in Figures 7 and 8.

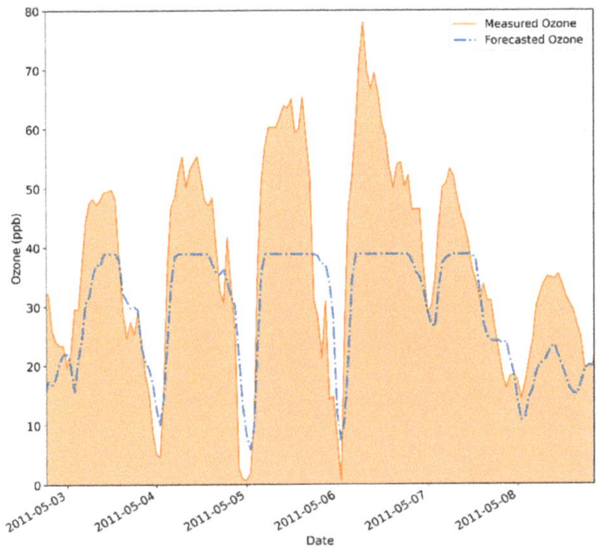


**Figure 6.** Time-series comparison between the predicted and measured ozone levels 6 h ahead for a specific period using z-score normalization, one ansatz repetition, and the topology depicted in Figure 4.

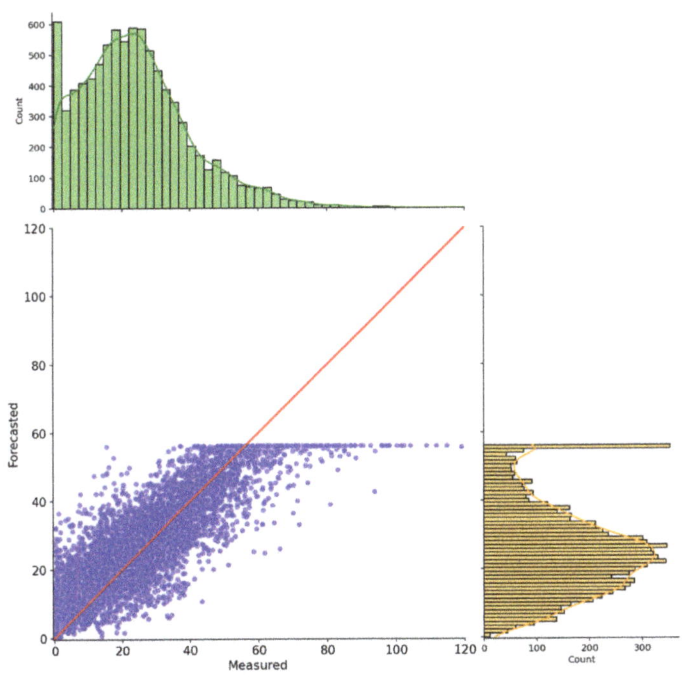


**Figure 7.** Scatter plot for the 6 h forecasting horizon using min–max normalization, one ansatz repetition, and the topology depicted in Figure 4.

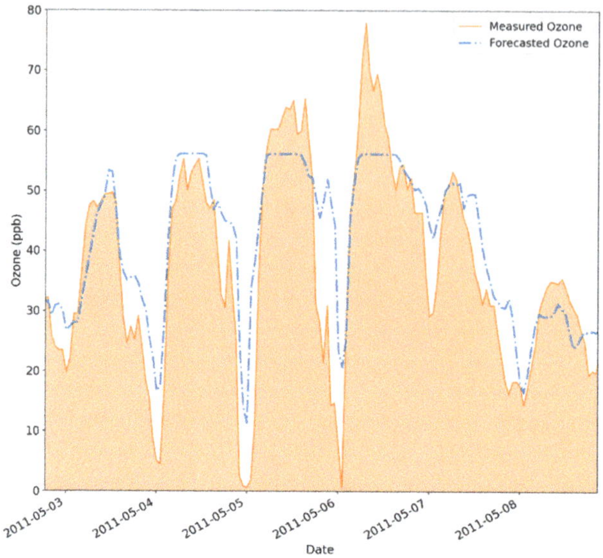


**Figure 8.** Time-series comparison between the predicted and measured ozone levels 6 h ahead for a specific period using min–max normalization, one ansatz repetition, and the topology depicted in Figure 4.

Figures 7 and 8 show similar behavior to the ones in Figures 5 and 6. Again, it is noticeable that saturation occurs, but this time at a higher threshold value, near 60 ppb. Overestimation for smaller values becomes more apparent, as indicated by the scattered points above the regression line. Additionally, using min–max as a normalization technique allowed the model to capture the first ozone peak present in Figure 8, showing some progress compared to the original model's implementation in Figure 6. This is reflected in the model's performance, which could yield better metric values compared to the ones presented in Table 5 for the 6 h forecasting scenario. This time, the model returned an RMSE of 8.31 ppb, an $R^2$ of 74.00%, and an FS of 56.07%. The min–max approach demonstrated that it was possible to push the saturated results upwards, thus improving the identification of higher ozone levels. However, it did not completely solve the saturation issue on its own.

Addressing this problem, the second modification investigated was the effect of two ansatz repetitions. Its outcomes were assessed using the original z-score normalization. Again, the 6 h forecasting horizon was selected for the reasons previously disclosed. Its results are presented in Figures 9 and 10, below.

Figures 9 and 10 again show the saturation problem occurring at the values predicted by the hybrid model. The limitation occurs at a threshold of around 40 ppb, similar to the one reported in Figures 5 and 6. Additionally, the marginal distributions in Figure 9 highlight the difference between the modeled and real ozone levels. There is a clear discrepancy in the extreme values predicted by the hybrid model, caused by underestimating the values in those regions.

Complementary to Figure 9, Figure 10 reveals that the lower values are better captured by the proposed model, which indicates better performance than its predecessors, as depicted in Figures 6 and 8. However, this approach cannot identify any of the peaks in Figure 10, contrary to the results presented in Figure 8. However, the configuration using two ansatz repetitions returned better metric results than the original ones in Table 5, with an RMSE of 8.80 ppb, an $R^2$ of 70.88%, and an FS of 53.52%. The improvement in the model's performance, even with the occurrence of the saturation phenomenon, indicates the importance of the ansatz repetition to the returned simulated values.

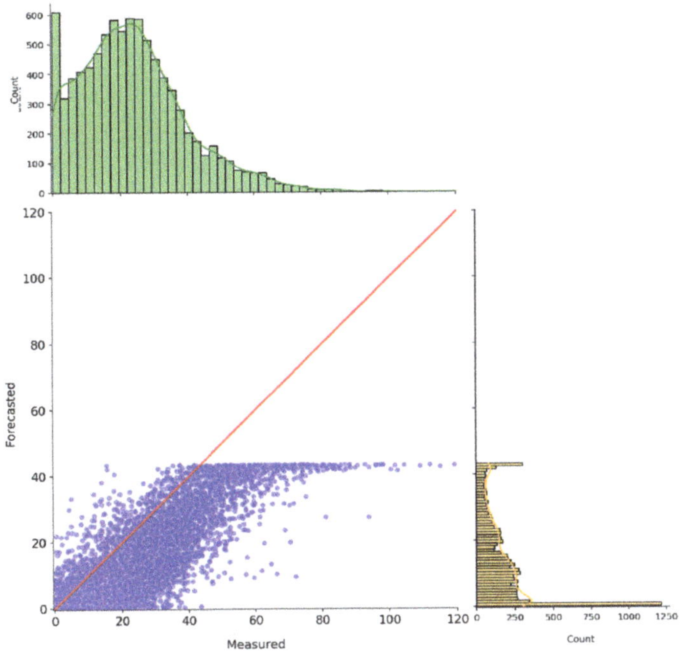


**Figure 9.** Scatter plot for the 6 h forecasting horizon using z-score normalization, two ansatz repetitions, and the topology depicted in Figure 4.

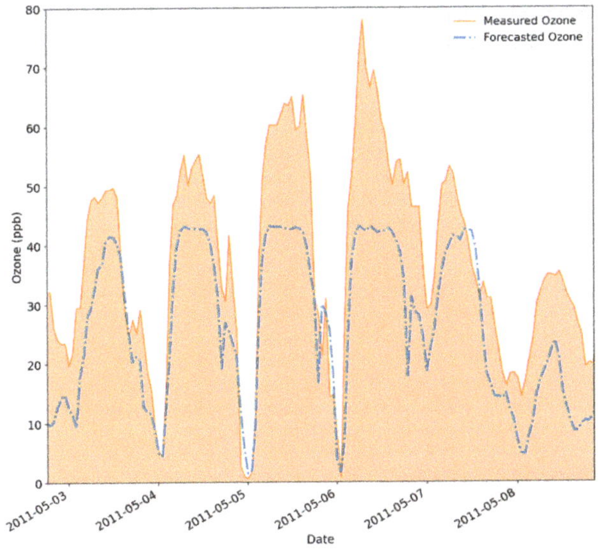


**Figure 10.** Time-series comparison between the predicted and measured ozone levels 6 h ahead for a specific period using z-score normalization, two ansatz repetitions, and the topology depicted in Figure 4.

By closely evaluating the individual results of min–max preprocessing and the ansatz repetition, it was decided that combining both strategies could further improve the model and solve the saturation problem. Indeed, this approach effectively addressed the model's limitation in identifying ozone peaks. Interestingly, predictions for 1 h ahead required three ansatz repetitions, while for 3 and 6 h time windows, two ansatz repetitions sufficed.

### 3.4. Ozone Forecasting Results for 1, 3, and 6 h Ahead

Finally, the results for the proposed forecasting horizons were achieved using the optimal configuration proposed for the hybrid approach, as described in Table 6. The outcomes are presented in Table 7.

**Table 6.** Final configuration for the hybrid GNN-SAGE-QNN model for each forecasting horizon.

| Forecasting Horizon | Normalization | Number of Ansatz Repetition |
|---|---|---|
| 1 h | Min–max | 3 |
| 3 h | Min–max | 2 |
| 6 h | Min–max | 2 |

**Table 7.** The outcomes for the evaluated forecasting horizons using the selected topology (Figure 4), min–max normalization, and an increased number of ansatz repetitions.

| Forecasting Horizon | RMSE (ppb) | $R^2$ (%) | FS (%) |
|---|---|---|---|
| 1 h | 3.96 | 94.12 | 31.01 |
| 3 h | 6.53 | 83.94 | 48.01 |
| 6 h | 8.01 | 75.62 | 57.46 |

The results for the predicted ozone levels at 1 h forecasting horizon are presented in Figures 11 and 12.

Figures 11 and 12 show how the inclusion of the proposed alterations managed to soar the model's performance, substantially improving the metrics values and its predictions, alike. The scatter points are visibly well-grouped around the regression line, and the marginal distributions are similar, indicating a good agreement between the predicted and measured values. With the current configuration, the RMSE, $R^2$, and FS achieved significant improvements of 28.00%, 5.51%, and 27.00%, respectively. Additionally, Figure 11 presents how well the model could capture the ozone levels in the dataset, as the marginal distributions suggest.

Assessing Figure 12, it is clear that the proposed hybrid architecture could adequately model all the ozone peaks in the dataset. The measured ozone trends are closely followed by the predicted values, again showing good accordance between the proposed methodology and ground truth numbers. This good performance can be observed for both small and large ozone concentrations, as indicated by the image, thus attesting its good performance for ozone level forecasting.

The following Figures 13 and 14 illustrate the results achieved by the model for a 3 h forecasting horizon.

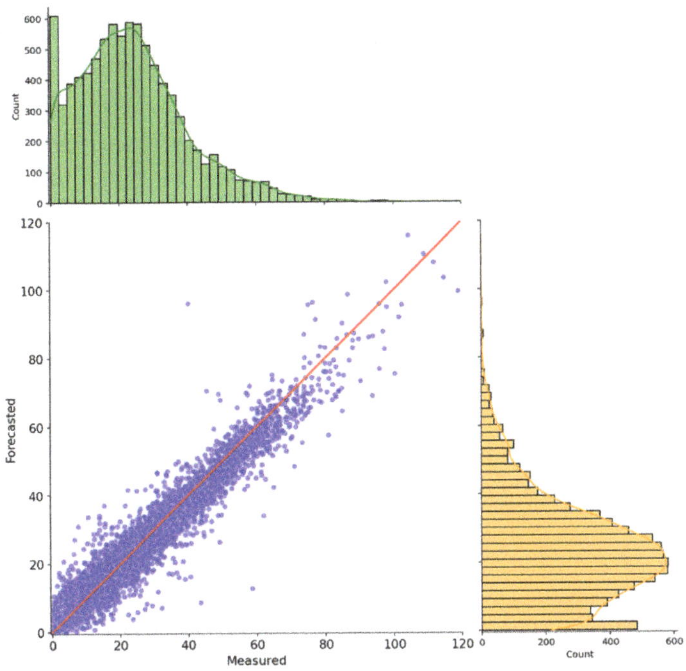


**Figure 11.** Scatter plot for the 1 h forecasting horizon using min–max normalization, three ansatz repetitions, and the topology depicted on Figure 4.

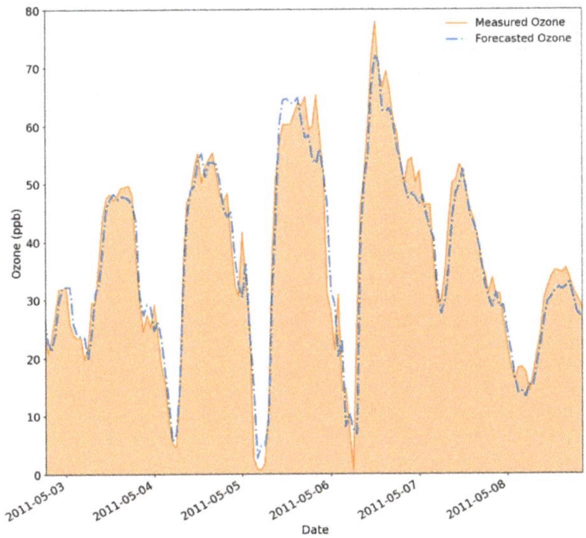


**Figure 12.** Time-series comparison between the predicted and measured ozone levels 1 h ahead for a specific period using min–max normalization, three ansatz repetitions, and the topology depicted on Figure 4.

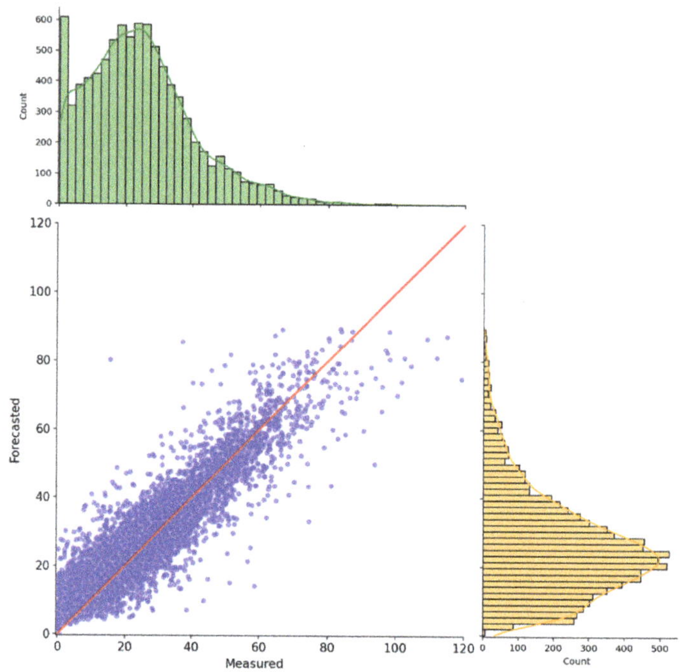


**Figure 13.** Scatter plot for the 3 h forecasting horizon using min–max normalization, two ansatz repetitions, and the topology depicted in Figure 4.

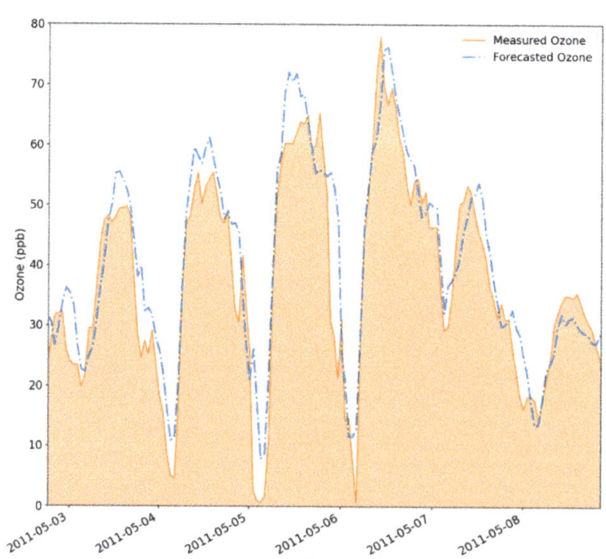


**Figure 14.** Time-series comparison between the predicted and measured ozone levels 1 h ahead for a specific period using min–max normalization, two ansatz repetitions, and the topology depicted in Figure 4.

Figure 13 also shows a good alignment between the predicted and the measured ozone data. Again, the points are scattered in a fashion that is close to the regression line. This time, however, the model exhibits some overestimation towards small values when verifying the marginal distribution graphs. Still, the metric results for this time window significantly improved over the original one shown in Table 5. The new configuration improves the RMSE by 19.38%, the coefficient of determination by 5.58%, and the FS by 12.35%.

Figure 14 corroborates the good efficiency of the proposed model. Using this enhanced configuration, the model can still capture significant ozone level peaks 3 h in advance. It is also possible to state that the hybrid configuration proposed in this study closely follows the real data trend, presenting some overestimation for the peaks.

Finally, the results for the 6 h forecasting horizon are presented in Figures 15 and 16.

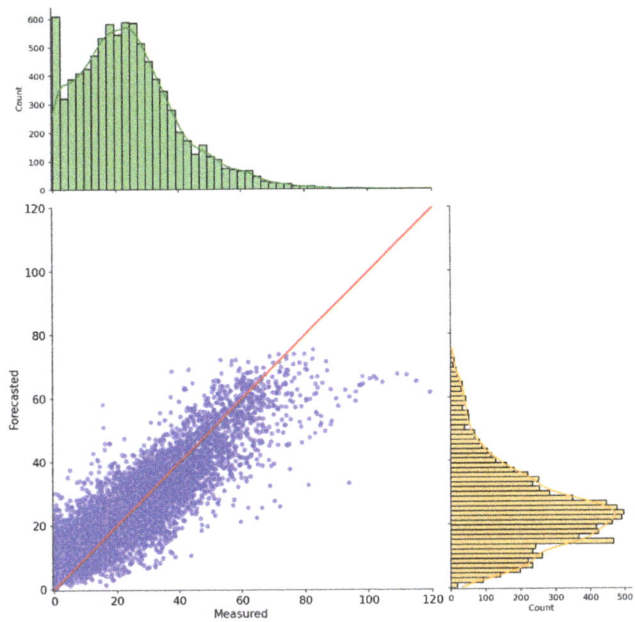


**Figure 15.** Scatter plot for the 6 h forecasting horizon using min–max normalization, two ansatz repetitions, and the topology depicted in Figure 4.

Figure 15 illustrates that for the 6 h forecasting horizon under study, the proposed hybrid GNN-SAGE-QNN model can predict the ozone level in Houston's atmosphere. From its depiction, the model presents some overestimation for the predicted value, similar to the 3 h forecasting horizon, as indicated by the marginal distribution picture. Still, this proved detrimental to the model's ability to predict larger ozone values. This is observed, again, by the marginal distributions, together with a closer examination of Figure 16. The highlight is that, once more, all the peaks described in that specific time window were adequately captured by the proposed hybrid approach.

The model's good performance is demonstrated once more by the value returned for the evaluation metrics. Assessing Tables 5 and 7, the gains for RMSE, $R^2$, and FS were 15.06%, 9.09%, and 7.30%. These values indicate the remarkable development of the proposed new configuration for the hybrid model compared to its initial structure.

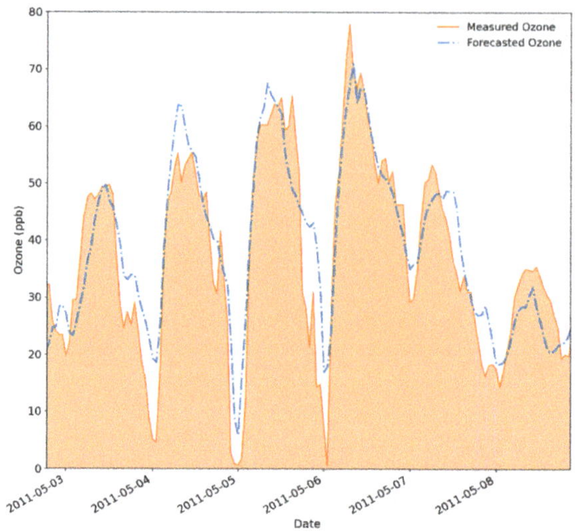


**Figure 16.** Time-series comparison between the predicted and measured ozone levels 6 h ahead for a specific period using min–max normalization, two ansatz repetitions, and the topology depicted in Figure 4.

# 4. Discussion

## 4.1. QML Results in the Current NISQ Era

The proposed hybrid approach, the GNN-SAGE-QNN model, has achieved good performance in estimating future ozone concentration levels. Given the current NISQ era, it is of utmost importance to validate the proposed methodology, whether using simulated quantum hardware or implementing the algorithms in real quantum hardware. Even though the current available quantum hardware lacks computing speed and processing power and is prone to noisy errors [124], validating these newer methodologies at this moment will pave the road for more robust approaches as we gain further understanding of the advantages and drawbacks of the quantum computing framework.

Therefore, even implementing QML models that fall short compared to their classical counterparts is important. This enables us to evaluate the viability of the implemented methodologies for specific situations, thus offering a deeper look into the working mechanisms of the proposed architectures. Thus, having this knowledge beforehand, as quantum technology becomes more accessible and more well-established, will allow us to leverage its exponential processing speed and achieve more accurate and precise outcomes.

## 4.2. Influence of Ansatz and Data Normalization

Although theory suggests that better generalization is achieved by increasing the ansatz repetitions in a parameterized quantum circuit, this phenomenon was not observed in the present study [86]. This discrepancy could be explained by the fact that the ansatz used in this investigation was validated for a different time-series problem [66] rather than ozone forecasting. Given that the ansatz is essential for properly generalizing the modeled Hilbert space by the QML model, the TwoLocal ansatz may not have been the best choice [86,89].

Nevertheless, it showed good outcomes, as observed in Table 7. As previously stated, the feature map and the ansatz are parameterized structures. This is because the quantum gates composing their frameworks will behave according to the data input from the previous layer. In this case, the classical data, normalized by the z-score approach, were mapped into the quantum framework by the feature map, and subsequently fed into ansatz structure. As the quantum gates are representations of qubits' rotations in a high-dimensional Hilbert space, we can fathom that the information is passed to the quantum gates as radians determining different rotations. Initially, using the z-score approach, the normalized values were not bound to a specific range, while min–max assured values within an interval between zero and one were included [112]. The use of values outside of the min–max range could impair the quantum structure in properly capturing the information within the dataset, as the conveyed normalized data may not allow a good representation of how the wave function describing the ozone forecasting problem behaved in its Hilbert Space, thus impacting its performance [125]. In the present study, this was observed in the form of saturation (Figures 5–10), from which the model could not learn to identify peak values for ozone concentrations.

Therefore, it is possible to conclude that the improvement behind an increasing number of ansatz repetitions depends significantly on a combination of factors, such as the data, the ansatz design, and the model's goal. Additionally, a combination between proper data normalization and the ansatz configuration promoted better modeling of the performance of the hybrid GNN-SAGE-QNN model by improving its representation of the quantum wave function on the Hilbert space. Ultimately, this has significantly boosted the results for ozone level predictions and boosted the results for ozone level predictions.

### *4.3. Comparison with Quantum Models Found in the Literature*

As we are investigating a new methodology combining quantum and classical schemes, it is hard to find similar published studies to compare it with. Some works have investigated the implementation of QML models for spatiotemporal scenarios. The previous work described showed promising results for a QNN application for renewable energies, considering solar energy forecasting, indicating the best outcomes for larger forecasting horizons. Another study conducted by Correa-Jullian et al. [97], applied the Quantum Support Vector Machine (QSVM) framework to another renewable energy scenario, this time considering the case of fault detection in wind turbines.

Another quantum application of an ML model can be found in the work by Farooq and colleagues [68]. They implemented the hybrid quantum–classical methodology for air quality forecasting based on SVM, achieving good accuracy. Their results showed that the QSVM approach was viable for this situation. The work developed by Zhang and colleagues [126] managed to use real quantum hardware to implement a quantum version of the Convolutional Long Short-Term Memory (ConvLSTM) DL model. Their methodology proved valuable for emission concentration forecasts by achieving lower error values and more accurate results. A comparable methodology can be found in the paper by Gohel and Joshi [65], where another quantum implementation of the LSTM architecture was proposed. Their work investigated the feasibility of this methodology for rain prediction. The QLSTM showed remarkable results for the proposed dataset.

Considering the case of hybrid methodologies, published results are still scarce, with few applications to be found. A hybrid method combining ML, the Internet of Things (IoT), and the quantum framework was proposed by [69]. Their implementation investigated the classification approach for air quality monitoring in industrial areas. Their model showed robust results for the proposed case, achieving improved accuracy and convergence speed. In Wang et al. [127], another hybrid architecture was proposed for air quality index prediction. The authors used a fuzzy-based model combined with the quantum structure to output more accurate results than the classical take. The DL model of LSTM and a QML model were combined [128] for solar irradiance forecasting. The combination of architectures reached better time-dependent information extraction from the dataset, thus improving the predictions for solar energy compared to classical ML models.

The proposed methodology in this study achieved similar conclusions to those found in the published works. Leveraging the benefits from both classical and quantum computing frameworks has been shown to advance the forecast of air quality conditions in terms of RMSE, coefficient of determination, and FS, as presented in Table 7. The validation of the viability of this model will lead to improved versions of similar methodologies in the future.

### 4.4. Comparison with Classical Models Found in the Literature

The performance of the proposed hybrid approach was evaluated against previously published classical models for the same task of ozone forecasting. We understand that a direct comparison between these different frameworks may not be entirely correct. However, it still offers a good insight into how the quantum-based methodology compares to the classical one. This is possible given the evaluation metrics, which allow assessing each model's accuracy, given their particularities [129]. Table 8 compiles the information for several proposed classical models for ozone forecasting.

**Table 8.** Values found for ozone forecasting for different predictive models in the literature.

| Model | Metric Value | References |
|---|---|---|
| GNN-SAGE | RMSE—$R^2$—FS<br>3.80 ppb—0.95%—33.70% for 1 h horizon<br>6.45 ppb—0.84%—48.70% for 3 h horizon<br>8.09 ppb—0.75%—57.10% for 6 h horizon | [52] |
| LSTM-based deep learning model | RMSE—$R^2$<br>7.50 ppb—83.69% for 1 h horizon | [130] |
| Attention-based deep learning model | RMSE<br>2.92 ppb for 1 h horizon | [49] |
| Transformer-based deep learning model | RMSE—$R^2$<br>3.99 ppb—92.20% for 1 h horizon<br>5.35 ppb—86.31% for 3 h horizon | [131] |
| Several classical ML models, where the SVM emerged as the best performing approach | RMSE—$R^2$<br>9.24 ppb—85.00% for 1 h horizon<br>14.01 ppb—65.50% for 3 h horizon<br>17.70 ppb—45.10% for 6 h horizon | [132] |

Comparing the results of the proposed model with the ones reported by [52], the original work where the GNN-SAGE structure was validated for ozone forecasting, it is possible to observe comparable performance by the hybrid quantum–classical approach. A close evaluation of the forecasting horizons of 1 and 3 h reveals that similar results were achieved by both models, with a difference of less than 1% for RMSE and $R^2$ metrics favoring the classical implementation. For the 6 h time frame, however, the hybrid approach surpassed its classical counterpart by a slight difference. This indicates that the hybrid model combining classical and quantum frameworks is suitable for longer forecasts. Interestingly, a similar conclusion was achieved by the results reported by [66]. The authors observed that the QML model improved its performance for longer prediction horizons, surpassing well-established classical ML implementations. The justification for this phenomenon is due to the highly complex levels of modeling achieved by the quantum configuration, which are not restricted by the classical information processing limitations (Section 2.1). This ultimately results in the identification of the complex relationships underlying the dataset, leading to improvements in the model's output for further predictive windows. Therefore, again, it is possible to attest to the good performance of quantum structures for forecasting time-series data further into the future, which is again observed in this study.

The work [130] developed a recurrent deep learning model based on LSTM and applied it to determine ozone levels in Los Angeles, USA. Their work achieved good results for ozone concentration prediction for 1 h ahead, reaching an RMSE of 7.50 ppb and an $R^2$ of 83.69%. Compared to the proposed approach, the hybrid methodology managed to overcome these values significantly, improving by 47.2% and 10.22% for RMSE and $R^2$, respectively. Even though the metrics reflect the performance of different applications, it is possible to observe that the hybrid model can perform better than classical deep learning models.

The works by [49,131] explored deep learning models based on attention. The former investigated a hybrid approach combining LSTM and attention structures for ozone forecasting in Beijing, China. For a forecasting horizon of 1 h, their proposed model achieved an excellent result of 2.92 ppb for the RMSE metric. This value was the best one among the studies compiled in Table 8. Nevertheless, the proposed methodology could return values within the same order of magnitude, thus being a competitive option. The second study used the transformer framework to predict ozone levels in Marrakech, Morocco. Their methodology returned very similar outcomes to those reported for the proposed GNN-SAGE-QNN. RMSE and the coefficient of determination were very similar for 1 and 3 h forecasting scenarios, indicating that the hybrid approach can offer competitive results compared to those yielded by the classical methodologies.

Finally, the work [132] investigated classical ML models, such as SVM, XGBoost, RF, and multilayer perceptron, to predict ozone concentrations in different locations across England. The highest-performing approach was identified as SVM, reaching significant metric values shown in Table 8. The GNN-SAGE-QNN model achieved better metric values for all time windows than our proposed approach. This could be explained by the fact that the quantum–classical model is more complex than simple ML methodologies, and is able to capture the spatiotemporal information within the data much more effectively. On top of that, differently from our methodology, the work by [132] did not include exogenous data as input variables. Consequently, future ozone concentration was forecasted based solely on its previous values. It is expected that the additional information from exogenous data could benefit the predictive model, allowing them to reach better values for both RMSE and the coefficient of determination, as reported in [52].

As a final remark, it is noticeable that the metrics values decrease for further leading times. This behavior is observed for the references [52,131,132], from Table 8. Such a phenomenon also occurs for the proposed methodology, as presented in Table 7. This is a known behavior in the time-series forecasting field, and has been reported in previous studies [30,119,133]. The reduced values indicate that further forecasting horizons are difficult to model, given the increased complexity of capturing the spatiotemporal information from these scenarios. This could be because the lack of additional information from the dataset, whether due to insufficient samples or independent variables, results in inferior results compared to those of short forecasting windows [100]. Nevertheless, it is important to disclose that even with the limitations of larger leading times, the proposed methodology was able to output remarkable results, proving to be a valuable tool for forecasting larger horizons and demonstrating the capacity of hybrid quantum–classical machine learning models.

## 5. Conclusions

This work proposed a new hybrid architecture combining quantum and classical ML models for air quality forecasting. To this end, historical data for Houston, Texas, ranging from 2011 to 2020, were used. With this information, ozone concentration in the atmosphere was predicted for three different forecasting horizons: 1, 3, and 6 h.

Preliminary results revealed that the initial hybrid graph-based DL model and the QNN implementation of the quantum counterpart (Figure 2) achieved suboptimal forecast error for the 1 h time frame regarding the FS metric. With further investigation of the proposed topology, a new configuration ensuring a smoother progression between the classical dense layer and the quantum one (Figure 4) was a better choice for the studied scenario. Further investigation also highlighted the importance of data preprocessing and the number of ansatzes in the QML structure. With this newer configuration, the hybrid model substantially improved its predictions for forecasting horizons of 1, 3, and 6 h, boosting its RMSE, $R^2$, and FS values (Table 7).

Despite the scarcity of publications in this field, a comparison with the existing literature for the QML standalone and hybrid approaches showed that the proposed methodology can achieve similar results. This deems the hybrid GNN-SAGE-QNN methodology a viable air quality forecasting tool. Additional comparisons with traditional DL and ML models published in this field revealed that the proposed methodology can provide competitive outcomes, surpassing some well-established ozone forecasting methodologies.

Several other configurations for the hybrid methodology can be evaluated for future work. In this case, different implementations could consider distinct quantum structure approaches. The proposed method can also be evaluated for different spatiotemporal practical applications, such as wind speed and renewable energy forecasting. Additionally, statistical tests such as the Wilcoxson signed-rank test can further investigate the significance of the improvements achieved by the hybrid methodology over classical benchmarks. Finally, implementing the GNN-SAGE-QNN model in the next generation of advanced quantum hardware would allow us to assess its full potential as an early warning system for real-time poor air quality zone mapping, where improved computational time and more accurate results are expected to be achieved.

The development of quantum-based air quality management technologies is anticipated to offer more precise and accurate results for real-time monitoring systems. This advancement will significantly improve the development of more accurate air quality zone mapping and allow the implementation of more reliable early alert systems. This will promote better decision-making strategies by the stakeholders and lawmakers and enhanced approaches for reducing the harmful effects of air pollutants on human health, thus promoting a more sustainable future.

**Author Contributions:** Conceptualization, J.V.G.T. and B.G.; methodology, P.A.C.R., J.V.G.T. and B.G.; software, V.O.S. and P.A.C.R.; validation, P.A.C.R., J.V.G.T. and B.G.; formal analysis, P.A.C.R.; investigation, P.A.C.R., J.V.G.T. and B.G.; resources, J.V.G.T. and B.G.; data curation, J.V.G.T. and B.G.; writing—original draft preparation, V.O.S. and P.A.C.R.; writing—review and editing, V.O.S., P.A.C.R., J.V.G.T. and B.G.; visualization, V.O.S. and P.A.C.R.; supervision, J.V.G.T. and B.G.; project administration, J.V.G.T. and B.G.; funding acquisition, B.G. and J.V.G.T. All authors have read and agreed to the published version of the manuscript.

**Funding:** This research study was funded by the Natural Sciences and Engineering Research Council of Canada (NSERC) Alliance, grant No. 401643, in association with Lakes Environmental Software Inc., and by the Conselho Nacional de Desenvolvimento Científico e Tecnológico—Brasil (CNPq), grant No. 303585/2022-6.

**Institutional Review Board Statement:** Not applicable.

**Informed Consent Statement:** Not applicable.

**Data Availability Statement:** The raw data supporting the conclusions of this article will be made available by the authors upon request.

**Conflicts of Interest:** The author Jesse Van Griensven Thé is employed by the company Lakes Environmental. The remaining authors declare that this research was conducted in the absence of any commercial or financial relationships that could be construed as potential conflicts of interest.

## References


1. Carneiro, F.O.M.; Moura, L.F.M.; Costa Rocha, P.A.; Pontes Lima, R.J.; Ismail, K.A.R. Application and Analysis of the Moving Mesh Algorithm AMI in a Small Scale HAWT: Validation with Field Test's Results against the Frozen Rotor Approach. *Energy* **2019**, *171*, 819–829. [CrossRef]
2. Vidal Bezerra, F.D.; Pinto Marinho, F.; Costa Rocha, P.A.; Oliveira Santos, V.; Van Griensven Thé, J.; Gharabaghi, B. Machine Learning Dynamic Ensemble Methods for Solar Irradiance and Wind Speed Predictions. *Atmosphere* **2023**, *14*, 1635. [CrossRef]
3. Verma, P.; Verma, R.; Mallet, M.; Sisodiya, S.; Zare, A.; Dwivedi, G.; Ristovski, Z. Assessment of Human and Meteorological Influences on PM10 Concentrations: Insights from Machine Learning Algorithms. *Atmos. Pollut. Res.* **2024**, *15*, 102123. [CrossRef]
4. Liao, M.; Braunstein, Z.; Rao, X. Sex Differences in Particulate Air Pollution-Related Cardiovascular Diseases: A Review of Human and Animal Evidence. *Sci. Total Environ.* **2023**, *884*, 163803. [CrossRef] [PubMed]
5. Sierra-Vargas, M.P.; Montero-Vargas, J.M.; Debray-García, Y.; Vizuet-de-Rueda, J.C.; Loaeza-Román, A.; Terán, L.M. Oxidative Stress and Air Pollution: Its Impact on Chronic Respiratory Diseases. *Int. J. Mol. Sci.* **2023**, *24*, 853. [CrossRef] [PubMed]
6. Lin, G.-Y.; Cheng, Y.-H.; Dejchanchaiwong, R. Insight into Secondary Inorganic Aerosol (SIA) Production Enhanced by Domestic Ozone Using a Machine Learning Technique. *Atmos. Environ.* **2024**, *316*, 120194. [CrossRef]
7. Xie, Y.; Zhou, M.; Hunt, K.M.R.; Mauzerall, D.L. Recent $PM_{2.5}$ Air Quality Improvements in India Benefited from Meteorological Variation. *Nat. Sustain.* **2024**, *7*, 983–993. [CrossRef]
8. Shindell, D.; Faluvegi, G.; Nagamoto, E.; Parsons, L.; Zhang, Y. Reductions in Premature Deaths from Heat and Particulate Matter Air Pollution in South Asia, China, and the United States under Decarbonization. *Proc. Natl. Acad. Sci. USA* **2024**, *121*, e2312832120. [CrossRef]
9. Ghaffarpasand, O.; Okure, D.; Green, P.; Sayyahi, S.; Adong, P.; Sserunjogi, R.; Bainomugisha, E.; Pope, F.D. The Impact of Urban Mobility on Air Pollution in Kampala, an Exemplar Sub-Saharan African City. *Atmos. Pollut. Res.* **2024**, *15*, 102057. [CrossRef]
10. Roy, R.; Braathen, N.A. *The Rising Cost of Ambient Air Pollution Thus Far in the 21st Century: Results from the BRIICS and the OECD Countries*; OECD: Paris, France, 2017.
11. Tran, H.M.; Tsai, F.-J.; Lee, Y.-L.; Chang, J.-H.; Chang, L.-T.; Chang, T.-Y.; Chung, K.F.; Kuo, H.-P.; Lee, K.-Y.; Chuang, K.-J.; et al. The Impact of Air Pollution on Respiratory Diseases in an Era of Climate Change: A Review of the Current Evidence. *Sci. Total Environ.* **2023**, *898*, 166340. [CrossRef]
12. Ma, M.; Liu, M.; Song, X.; Liu, M.; Fan, W.; Wang, Y.; Xing, H.; Meng, F.; Lv, Y. Spatiotemporal Patterns and Quantitative Analysis of Influencing Factors of $PM_{2.5}$ and $O_3$ Pollution in the North China Plain. *Atmos. Pollut. Res.* **2024**, *15*, 101950. [CrossRef]
13. Steinebach, Y. Instrument Choice, Implementation Structures, and the Effectiveness of Environmental Policies: A Cross-National Analysis. *Regul. Gov.* **2022**, *16*, 225–242. [CrossRef]
14. Tang, J.-H.; Pan, S.-R.; Li, L.; Chan, P.-W. A Machine Learning-Based Method for Identifying the Meteorological Field Potentially Inducing Ozone Pollution. *Atmos. Environ.* **2023**, *312*, 120047. [CrossRef]
15. Zafra-Pérez, A.; Medina-García, J.; Boente, C.; Gómez-Galán, J.A.; Sánchez De La Campa, A.; De La Rosa, J.D. Designing a Low-Cost Wireless Sensor Network for Particulate Matter Monitoring: Implementation, Calibration, and Field-Test. *Atmos. Pollut. Res.* **2024**, *15*, 102208. [CrossRef]
16. Sayeed, A.; Choi, Y.; Jung, J.; Lops, Y.; Eslami, E.; Salman, A.K. A Deep Convolutional Neural Network Model for Improving WRF Simulations. *IEEE Trans. Neural Netw. Learn. Syst.* **2021**, *34*, 750–760. [CrossRef]
17. Li, Z.; Wang, W.; He, Q.; Chen, X.; Huang, J.; Zhang, M. Estimating Ground-Level High-Resolution Ozone Concentration across China Using a Stacked Machine-Learning Method. *Atmos. Pollut. Res.* **2024**, *15*, 102114. [CrossRef]
18. Liu, Y.; Geng, X.; Smargiassi, A.; Fournier, M.; Gamage, S.M.; Zalzal, J.; Yamanouchi, S.; Torbatian, S.; Minet, L.; Hatzopoulou, M.; et al. Changes in Industrial Air Pollution and the Onset of Childhood Asthma in Quebec, Canada. *Environ. Res.* **2024**, *243*, 117831. [CrossRef] [PubMed]
19. Tang, B.; Stanier, C.O.; Carmichael, G.R.; Gao, M. Ozone, Nitrogen Dioxide, and $PM_{2.5}$ Estimation from Observation-Model Machine Learning Fusion over S. Korea: Influence of Observation Density, Chemical Transport Model Resolution, and Geostationary Remotely Sensed AOD. *Atmos. Environ.* **2024**, *331*, 120603. [CrossRef]
20. Yang, W.; Wu, Q.; Li, J.; Chen, X.; Du, H.; Wang, Z.; Li, D.; Tang, X.; Sun, Y.; Ye, Z.; et al. Predictions of Air Quality and Challenges for Eliminating Air Pollution during the 2022 Olympic Winter Games. *Atmos. Res.* **2024**, *300*, 107225. [CrossRef]
21. Wang, L.; Zhao, Y.; Shi, J.; Ma, J.; Liu, X.; Han, D.; Gao, H.; Huang, T. Predicting Ozone Formation in Petrochemical Industrialized Lanzhou City by Interpretable Ensemble Machine Learning. *Environ. Pollut.* **2022**, *318*, 120798. [CrossRef] [PubMed]

22. Rocha, P.A.C.; Santos, V.O. Global Horizontal and Direct Normal Solar Irradiance Modeling by the Machine Learning Methods XGBoost and Deep Neural Networks with CNN-LSTM Layers: A Case Study Using the GOES-16 Satellite Imagery. *Int. J. Energy Environ. Eng.* **2022**, *13*, 1271–1286. [CrossRef]
23. Friberg, M.D.; Zhai, X.; Holmes, H.A.; Chang, H.H.; Strickland, M.J.; Sarnat, S.E.; Tolbert, P.E.; Russell, A.G.; Mulholland, J.A. Method for Fusing Observational Data and Chemical Transport Model Simulations To Estimate Spatiotemporally Resolved Ambient Air Pollution. *Environ. Sci. Technol.* **2016**, *50*, 3695–3705. [CrossRef]
24. Goodfellow, I.; Bengio, Y.; Courville, A. *Deep Learning*; MIT Press: Cambridge, MA, USA, 2016; ISBN 978-0-262-03561-3.
25. LeCun, Y.; Bengio, Y.; Hinton, G. Deep Learning. *Nature* **2015**, *521*, 436–444. [CrossRef] [PubMed]
26. Salah Eddine, S.; Drissi, L.B.; Mejjad, N.; Mabrouki, J.; Romanov, A.A. Machine Learning Models Application for Spatiotemporal Patterns of Particulate Matter Prediction and Forecasting over Morocco in North of Africa. *Atmos. Pollut. Res.* **2024**, *15*, 102239. [CrossRef]
27. Luo, R.; Wang, J.; Gates, I. Estimating Air Methane and Total Hydrocarbon Concentrations in Alberta, Canada Using Machine Learning. *Atmos. Pollut. Res.* **2024**, *15*, 101984. [CrossRef]
28. De La Cruz Libardi, A.; Masselot, P.; Schneider, R.; Nightingale, E.; Milojevic, A.; Vanoli, J.; Mistry, M.N.; Gasparrini, A. High Resolution Mapping of Nitrogen Dioxide and Particulate Matter in Great Britain (2003–2021) with Multi-Stage Data Reconstruction and Ensemble Machine Learning Methods. *Atmos. Pollut. Res.* **2024**, *15*, 102284. [CrossRef] [PubMed]
29. Nouayti, A.; Berriban, I.; Chham, E.; Azahra, M.; Satti, H.; Drissi El-Bouzaidi, M.; Yerrou, R.; Arectout, A.; Ziani, H.; El Bardouni, T.; et al. Utilizing Innovative Input Data and ANN Modeling to Predict Atmospheric Gross Beta Radioactivity in Spain. *Atmos. Pollut. Res.* **2024**, *15*, 102264. [CrossRef]
30. Costa Rocha, P.A.; Oliveira Santos, V.; Scott, J.; Van Griensven Thé, J.; Gharabaghi, B. Application of Graph Neural Networks to Forecast Urban Flood Events: The Case Study of the 2013 Flood of the Bow River, Calgary, Canada. *Int. J. River Basin Manag.* **2024**, 1–18. [CrossRef]
31. Ren, K.; Chen, K.; Jin, C.; Li, X.; Yu, Y.; Lin, Y. TEMDI: A Temporal Enhanced Multisource Data Integration Model for Accurate PM2.5 Concentration Forecasting. *Atmos. Pollut. Res.* **2024**, *15*, 102269. [CrossRef]
32. Lei, T.M.T.; Ng, S.C.W.; Siu, S.W.I. Application of ANN, XGBoost, and Other ML Methods to Forecast Air Quality in Macau. *Sustainability* **2023**, *15*, 5341. [CrossRef]
33. Al-Eidi, S.; Amsaad, F.; Darwish, O.; Tashtoush, Y.; Alqahtani, A.; Niveshitha, N. Comparative Analysis Study for Air Quality Prediction in Smart Cities Using Regression Techniques. *IEEE Access* **2023**, *11*, 115140–115149. [CrossRef]
34. Ketu, S. Spatial Air Quality Index and Air Pollutant Concentration Prediction Using Linear Regression Based Recursive Feature Elimination with Random Forest Regression (RFERF): A Case Study in India. *Nat. Hazards* **2022**, *114*, 2109–2138. [CrossRef]
35. Zhou, Z.; Qiu, C.; Zhang, Y. A Comparative Analysis of Linear Regression, Neural Networks and Random Forest Regression for Predicting Air Ozone Employing Soft Sensor Models. *Sci. Rep.* **2023**, *13*, 22420. [CrossRef]
36. He, Z.; Guo, Q.; Wang, Z.; Li, X. Prediction of Monthly PM2.5 Concentration in Liaocheng in China Employing Artificial Neural Network. *Atmosphere* **2022**, *13*, 1221. [CrossRef]
37. Ren, L.; An, F.; Su, M.; Liu, J. Exposure Assessment of Traffic-Related Air Pollution Based on CFD and BP Neural Network and Artificial Intelligence Prediction of Optimal Route in an Urban Area. *Buildings* **2022**, *12*, 1227. [CrossRef]
38. Maltare, N.N.; Vahora, S. Air Quality Index Prediction Using Machine Learning for Ahmedabad City. *Digit. Chem. Eng.* **2023**, *7*, 100093. [CrossRef]
39. Imam, M.; Adam, S.; Dev, S.; Nesa, N. Air Quality Monitoring Using Statistical Learning Models for Sustainable Environment. *Intell. Syst. Appl.* **2024**, *22*, 200333. [CrossRef]
40. Zhen, L.; Chen, B.; Wang, L.; Yang, L.; Xu, W.; Huang, R.-J. Data Imbalance Causes Underestimation of High Ozone Pollution in Machine Learning Models: A Weighted Support Vector Regression Solution. *Atmos. Environ.* **2025**, *343*, 120952. [CrossRef]
41. Salman, A.K.; Choi, Y.; Singh, D.; Kayastha, S.G.; Dimri, R.; Park, J. Temporal CNN-Based 72-h Ozone Forecasting in South Korea: Explainability and Uncertainty Quantification. *Atmos. Environ.* **2025**, *343*, 120987. [CrossRef]
42. Aarthi, C.; Ramya, V.J.; Falkowski-Gilski, P.; Divakarachari, P.B. Balanced Spider Monkey Optimization with Bi-LSTM for Sustainable Air Quality Prediction. *Sustainability* **2023**, *15*, 1637. [CrossRef]
43. Nguyen, H.A.D.; Ha, Q.P.; Duc, H.; Azzi, M.; Jiang, N.; Barthelemy, X.; Riley, M. Long Short-Term Memory Bayesian Neural Network for Air Pollution Forecast. *IEEE Access* **2023**, *11*, 35710–35725. [CrossRef]
44. Jia, T.; Cheng, G.; Chen, Z.; Yang, J.; Li, Y. Forecasting Urban Air Pollution Using Multi-Site Spatiotemporal Data Fusion Method (Geo-BiLSTMA). *Atmos. Pollut. Res.* **2024**, *15*, 102107. [CrossRef]
45. Guo, Z.; Yang, C.; Wang, D.; Liu, H. A Novel Deep Learning Model Integrating CNN and GRU to Predict Particulate Matter Concentrations. *Process Saf. Environ. Prot.* **2023**, *173*, 604–613. [CrossRef]
46. Lu, Y.; Li, K. Multistation Collaborative Prediction of Air Pollutants Based on the CNN-BiLSTM Model. *Environ. Sci. Pollut. Res.* **2023**, *30*, 92417–92435. [CrossRef] [PubMed]

47. Rabie, R.; Asghari, M.; Nosrati, H.; Emami Niri, M.; Karimi, S. Spatially Resolved Air Quality Index Prediction in Megacities with a CNN-Bi-LSTM Hybrid Framework. *Sustain. Cities Soc.* **2024**, *109*, 105537. [CrossRef]
48. Wang, Z.; Wu, F.; Yang, Y. Air Pollution Measurement Based on Hybrid Convolutional Neural Network with Spatial-and-Channel Attention Mechanism. *Expert Syst. Appl.* **2023**, *233*, 120921. [CrossRef]
49. Li, W.; Jiang, X. Prediction of Air Pollutant Concentrations Based on TCN-BiLSTM-DMAttention with STL Decomposition. *Sci. Rep.* **2023**, *13*, 4665. [CrossRef] [PubMed]
50. Elbaz, K.; Shaban, W.M.; Zhou, A.; Shen, S.-L. Real Time Image-Based Air Quality Forecasts Using a 3D-CNN Approach with an Attention Mechanism. *Chemosphere* **2023**, *333*, 138867. [CrossRef] [PubMed]
51. Ren, D.; Hu, Q.; Zhang, T. EKLT: Kolmogorov-Arnold Attention-Driven LSTM with Transformer Model for River Water Level Prediction. *J. Hydrol.* **2025**, *649*, 132430. [CrossRef]
52. Oliveira Santos, V.; Costa Rocha, P.A.; Scott, J.; Van Griensven Thé, J.; Gharabaghi, B. Spatiotemporal Air Pollution Forecasting in Houston-TX: A Case Study for Ozone Using Deep Graph Neural Networks. *Atmosphere* **2023**, *14*, 308. [CrossRef]
53. Jin, X.-B.; Wang, Z.-Y.; Kong, J.-L.; Bai, Y.-T.; Su, T.-L.; Ma, H.-J.; Chakrabarti, P. Deep Spatio-Temporal Graph Network with Self-Optimization for Air Quality Prediction. *Entropy* **2023**, *25*, 247. [CrossRef] [PubMed]
54. Wang, X.; Zhang, S.; Chen, Y.; He, L.; Ren, Y.; Zhang, Z.; Li, J.; Zhang, S. Air Quality Forecasting Using a Spatiotemporal Hybrid Deep Learning Model Based on VMD–GAT–BiLSTM. *Sci. Rep.* **2024**, *14*, 17841. [CrossRef] [PubMed]
55. Biamonte, J.; Wittek, P.; Pancotti, N.; Rebentrost, P.; Wiebe, N.; Lloyd, S. Quantum Machine Learning. *Nature* **2017**, *549*, 195–202. [CrossRef] [PubMed]
56. Sachdeva, N.; Harnett, G.S.; Maity, S.; Marsh, S.; Wang, Y.; Winick, A.; Dougherty, R.; Canuto, D.; Chong, Y.Q.; Hush, M.; et al. Quantum Optimization Using a 127-Qubit Gate-Model IBM Quantum Computer Can Outperform Quantum Annealers for Nontrivial Binary Optimization Problems. *arXiv* **2024**, arXiv:2406.01743.
57. Klusch, M.; Lässig, J.; Müssig, D.; Macaluso, A.; Wilhelm, F.K. Quantum Artificial Intelligence: A Brief Survey. *KI Künstl. Intell.* **2024**. [CrossRef]
58. Cerezo, M.; Verdon, G.; Huang, H.-Y.; Cincio, L.; Coles, P.J. Challenges and Opportunities in Quantum Machine Learning. *Nat. Comput. Sci.* **2022**, *2*, 567–576. [CrossRef]
59. Thakkar, S.; Kazdaghli, S.; Mathur, N.; Kerenidis, I.; Ferreira–Martins, A.J.; Brito, S. Improved Financial Forecasting via Quantum Machine Learning. *Quantum Mach. Intell.* **2024**, *6*, 27. [CrossRef]
60. Bhasin, N.K.; Kadyan, S.; Santosh, K.; HP, R.; Changala, R.; Bala, B.K. Enhancing Quantum Machine Learning Algorithms for Optimized Financial Portfolio Management. In Proceedings of the 2024 Third International Conference on Intelligent Techniques in Control, Optimization and Signal Processing (INCOS), Tamil Nadu, India, 14–16 March 2024; pp. 1–7.
61. Avramouli, M.; Savvas, I.K.; Vasilaki, A.; Garani, G. Unlocking the Potential of Quantum Machine Learning to Advance Drug Discovery. *Electronics* **2023**, *12*, 2402. [CrossRef]
62. Sagingalieva, A.; Kordzanganeh, M.; Kenbayev, N.; Kosichkina, D.; Tomashuk, T.; Melnikov, A. Hybrid Quantum Neural Network for Drug Response Prediction. *Cancers* **2023**, *15*, 2705. [CrossRef] [PubMed]
63. Lourenço, M.P.; Herrera, L.B.; Hostaš, J.; Calaminici, P.; Köster, A.M.; Tchagang, A.; Salahub, D.R. QMLMaterial–A Quantum Machine Learning Software for Material Design and Discovery. *J. Chem. Theory Comput.* **2023**, *19*, 5999–6010. [CrossRef] [PubMed]
64. Rosyid, M.R.; Mawaddah, L.; Santosa, A.P.; Akrom, M.; Rustad, S.; Dipojono, H.K. Implementation of Quantum Machine Learning in Predicting Corrosion Inhibition Efficiency of Expired Drugs. *Mater. Today Commun.* **2024**, *40*, 109830. [CrossRef]
65. Gohel, P.; Joshi, M. Quantum Time Series Forecasting. In Proceedings of the Sixteenth International Conference on Machine Vision (ICMV 2023), Yerevan, Armenia, 15–18 November 2024; Osten, W., Ed.; SPIE: Yerevan, Armenia, 2024; p. 39.
66. Oliveira Santos, V.; Marinho, F.P.; Costa Rocha, P.A.; Thé, J.V.G.; Gharabaghi, B. Application of Quantum Neural Network for Solar Irradiance Forecasting: A Case Study Using the Folsom Dataset, California. *Energies* **2024**, *17*, 3580. [CrossRef]
67. Dong, Y.; Li, F.; Zhu, T.; Yan, R. Air Quality Prediction Based on Quantum Activation Function Optimized Hybrid Quantum Classical Neural Network. *Front. Phys.* **2024**, *12*, 1412664. [CrossRef]
68. Farooq, O.; Shahid, M.; Arshad, S.; Altaf, A.; Iqbal, F.; Vera, Y.A.M.; Flores, M.A.L.; Ashraf, I. An Enhanced Approach for Predicting Air Pollution Using Quantum Support Vector Machine. *Sci. Rep.* **2024**, *14*, 19521. [CrossRef] [PubMed]
69. Tsoukalas, M.Z.; Stratogiannis, D.G.; Dritsas, L.; Papadakis, A. Hybrid Quantum and Classical Machine Learning Classification for IoT-Based Air Quality Monitoring. In Proceedings of the 2024 10th International Conference on Automation, Robotics and Applications (ICARA), Athens, Greece, 22–24 February 2024; pp. 546–550.
70. Khan, M.A.-Z.; Al-Karaki, J.; Omar, M. Predicting Water Quality Using Quantum Machine Learning: The Case of the Umgeni Catchment (U20A) Study Region. *arXiv* **2024**, arXiv:2411.18141v1.
71. Nammouchi, A.; Kassler, A.; Theorachis, A. Quantum Machine Learning in Climate Change and Sustainability: A Review. *arXiv* **2023**, arXiv:2310.09162. [CrossRef]
72. Rahman, S.M.; Alkhalaf, O.H.; Alam, M.S.; Tiwari, S.P.; Shafiullah, M.; Al-Judaibi, S.M.; Al-Ismail, F.S. Climate Change Through Quantum Lens: Computing and Machine Learning. *Earth Syst. Environ.* **2024**, *8*, 705–722. [CrossRef]

73. Feynman, R.P. Simulating Physics with Computers. *Int. J. Theor. Phys.* **1982**, *21*, 467–488. [CrossRef]
74. Berberian, S.K. *Introduction to Hilbert Space*; American Mathematical Soc.: Providence, RI, USA, 1999; ISBN 978-0-8218-1912-8.
75. Halmos, P.R. *Introduction to Hilbert Space and the Theory of Spectral Multiplicity: Second Edition*; Courier Dover Publications: Garden City, NY, USA, 2017; ISBN 978-0-486-81733-0.
76. Fleisch, D.A. *A Student's Guide to the Schrödinger Equation*; Cambridge University Press: Cambridge, UK, 2020; ISBN 978-1-108-83473-5.
77. Nielsen, M.A.; Chuang, I.L. *Quantum Computation and Quantum Information*, 10th anniversary ed.; Cambridge University Press: Cambridge, UK; New York, NY, USA, 2010; ISBN 978-1-107-00217-3.
78. Kaye, P.; Laflamme, R.; Mosca, M. *An Introduction to Quantum Computing*; Oxford University Press: Oxford, UK, 2007; ISBN 978-0-19-857000-4.
79. Einstein, A.; Podolsky, B.; Rosen, N. Can Quantum-Mechanical Description of Physical Reality Be Considered Complete? *Phys. Rev.* **1935**, *47*, 777–780. [CrossRef]
80. Shor, P.W. Polynomial-Time Algorithms for Prime Factorization and Discrete Logarithms on a Quantum Computer. *SIAM J. Comput.* **1997**, *26*, 1484–1509. [CrossRef]
81. Shankar, R. *Principles of Quantum Mechanics*; Springer: Berlin/Heidelberg, Germany, 1994; ISBN 978-0-306-44790-7.
82. Grover, L.K. A Fast Quantum Mechanical Algorithm for Database Search. In Proceedings of the Twenty-Eighth Annual ACM Symposium on Theory of Computing—STOC '96, Philadelphia, PA, USA, 22–24 May 1996; pp. 212–219.
83. Sutor, R.S. *Dancing with Qubits: How Quantum Computing Works and How It May Change the World*; Expert insight; Packt: Birmingham, UK; Mumbai, India, 2019; ISBN 978-1-83882-736-6.
84. Lee, D.P.Y.; Ji, D.H.; Cheng, D.R. *Quantum Computing and Information: A Scaffolding Approach*; Polaris QCI Publishing: Albany, NY, USA, 2024; ISBN 978-1-961880-03-0.
85. Griffiths, D.J. *Introduction to Quantum Mechanics*; Cambridge University Press: Cambridge, UK, 2017; ISBN 978-1-107-17986-8.
86. Schuld, M.; Petruccione, F. *Machine Learning with Quantum Computers*; Springer Nature: Berlin/Heidelberg, Germany, 2021; ISBN 978-3-030-83098-4.
87. Schuld, M.; Killoran, N. Quantum Machine Learning in Feature Hilbert Spaces. *Phys. Rev. Lett.* **2019**, *122*, 040504. [CrossRef] [PubMed]
88. Mengoni, R.; Di Pierro, A. Kernel Methods in Quantum Machine Learning. *Quantum Mach. Intell.* **2019**, *1*, 65–71. [CrossRef]
89. Combarro, E.F.; Gonzalez-Castillo, S.; Meglio, A.D. *A Practical Guide to Quantum Machine Learning and Quantum Optimization: Hands-on Approach to Modern Quantum Algorithms*; Packt Publishing Ltd.: Birmingham, UK, 2023; ISBN 978-1-80461-830-1.
90. Sushmit, M.M.; Mahbubul, I.M. Forecasting Solar Irradiance with Hybrid Classical–Quantum Models: A Comprehensive Evaluation of Deep Learning and Quantum-Enhanced Techniques. *Energy Convers. Manag.* **2023**, *294*, 117555. [CrossRef]
91. IBM Quantum Documentation TwoLocal. Available online: https://docs.quantum.ibm.com/api/qiskit/qiskit.circuit.library.TwoLocal (accessed on 12 June 2024).
92. Dalvand, Z.; Hajarian, M. Solving Generalized Inverse Eigenvalue Problems via L-BFGS-B Method. *Inverse Probl. Sci. Eng.* **2020**, *28*, 1719–1746. [CrossRef]
93. Wilson, M.; Stromswold, R.; Wudarski, F.; Hadfield, S.; Tubman, N.M.; Rieffel, E.G. Optimizing Quantum Heuristics with Meta-Learning. *Quantum Mach. Intell.* **2021**, *3*, 13. [CrossRef]
94. Zheng, L.; Li, D.; Xu, J.; Xia, Z.; Hao, H.; Chen, Z. A Twenty-Years Remote Sensing Study Reveals Changes to Alpine Pastures under Asymmetric Climate Warming. *ISPRS J. Photogramm. Remote Sens.* **2022**, *190*, 69–78. [CrossRef]
95. Bergholm, V.; Izaac, J.; Schuld, M.; Gogolin, C.; Ahmed, S.; Ajith, V.; Alam, M.S.; Alonso-Linaje, G.; AkashNarayanan, B.; Asadi, A.; et al. PennyLane: Automatic Differentiation of Hybrid Quantum-Classical Computations. *arXiv* **2022**, arXiv:1811.04968.
96. Javadi-Abhari, A.; Treinish, M.; Krsulich, K.; Wood, C.J.; Lishman, J.; Gacon, J.; Martiel, S.; Nation, P.D.; Bishop, L.S.; Cross, A.W.; et al. Quantum Computing with Qiskit. *arXiv* **2024**, arXiv:2405.08810.
97. Correa-Jullian, C.; Cofre-Martel, S.; San Martin, G.; Lopez Droguett, E.; De Novaes Pires Leite, G.; Costa, A. Exploring Quantum Machine Learning and Feature Reduction Techniques for Wind Turbine Pitch Fault Detection. *Energies* **2022**, *15*, 2792. [CrossRef]
98. Jing, H.; Li, Y.; Brandsema, M.J.; Chen, Y.; Yue, M. HHL Algorithm with Mapping Function and Enhanced Sampling for Model Predictive Control in Microgrids. *Appl. Energy* **2024**, *361*, 122878. [CrossRef]
99. Labonne, M. *Hands-On Graph Neural Networks Using Python: Practical Techniques and Architectures for Building Powerful Graph and Deep Learning Apps with PyTorch*; Packt Publishing Ltd.: Birmingham, UK, 2023; ISBN 1-80461-070-4.
100. Oliveira Santos, V.; Costa Rocha, P.A.; Scott, J.; Van Griensven Thé, J.; Gharabaghi, B. A New Graph-Based Deep Learning Model to Predict Flooding with Validation on a Case Study on the Humber River. *Water* **2023**, *15*, 1827. [CrossRef]
101. Oliveira Santos, V.; Costa Rocha, P.A.; Thé, J.V.G.; Gharabaghi, B. Graph-Based Deep Learning Model for Forecasting Chloride Concentration in Urban Streams to Protect Salt-Vulnerable Areas. *Environments* **2023**, *10*, 157. [CrossRef]

102. Oliveira Santos, V.; Costa Rocha, P.A.; Scott, J.; Van Griensven Thé, J.; Gharabaghi, B. Spatiotemporal Analysis of Bidimensional Wind Speed Forecasting: Development and Thorough Assessment of LSTM and Ensemble Graph Neural Networks on the Dutch Database. *Energy* **2023**, *278*, 127852. [CrossRef]
103. Owusu-Mfum, S.; Hudson, M.D.; Osborne, P.E.; Roberts, T.J.; Zapata-Restrepo, L.M.; Williams, I.D. Atmospheric Pollution in Port Cities. *Atmosphere* **2023**, *14*, 1135. [CrossRef]
104. Pan, S.; Roy, A.; Choi, Y.; Eslami, E.; Thomas, S.; Jiang, X.; Gao, H.O. Potential Impacts of Electric Vehicles on Air Quality and Health Endpoints in the Greater Houston Area in 2040. *Atmos. Environ.* **2019**, *207*, 38–51. [CrossRef]
105. Sadeghi, B.; Choi, Y.; Yoon, S.; Flynn, J.; Kotsakis, A.; Lee, S. The Characterization of Fine Particulate Matter Downwind of Houston: Using Integrated Factor Analysis to Identify Anthropogenic and Natural Sources. *Environ. Pollut.* **2020**, *262*, 114345. [CrossRef] [PubMed]
106. Beck, H.E.; Zimmermann, N.E.; McVicar, T.R.; Vergopolan, N.; Berg, A.; Wood, E.F. Present and Future Köppen-Geiger Climate Classification Maps at 1-Km Resolution. *Sci. Data* **2018**, *5*, 180214. [CrossRef] [PubMed]
107. Kyropoulou, M. Thermal Comfort and Indoor Air Quality in Higher Education: A Case Study in Houston, TX, during Mid-Season. *J. Phys. Conf. Ser.* **2023**, *2600*, 102022. [CrossRef]
108. EPA, U. Green Book. Available online: https://www3.epa.gov/airquality/greenbook/jnc.html (accessed on 16 December 2022).
109. Liao, D.; Wang, L.; Wang, Y.; Huang, H.; Zhuang, Z.; Choi, S.-D.; Hong, Y. Effects of Heat Waves on Ozone Pollution in a Coastal Industrial City: Meteorological Impacts and Photochemical Mechanisms. *Atmos. Pollut. Res.* **2024**, *15*, 102280. [CrossRef]
110. Open Street Map OpenStreetMap. Available online: https://www.openstreetmap.org (accessed on 27 February 2024).
111. QGIS Development Team QGIS Geographic Information System 2024, Accessed Public-Domain Spatial Visualization and Decision-Making Tools. Available online: https://qgis.org/ (accessed on 1 June 2024).
112. Géron, A. *Hands-On Machine Learning with Scikit-Learn, Keras, and TensorFlow*; O'Reilly Media, Inc.: Sebastopol, CA, USA, 2022; ISBN 978-1-0981-2246-1.
113. Zhang, A.; Lipton, Z.C.; Li, M.; Smola, A.J. *Dive into Deep Learning*; Cambridge University Press: Cambridge, UK, 2023; ISBN 978-1-009-38943-3.
114. Paszke, A.; Gross, S.; Massa, F.; Lerer, A.; Bradbury, J.; Chanan, G.; Killeen, T.; Lin, Z.; Gimelshein, N.; Antiga, L.; et al. PyTorch: An Imperative Style, High-Performance Deep Learning Library. In *Advances in Neural Information Processing Systems*; Curran Associates, Inc.: Red Hook, NY, USA, 2019; Volume 32.
115. Baïle, R.; Muzy, J.F.; Poggi, P. Short-Term Forecasting of Surface Layer Wind Speed Using a Continuous Random Cascade Model. *Wind Energy* **2011**, *14*, 719–734. [CrossRef]
116. Hanifi, S.; Liu, X.; Lin, Z.; Lotfian, S. A Critical Review of Wind Power Forecasting Methods—Past, Present and Future. *Energies* **2020**, *13*, 3764. [CrossRef]
117. Yang, D.; Kleissl, J.; Gueymard, C.A.; Pedro, H.T.C.; Coimbra, C.F.M. History and Trends in Solar Irradiance and PV Power Forecasting: A Preliminary Assessment and Review Using Text Mining. *Sol. Energy* **2018**, *168*, 60–101. [CrossRef]
118. Chicco, D.; Warrens, M.J.; Jurman, G. The Coefficient of Determination R-Squared Is More Informative than SMAPE, MAE, MAPE, MSE and RMSE in Regression Analysis Evaluation. *PeerJ Comput. Sci.* **2021**, *7*, e623. [CrossRef]
119. Ding, Y.; Zhu, Y.; Feng, J.; Zhang, P.; Cheng, Z. Interpretable Spatio-Temporal Attention LSTM Model for Flood Forecasting. *Neurocomputing* **2020**, *403*, 348–359. [CrossRef]
120. Dazzi, S.; Vacondio, R.; Mignosa, P. Flood Stage Forecasting Using Machine-Learning Methods: A Case Study on the Parma River (Italy). *Water* **2021**, *13*, 1612. [CrossRef]
121. Brunton, S.L.; Kutz, J.N. *Data-Driven Science and Engineering: Machine Learning, Dynamical Systems, and Control*; Cambridge University Press: Cambridge, UK, 2022; ISBN 978-1-009-09848-9.
122. Ghojogh, B.; Crowley, M.; Karray, F.; Ghodsi, A. *Elements of Dimensionality Reduction and Manifold Learning*; Springer International Publishing: Cham, Switzerland, 2023; ISBN 978-3-031-10601-9.
123. James, G.; Witten, D.; Hastie, T.; Tibshirani, R.; Taylor, J. *An Introduction to Statistical Learning: With Applications in Python*; Springer International Publishing: Berlin/Heidelberg, Germany, 2023; ISBN 978-3-031-38746-3.
124. Tennie, F.; Palmer, T.N. Quantum Computers for Weather and Climate Prediction: The Good, the Bad, and the Noisy. *Bull. Am. Meteorol. Soc.* **2023**, *104*, E488–E500. [CrossRef]
125. Lim, H.; Kang, D.H.; Kim, J.; Pellow-Jarman, A.; McFarthing, S.; Pellow-Jarman, R.; Jeon, H.-N.; Oh, B.; Rhee, J.-K.K.; No, K.T. Fragment Molecular Orbital-Based Variational Quantum Eigensolver for Quantum Chemistry in the Age of Quantum Computing. *Sci. Rep.* **2024**, *14*, 2422. [CrossRef] [PubMed]
126. Zheng, K.; Van Griensven, J.; Fraser, R. A Quantum Machine Learning Approach to Spatiotemporal Emission Modelling. *Atmosphere* **2023**, *14*, 944. [CrossRef]
127. Wang, J.; Li, H.; Yang, H.; Wang, Y. Intelligent Multivariable Air-Quality Forecasting System Based on Feature Selection and Modified Evolving Interval Type-2 Quantum Fuzzy Neural Network. *Environ. Pollut.* **2021**, *274*, 116429. [CrossRef]

128. Yu, Y.; Hu, G.; Liu, C.; Xiong, J.; Wu, Z. Prediction of Solar Irradiance One Hour Ahead Based on Quantum Long Short-Term Memory Network. *IEEE Trans. Quantum Eng.* **2023**, *4*, 1–15. [CrossRef]
129. Nevo, S.; Morin, E.; Gerzi Rosenthal, A.; Metzger, A.; Barshai, C.; Weitzner, D.; Voloshin, D.; Kratzert, F.; Elidan, G.; Dror, G.; et al. Flood Forecasting with Machine Learning Models in an Operational Framework. *Hydrol. Earth Syst. Sci.* **2022**, *26*, 4013–4032. [CrossRef]
130. Hong, F.; Ji, C.; Rao, J.; Chen, C.; Sun, W. Hourly Ozone Level Prediction Based on the Characterization of Its Periodic Behavior via Deep Learning. *Process Saf. Environ. Prot.* **2023**, *174*, 28–38. [CrossRef]
131. Bekkar, A.; Hssina, B.; Douzi, S.; Douzi, K. Transformer-Based Model for Multi-Horizon Forecasting Ozone in Marrakech City, Morocco. In Proceedings of the 2023 14th International Conference on Intelligent Systems: Theories and Applications (SITA), Casablanca, Morocco, 22–23 November 2023; pp. 1–8.
132. Wood, D.A. Ozone Air Concentration Trend Attributes Assist Hours-Ahead Forecasts from Univariate Recorded Data Avoiding Exogenous Data Inputs. *Urban Clim.* **2023**, *47*, 101382. [CrossRef]
133. Zavala-Romero, O.; Segura-Chavez, P.A.; Camacho-Gonzalez, P.; Zavala-Hidalgo, J.; Garcia, A.R.; Oropeza-Alfaro, P.; Romero-Centeno, R.; Gomez-Ramos, O. Operational Ozone Forecasting System in Mexico City: A Machine Learning Framework Integrating Forecasted Weather and Historical Ozone Data. *Atmos. Environ.* **2025**, *344*, 121017. [CrossRef]

atmosphere

MDPI

Article

# Air Pollutant Concentration Forecasting with WTMP: Wavelet Transform-Based Multilayer Perceptron

**Xiaoling Wang [1], Liangzhao Tao [1,*], Mingliang Fu [2,*] and Qi Wang [3]**

1 School of Information and Electronic Engineering, Zhejiang Gongshang University, Hangzhou 310018, China
2 State Environmental Protection Key Laboratory of Vehicle Emission Control and Simulation, Chinese Research Academy of Environmental Sciences, Beijing 100012, China
3 College of Life and Environmental Sciences, Wenzhou University, Wenzhou 325000, China
* Correspondence: 22020090021@pop.zjgsu.edu.cn (L.T.); fumingliang160@163.com (M.F.)

**Abstract:** Atmospheric pollutants' real-time changes and the internal interactions among various data make it challenging to efficiently predict concentration variations. In order to extract more information from the time series of pollutants and improve the accuracy of prediction models, we propose a type of Multilayer Perceptron model based on wavelet decomposition, named Wavelet Transform-based Multilayer Perceptron (WTMP) model. This model decomposes pollutant data through overlapping discrete wavelet transforms to extract non-stationarity and nonlinear dependencies in the time series. It combines the decomposed data with static covariate information such as data collection time and inputs them into an improved Multilayer Perceptron (MLP) model, reconstructing and outputting the prediction results. Finally, the model is validated using atmospheric pollutant data collected at a specific location in Ruian City, Zhejiang Province, China. The results indicate that the model performs well with minimal prediction errors.

**Keywords:** atmospheric pollutants; wavelet decomposition; multi-layer perceptron model

## 1. Introduction

In recent years, the escalating issue of environmental pollution has brought the subject of air pollution to the forefront of global concern [1], especially at the United Nations Climate Change Conferences. The primary atmospheric pollutants include six categories: nitrogen oxides, sulfur oxides, carbon oxides, ozone, PM10, and PM2.5 Specifically, PM2.5 refers to particulate matter with a diameter less than 2.5 microns, which are so minuscule and lightweight that they can linger in the atmosphere for extended periods. When inhaled, these particles can bypass the nasal cavity, reaching the throat and even the lungs [2], potentially leading to conditions such as bronchitis, asthma, cardiovascular diseases, and lung cancer, thereby posing a significant risk to human health. In 2018, research by Heft-Neal et al. [3] suggested that PM2.5 levels above the minimum exposure threshold are a primary cause of 22% of infant mortality across 30 countries. Additionally, various atmospheric pollutants can irritate the respiratory tract, predisposing individuals to respiratory illnesses. Prolonged exposure to polluted environments may result in impaired vascular endothelial function and an array of cardiovascular diseases. Fine particulate matter in the atmosphere can also impact the nervous system, elevating the risk of neurological disorders such as Alzheimer's and Parkinson's diseases.

Given the profound impact of air pollution on living environments and human health, research into air pollution forecasting has become increasingly vital [4]. It is recognized as a crucial aspect of environmental protection and a key instrument in guiding scientific decision-making for air pollution alerts and control measures. Numerous cities have established air quality monitoring stations to provide real-time surveillance of PM2.5 and other atmospheric pollutants.

The composition and concentration of various gas pollutants form a time series, characterized by chronological order. By analyzing historical data, it is possible to identify dynamic patterns and predict future values. This capability is crucial for forecasting problems involving time-dependent components. This paper focuses on analyzing and extracting patterns from the concentration data of air pollutants within a specific time period to predict future air quality levels. Researchers have discovered that, in addition to other atmospheric pollutants, climatic factors such as wind speed, wind direction, and relative humidity significantly impact the prediction of future concentrations [5]. Therefore, it is essential to extract these complex features, determine their correlations, and construct a time series model to foresee potential future changes.

With the continuous advancement of science and technology and the widespread use of various sensors, air quality monitoring networks have been established nationwide. These networks, equipped with sensors covering multiple monitoring points, facilitate the collection of time series data, making it increasingly easier to obtain various data. Consequently, there is a higher demand for accurate results from time series predictions. Multivariate analysis and precise forecasting of time series data are becoming increasingly important in time prediction. Currently, time series forecasting techniques are a significant research focus in many fields, including weather prediction, disease prevention, pollutant detection, and stock market forecasting [6]. They are also relevant in environmental management and forecasting, regional development planning, and other areas.

In the evolution of air quality prediction, numerical modeling methods have seen significant advancements. Early numerical models primarily relied on simple physical and chemical processes to forecast the dispersion of pollutants. For instance, in 2004, Arystanbekova K et al. [7] proposed the Gaussian plume model based on the Gaussian distribution characteristics of atmospheric pollutant concentrations. While these models offered an initial understanding of pollutant behavior, they often had limited capabilities when dealing with complex terrain and meteorological conditions.

With the increase in computational power, modern numerical models such as climate models and regional air quality models (like the Community Multiscale Air Quality (CMAQ) model and the Weather Research and Forecasting (WRF)-Chem model) have incorporated more complex chemical reactions and meteorological factors. These models can more accurately simulate the generation, transformation, and transport of pollutants, providing high-resolution forecasts. However, they typically require a substantial amount of input data and come with high computational costs, and in practical applications, they may face challenges in data acquisition and model parameterization.

In contrast, data-driven models, such as machine learning methods, train on historical observational data and can quickly generate predictions, often performing well when there is an abundance of data. However, they lack an in-depth understanding of physical processes, which may lead to suboptimal performance in the face of extreme weather or new pollutants.

With the continuous development of computer technology, traditional time series forecasting methods, often used for short-term univariate predictions, struggle with nonlinear relationships and multivariate, complex data. Consequently, machine learning and data mining techniques have been introduced to address these challenges [8]. These methods analyze data patterns, extract features, and capture correlations through training to predict target events. Machine learning-based models excel at fitting nonlinear time series data but often overlook historical temporal dependencies, resulting in high computational complexity and potential local optimization issues. In contrast, deep learning methods are highly effective in large-scale data feature extraction and model construction [9]. Deep neural networks, with their multiple hidden layers, can identify complex nonlinear patterns and require large datasets for accurate predictions.Recurrent Neural Network (RNN) models address the temporal characteristics of time series data [10], offering short-term memory capabilities. They are widely used in fields such as financial forecasting and power load prediction. Long Short-Term Memory (LSTM) networks, a variant of RNNs, were proposed by Hochreiter et al. [11] to address gradient vanishing in long-term dependencies. For example, Li and colleagues developed an air quality prediction model using LSTM networks [12], which outperformed ARIMA [13] and SVR [14] models. Following this, Wu and colleagues [15] integrated LSTM with quadratic decomposition and optimization techniques to create a hybrid forecasting model. Additionally, Tao and colleagues proposed a model combining 1D-CNNs and Bi-GRU [9] for predicting the Air Pollution Index. Building on the success of CNNs in image analysis, Dong and colleagues hypothesized its potential for deciphering the spatial patterns of air pollutants. Wu and colleagues enhanced prediction accuracy by applying CNN to multi-site air quality data [16].

To overcome the limitations of single networks, various combinations of deep learning networks have been explored. By integrating CNN's feature extraction capabilities with LSTM's proficiency in handling time series data, Kuo and colleagues [17] introduced the CNN-LSTM model for PM2.5 forecasting. Similarly, Pak and colleagues developed a CNN-LSTM model for ozone prediction [18], significantly improving predictive capabilities. In recent decades, the complexity of factors influencing air pollution predictions, combined with the widespread use of various sensors, has led to the establishment of air quality monitoring networks across the country. This has enabled the collection of comprehensive time series data, providing valuable insights for developing more effective and accurate industrial pollutant index prediction models. Qi and colleagues combined LSTM networks with Graph Convolutional Networks (GCN) to predict PM2.5 concentrations in the air [19]. They proposed a hybrid GC-LSTM model and validated its effectiveness and feasibility by evaluating metrics such as recall rate, false alarm rate, and correlation coefficient matrices.

In the domain of sequence modeling such as language and vision, Transformers [20] have emerged as the most successful deep learning architecture, which utilize multi-head attention and positional encoding to significantly enhance the model's ability to capture complex relationships within data [21]. By leveraging multiple attention heads, the Transformer can focus on different parts of the input sequence simultaneously, leading to a richer representation of the data [22]. The inclusion of positional encoding allows the model to understand the order of the sequence, which is crucial for tasks involving time series data where temporal relationships are important.

An advanced Transformer-based model, known as Informer [23], is specifically designed for long-sequence prediction. It introduces three key innovations to improve efficiency and reduce memory consumption in long-sequence forecasting: ProbSparse self-attention, which selects the most relevant attention scores to focus on, thereby reducing computational complexity; self-attention distillation, which distills the attention maps to retain the most important information while discarding redundancy; and generative decoding, which enhances the model's ability to generate long sequences by iteratively predicting future values.

The model, Autoformer [24], proposes a depth-wise decomposition architecture and a self-correlation mechanism to more effectively forecast time series data. The depth-wise decomposition architecture separates trend and seasonal components of the time series, allowing the model to capture both long-term trends and short-term fluctuations more accurately. The self-correlation mechanism leverages the inherent autocorrelation in time series data to improve prediction accuracy by identifying and utilizing repeating patterns within the data.

The model incorporates a frequency-enhanced structure, known as FEDFormer [25], to enhance the accuracy of time series forecasting by better capturing frequency components. This structure allows the model to better handle periodic variations in time series data, excelling in capturing both long-term and short-term patterns. Pyraformer [26] uses a pyramidal self-attention mechanism, which has linear complexity and can attend to different granularities of time information. This multi-granularity attention allows Pyraformer to flexibly handle complex time series data while maintaining computational efficiency.

Recently, the DLinear model [27] has been introduced as a straightforward linear model that, contrary to expectations, surpasses the performance of numerous Transformer-based models. DLinear adeptly acquires a direct linear relationship between the context and the forecast horizon, highlighting the limitations of sub-quadratic approximations used in self-attention mechanisms. This model focuses on the simplicity and interpretability of linear relationships, demonstrating that in many cases, complex attention mechanisms may not be necessary for achieving high performance in time series forecasting. By leveraging the strengths of linear modeling, DLinear provides a robust and efficient alternative to more complex deep learning models. However, it has notable deficiencies, as it is ill-suited for modeling non-linear dependencies that may exist within the time-series sequence and between time-independent covariates. Consequently, for time-series forecasting tasks that require capturing intricate and non-linear interactions, more sophisticated models might be necessary to achieve optimal performance.

In this paper, we introduce a concise and efficient deep learning framework designed to significantly improve long-term predictions of atmospheric pollution concentrations compared to traditional neural network models. Our model, based on the Multi-Layer Perceptron (MLP), maintains simplicity by excluding complex self-attention or convolutional mechanisms. This design leads to linear computational growth as the context and forecast horizon lengths expand, making it more suitable for long-term predictions of pollutant concentrations compared to many Transformer-based models.

We propose the Wavelet Transform-based Multilayer Perceptron (WTMP) model architecture for long-term air pollution concentration forecasting. This model combines the signal decomposition algorithm of Maximal Overlap Discrete Wavelet Transform (MODWT) with an enhanced MLP structure. When applied to the air pollutant dataset from the Ruian Economic Development Zone in Zhejiang Province, our model demonstrates superior or similar performance compared to existing neural network-based benchmarks, achieving a Mean Squared Error reduction of over 15%. Additionally, WTMP is more than five times faster in both inference and training compared to the best Transformer-based model.

The effectiveness and feasibility of the proposed model were validated through a comprehensive set of experiments. Our findings reveal that the WTMP model has significant potential in accurately predicting long-term variations in atmospheric pollution concentrations, providing valuable insights for air quality monitoring and management. This innovative approach opens new avenues for future research in this field.

The remaining parts of this paper are arranged as follows. Section 2 discusses the relevant deep learning models. Section 3 presents the details of the proposed WTMP model. Section 4 provides the experiments and result analysis. Finally, Section 5 concludes the paper.

## 2. Background and Related Works

Models for long-term forecasting can be broadly categorized into multivariate and univariate models. Multivariate models leverage the historical data of multiple interrelated time series variables to predict their future behavior. Recent neural network-based multivariate models have gained traction, such as LongTrans [28], Informer [23], Autoformer [24], FEDFormer [25], and Pyraformer [26].

In contrast, univariate models rely solely on the historical data of a single time series and its covariates to predict its future. In this case, the historical data of other time series is not considered during prediction. Univariate models can be further divided into local and global models. Local univariate models are trained and inferred independently for each time series. Classical models such as Autoregressive (AR), Autoregressive Integrated Moving Average (ARIMA), exponential smoothing models, and the Box-Jenkins methodology fall into this category. Global univariate models use a shared model that is trained across the entire dataset and can be applied to different time series like DLinear model.

It is important to note that both multivariate and univariate models can be fairly compared in long-term forecasting tasks as long as they are evaluated on the same test set. In Section 4, we will follow this evaluation protocol to compare the performance of different models.

### 2.1. Residual Block

Throughout the model architecture, residual connection blocks are frequently employed (as shown in Figure 1). Each residual block is composed of a Multi-Layer Perceptron (MLP) with a hidden layer and employs ReLU activation functions to introduce non-linearity, thereby enhancing the model's capacity to capture complex features. A critical aspect of each residual block is its skip connection, which directly maps the input to the output, allowing gradients to flow uninterrupted through the network. This feature significantly improves the training efficiency of deep networks. Both the dropout and the layer norm can be turned on or off by a model level hyperparameter [29].

Our residual block design presents several advantages. Compared to architectures with complex self-attention mechanisms [20], our MLP-based structure is computationally more efficient, reducing both parameter count and computational complexity. The skip connection also alleviates the vanishing gradient problem, supporting the training of deeper networks. Furthermore, the incorporation of dropout in the skip connection's linear layer enhances the model's generalization and reduces overfitting.

Additionally, the design of the residual block allows for easy integration into various network architectures, making it versatile for different data types and tasks. Compared to residual blocks in other models [30], our design achieves a balance between simplicity and efficiency. Notably, in long-sequence time series forecasting tasks, our residual blocks have shown exceptional performance, as demonstrated by the experimental results.

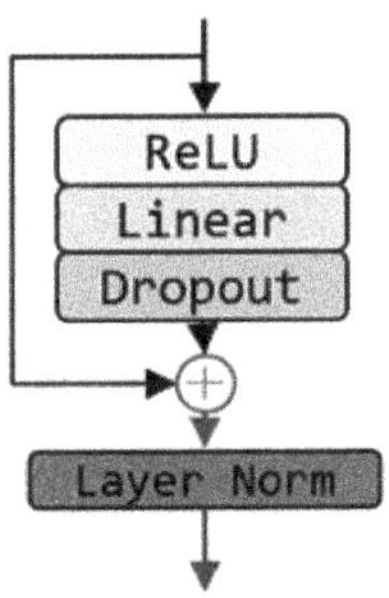


**Figure 1.** Residual Block.

### 2.2. Wavelet Decomposition

A wavelet is a wave-like oscillation localized in time, capable of capturing both the temporal and frequency characteristics of a signal [31]. This contrasts with the Fourier Transform, which solely captures the frequency domain. Wavelets can be categorized into two types: (a) Continuous Wavelet Transforms (CWT), which involve an infinite number of wavelets (i.e., every possible wavelet), and (b) Discrete Wavelet Transforms (DWT), which utilize a finite number of wavelets at specific times and locations. Our research focuses on the DWT, which represents a signal using an orthonormal basis representation. This method is widely employed in various fields, such as hydrology, epidemiology, and geophysics. The decomposition provided by DWT yields a set of time series, where each series has a coefficient describing the signal's temporal evolution within a specific frequency band [32]. The DWT operates on a discrete dyadic (octave) grid for the scale parameter $i$ and shift parameter $k$. The equation for the forward DWT is given by [31]:

$$c(i,k) = \sum_t f(t)\Psi^*_{i,k}(t), \quad \text{where} \quad \Psi^*_{i,k}(t) = 2^{\frac{i}{2}}\Psi(2^i t - k). \tag{1}$$

The inverse DWT is defined as:

$$f(t) = \sum_k \sum_i c(i,k)\Psi_{i,k}(t), \tag{2}$$

where $f(\cdot)$ denotes the function and $\Psi$ is the mother wavelet.

### 2.3. MODWT Algorithm

The Maximal Overlap Discrete Wavelet Transform (MODWT) is a variant of the Discrete Wavelet Transform (DWT) with several notable features that address the drawbacks and limitations of the DWT [33]. Both methods use a pair of filters to split the frequency range of the input data into low- and high-frequency components, referred to as scaling and wavelet coefficients, respectively. Unlike the DWT, the MODWT does not include a downsampling step in its decomposition process. As a result, it is generally interpreted that the computational load of the DWT is less than that of the MODWT. However, in matrix operations, all coefficients are calculated simultaneously, making it a single operation. For instance, considering an input signal with $N$ samples, in the DWT, the $N/2$ wavelet coefficients are computed through the multiplication of an $N/2 \times N$ transformation matrix and an $N \times 1$ input signal. By contrast, the $N$ wavelet coefficients of the MODWT are calculated using an $N \times N$ transformation matrix multiplied by an $N \times 1$ input signal.Additionally, in real-time applications, most of the processing time is spent on the computation of wavelet coefficients, which is significantly lower than the sampling frequency. Moreover, the MODWT offers advantages such as the ability to handle any sample size and signal shift invariance.

The Haar wavelet, favored for its computational simplicity, strong locality, and interpretability, stands out as the wavelet basis of choice for time series analysis [31]. While the db4, coif4, and fk4 wavelets may capture certain signal features more intricately, our ablation study in Section 4.3 revealed that their overall impact on atmospheric pollutant monitoring results is negligible, particularly with the coif4 wavelet performing similarly to the Haar wavelet. Given that the Haar wavelet is the simplest and most widely used option, offering superior computational efficiency, ease of use, and versatility, we concluded that it is the preferred choice for our monitoring and analysis efforts [34]. Additionally, the Haar wavelet's filter length of just 2 offers a significant advantage in terms of computational efficiency during the decomposition and reconstruction processes. Figure 2 is an example of using MODWT decomposition.

Below, we define the detail coefficients (denoted by $\tilde{s}_{j,l}$) and scaling coefficients (denoted by $\tilde{w}_{j,l}$) generated by an MODWT algorithm utilizing the Haar filter, which is among the most prevalent wavelet filters in time series literature.

$$\tilde{d}_{j,l} = \frac{d_{j,l}}{2^{j/2}}, \quad \tilde{s}_{j,l} = \frac{s_{j,l}}{2^{j/2}} \tag{3}$$

The wavelet and scaling filters, denoted as $d_{(j,l)}$ and $s_{(j,l)}$, respectively, convolve the coefficients of the $j$-th level (where $l = 0, 1, \ldots, L-1$) from the original time series $\{Y_t : t = 0, 1, \ldots, N-1\}$ to generate the detailed coefficients ($D_{(\tilde{j,t})}$) and scaling coefficients ($S_{(\tilde{j,t})}$) [31] as follows:

$$\tilde{D}_{j,t} = \sum_{l=0}^{L_j-1} \tilde{d}_{j,l} Y_{(t-l)modN}, \quad \tilde{S}_{j,t} = \sum_{l=0}^{L_j-1} \tilde{s}_{j,l} Y_{(t-l)modN} \tag{4}$$

where ($D_{(\tilde{j,t})}$). The length of the decomposition into different scale detail coefficients is consistent with the length of the original signal.

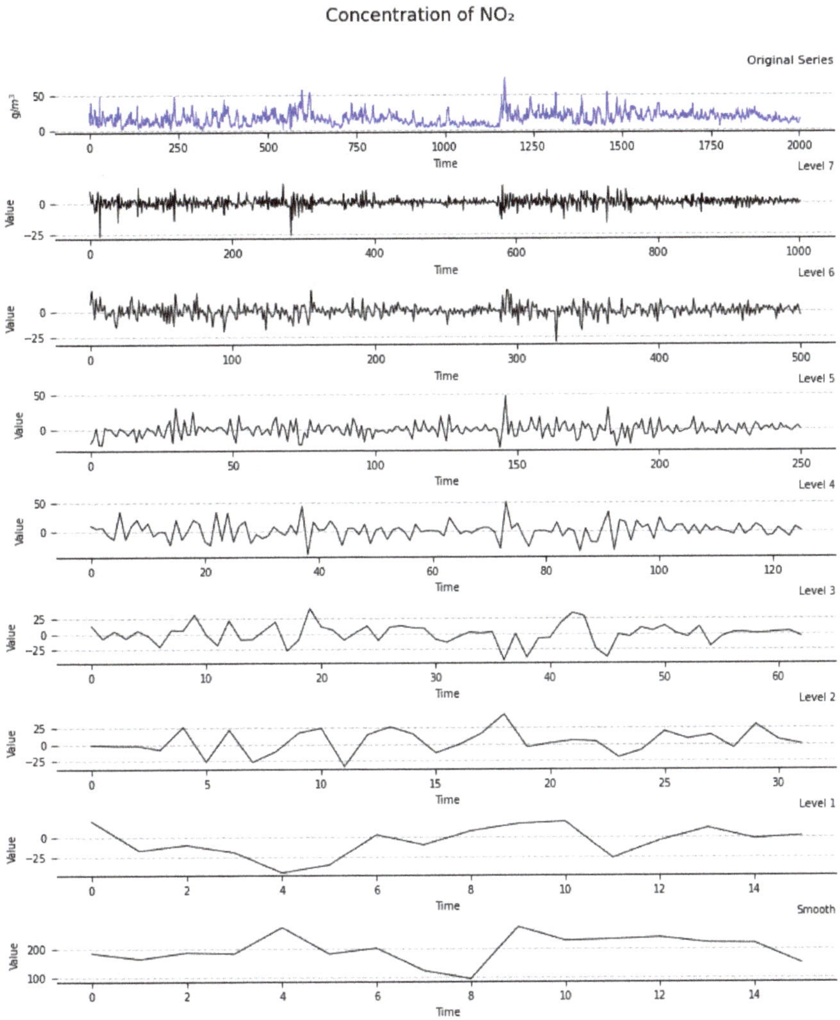


**Figure 2.** MODWT decomposition of the $NO_2$ dataset. The first curve represents the original time series; the last represents the smooth series, i.e., the scaling coefficient of the time series after computing the MODWT algorithm. The other seven curves (in between) represent the different level coefficient calculated by the MODWT algorithm. The filter used for this decomposition is the Haar filter.

## *2.4. Decomposition Level Selection*

In time series forecasting, selecting the number of MODWT decomposition levels is a critical step that determines the extent of frequency components that can be captured. Setting the number of decomposition levels $J+1$ to $\lfloor \log_e N \rfloor$ is a method based on both theory and experience [31]. It ensures that the number of decomposition levels is proportional to the length of the time series $N$, preventing an excess that could increase noise or a deficiency that might miss vital information. This approach provides the capability for multi-scale analysis, allowing the model to extract features from different temporal scales, thereby better understanding and forecasting the behavior of the time series. Additionally, this setting is flexible and can adapt to time series data of varying lengths, achieving a good balance between theory and practice, and supporting the enhancement of the predictive model's performance.

### 2.5. Covariate Information

Covariates are commonly used in statistics and data analysis to refer to variables that may influence the outcome of a primary dependent variable. While these covariates are not the main focus of the study, they may be associated with the primary variable and therefore need to be accounted for and controlled in the analysis. Covariates can be categorized into static and dynamic types [29]. Dynamic covariates change over time, such as wind speed, wind direction, or temperature. In contrast, static covariates are variables that do not change or can be determined in advance, such as the date.

We use $x_i^t \in \mathbb{R}^r$ to denote the $r$-dimensional covariates of time-series $i$ at time $t$. These covariates can include variables specific to the time-series, such as the day of the week and holiday information. Although these variables vary over time, they are known in advance and are thus more appropriately treated as static covariates. In many applications, incorporating such covariates has been shown to significantly improve forecasting accuracy, and a well-designed model architecture should effectively integrate them.

### 2.6. Problem Setting

Let $i : j$ denote the data set $\{i, i+1, \ldots, j\}$. We define the input of pollutant data at time $t$ as $y^t_{(1:L)}$, and $x^t_{(1:L+H)}$ as the input time data at time $t$, which includes both known sequences and forecasting sequence time information. The predicted pollutant results at time $t$ are represented by $\tilde{y}^t_{(L+1:L+H)}$. Here, $L$ and $H$ represent the lengths of the input and output sequences, respectively. The dimension of $x$ is $r$, and the feature dimensions of $y$ and $\tilde{y}$ are denoted by $d_y$ and $d_{\tilde{y}}$, respectively, representing the number of pollutant species for input and output.Let the predictive model function be denoted as $f$.

The prediction process is then defined as:

$$f : \left( \{y^t_{(1:L)}\}_{t=1}^N, \{x^t_{(1:L+H)}\}_{t=1}^N \right) \rightarrow \{\tilde{y}^t_{(L+1:L+H)}\}_{t=1}^N. \quad (5)$$

## 3. Methodology

In the current section, we provide an in-depth description of our model. With reference to Figure 3, we propose an enhanced MLP network model that synergizes historical data with both time series and wavelet characteristics to modify the DLinear model. To address the inherent discontinuities in the temporal monitoring of atmospheric pollutant concentrations, we leverage a sophisticated GAN [35] for effective data interpolation. Thereafter, the refined dataset is integrated into our model framework, which employs a modwt decomposition, enriched with temporal covariate data, to yield the forecasted results.

### 3.1. Data Description

The data is sourced from the environmental monitoring station located at coordinates 27.778203, 120.655098 in the Economic Development Zone of Ruian City. Ruian Economic Development Zone is located in the southeast coast of Zhejiang Province, with a terrain dominated by hills and mountains. The land is higher in the west and lower in the east, with the western part being a region of medium and low hills and mountains, generally above 600 m in altitude. The industry in the area is well-developed, forming industries including automotive parts, polymer materials, and mechanical electronics. The station monitors various atmospheric pollutants including nitrogen oxides, sulfur oxides, carbon oxides, ozone, PM10, PM2.5, as well as meteorological indicators such as wind speed, wind direction, temperature, and relative humidity. Data indicators with missing values accounting for more than 10% of the total samples are excluded, resulting in a dataset comprising 104 data indicators. Both the original and prediction sequences in the experiments are set to have dimensions of 104. The experimental dataset consists of historical monitoring data from 1 January 2021 to 30 June 2022, with a sampling interval of 1 h, totaling 13,095 data points.

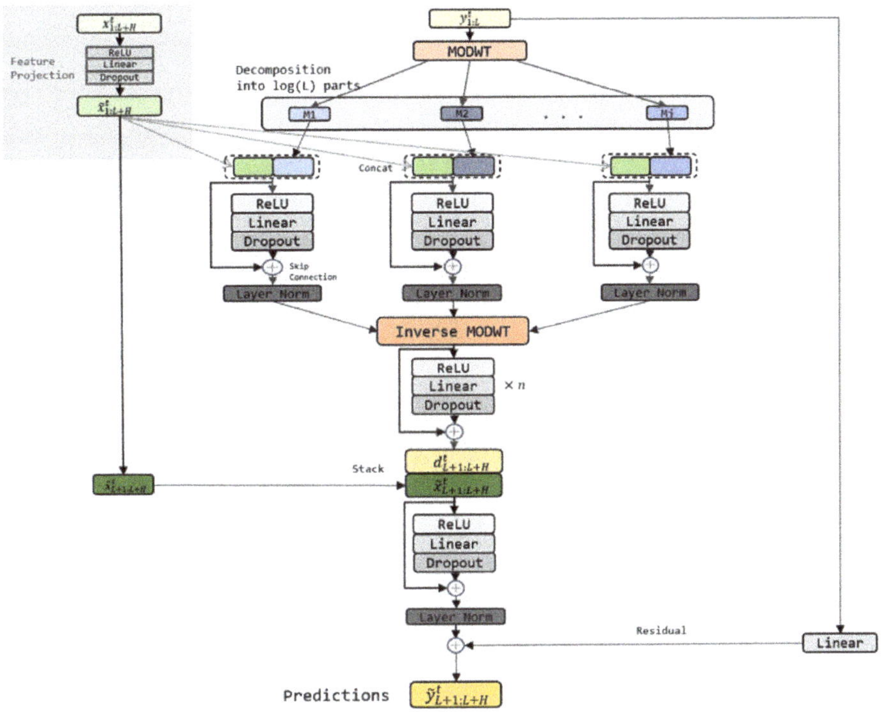


**Figure 3.** The architecture of the WTMP model.

The data is sourced from the environmental monitoring station in the Economic Development Zone of Ruian City. The station monitors various atmospheric pollutants including nitrogen oxides, sulfur oxides, carbon oxides, ozone, PM10, PM2.5, as well as meteorological indicators such as wind speed, wind direction, temperature, and relative humidity. During the data preprocessing phase, the collected raw data went through multiple steps. Initially, data entries with more than ten percent missing values were removed, followed by the elimination of irrelevant attributes. Table 1 provides an overview of the attributes and their corresponding symbols in each dataset after the completion of the preprocessing stage.

**Table 1.** Attribute Comparison Table.

| Name | Unit | Name | Unit |
|---|---|---|---|
| Time | year-month-day-hour | $SO_2$ | $\mu g/m^3$ |
| $NO_2$ | $\mu g/m^3$ | $O_3$ | $\mu g/m^3$ |
| $PM_{2.5}$ | $\mu g/m^3$ | $PM_{10}$ | $\mu g/m^3$ |
| Wind speed | m/s | Wind direction | ° |
| Temperature | °C | Isobutane | ppb |
| Ethane | ppb | Propane | ppb |
| Butane | ppb | Isopentane | ppb |
| 2,3-Dimethylbutane | ppb | 2-Methylpentane | ppb |

**Table 1.** *Cont.*

| Name | Unit | Name | Unit |
|---|---|---|---|
| 3-Methylpentane | ppb | n-Hexane | ppb |
| Methylcyclopentane | ppb | 2,4-Dimethylpentane | ppb |
| Cyclohexane | ppb | 2-Methylhexane | ppb |
| 2,3-Dimethylpentane | ppb | 3-Methylhexane | ppb |
| 2,2,4-Trimethylpentane | ppb | n-Heptane | ppb |
| Methylcyclohexane | ppb | 2,3,4-Trimethylpentane | ppb |
| 2-Methylheptane | ppb | 3-Methylheptane | ppb |
| n-Octane | ppb | n-Nonane | ppb |
| n-Decane | ppb | n-Undecane | ppb |
| n-Dodecane | ppb | 1-Hexene | ppb |
| Ethylene | ppb | Propylene | ppb |

### 3.2. Data Processing

In the actual training phase of the WTMP model, we need time series data that are as continuous as possible, and each record must have complete annotations. However, due to the missing data in the time series, we need to adopt appropriate methods to initialize the data and obtain the covariate information within the data. The data processing methods in this paper include data imputation.

Accurately imputing missing values in datasets is critical for subsequent analyses. To address this challenge, Yoon et al. introduced the Generative Adversarial Imputation Network (GAIN) [35]. This model leverages the framework of Generative Adversarial Networks (GANs), where a generator and a discriminator are trained adversarially to produce plausible imputations for missing data.

The GAIN model's principal virtue is its adeptness at capitalizing on the comprehensive data distribution to deduce missing values, yielding more authentic and credible imputations. As delineated in Figure 4, GAIN distinguishes itself from conventional imputation strategies—like mean and regression imputation—and even from sophisticated methodologies such as MissForest [36] and Autoencoders [37], by adeptly leveraging the global data distribution. This proficiency enables GAIN to craft imputations that transcend mere reliance on localized correlations, instead embracing the holistic attributes of the dataset.

Our extensive array of experiments, conducted across a tapestry of datasets with missing data ranging from a mere 2% to a substantial 20%, consistently underscored the GAIN algorithm's superior efficacy across the continuum of data incompleteness. This distinctive edge endows GAIN with a unique advantage for multidimensional data applications, particularly excelling in the domain of atmospheric pollutant forecasting.

Our experiments with atmospheric pollutant datasets not only confirm the superior performance of the GAIN algorithm but also highlight its adaptability and precision in environmental data analysis. These features make GAIN an essential tool for environmental scientists and policymakers, playing a significant role in developing effective strategies to address atmospheric pollution.

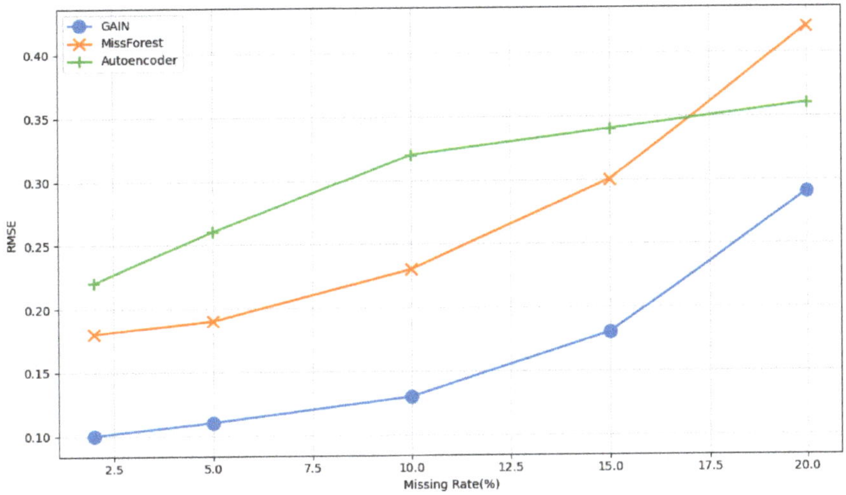


**Figure 4.** RMSE performance in various missing rates.

### 3.3. Model Architecture

For the input pollutant raw data $y^t_{1:L}$, it is initially decomposed via the MODWT to capture more explicit information about the observed time series. ethm will provide us with $J$ higher-frequency detail coefficients and a smooth (trend component) of the available time series. In Figure 2, we present a visual representation of the MODWT decomposition applied to the $NO_2$ concentration series in atmospheric pollution data, utilizing a Haar filter. The MODWT decomposition can be represented as:

$$F_{\text{modwt}}(y^t_{1:L}) = \sum_{j=1}^{J} M_j + S \tag{6}$$

Here, $M_j$ represents the $j$-th level details, and $S$ is the smooth component of the decomposed time series.The MRA-based MODWT decomposed series can handle nonstationarity and seasonality in the time series datasets and capture the overall trend of the series.

As depicted in the top-left of Figure 3, for the input static covariates $x^t_{1:L+H}$, which include information such as time, date, holidays, known in advance for both the input sequence (of length $L$) and the prediction sequence (of length $H$) [38], we utilize a residual block to map the dimensionality from $r$ to $\tilde{r}$ ($\tilde{r} \ll r$) , resulting in the mapped sequence $\tilde{x}^t_{1:L+H}$ :

$$\tilde{x}^t_{1:L+H} = \text{ResidualBlock}(x^t_{1:L+H}) \tag{7}$$

This essentially performs dimensionality reduction, preventing the flattening of static covariates into a large one-dimensional vector of size $(L + H)r$. This reduction diminishes the weight of pollutant data during training. After dimensionality reduction, the flattened vector size is only $(L + H)\tilde{r}$.

For the sequence of covariates that has been dimensionality reduced, $\tilde{x}^t_{(1:L+H)}$, in order to incorporate covariate information into the sequence, $\tilde{x}^t_{(1:L+H)}$ is concatenated with the wavelet-decomposed $M_j$ and $S$. The concatenation method involves flattening the two sequences into a one-dimensional connection, resulting in $J + 1$ vectors of size $(L + H)\tilde{r} + L \cdot d_y$. Assuming the output of the residual block is denoted as $h$, then through the residual block, we obtain vectors $\tilde{M}_j$ and $\tilde{S}$ of length $h$:

$$\tilde{M}_j = \text{ResidualBlock}(M_j; \tilde{x}^t_{1:L}), \quad j = 0, 1, \ldots, J \tag{8}$$

$$\tilde{S} = \text{ResidualBlock}(S; \tilde{x}^t_{1:L}) \tag{9}$$

Next, we perform an inverse wavelet transform on $\tilde{M}_j$ and $\tilde{S}$ to obtain one-dimensional vectors of length $h$. These vectors are then passed through $n$ residual blocks to generate a one-dimensional prediction component $g^t$ of length $H \cdot p$, where $H$ represents the output sequence length, $p$ denotes the output dimension, and $n$ is the number of residual blocks, a hyperparameter.

$$g^t = \text{ResidualBlock}\big(F_{\text{imodwt}}(\tilde{M}_1, \ldots, \tilde{M}_J, \tilde{S})\big) \in \mathbb{R}^{p \cdot H} \tag{10}$$

We reshape $g^t$ to obtain the prediction component $d^t_{(L+1):(L+H)}$ in $\mathbb{R}^{H \times p}$:

$$d^t_{(L+1):(L+H)} = \text{Reshape}(g^t) \in \mathbb{R}^{H \times p} \tag{11}$$

Then, we concatenate the prediction component $d^t_{(L+1):(L+H)}$ with the covariates $\tilde{x}^t_{(L+1):(L+H)}$, projecting the transformed covariates onto the prediction component. This results in a vector of dimension $H \times (p + \tilde{r})$, which is passed through another residual block to generate the predicted result $\tilde{d}^t_{(L+1):(L+H)}$:

$$\tilde{d}^t_{(L+1):(L+H)} = \text{ResidualBlock}\left(d^t_{(L+1):(L+H)}; \tilde{x}^t_{(L+1):(L+H)}\right) \in \mathbb{R}^{H \times \tilde{\delta}_y} \tag{12}$$

To strengthen the influence of covariates, particularly for prediction tasks like atmospheric pollutant prediction where date and holiday features have an impact on actual values, we reintroduce the covariates of the future prediction sequence onto the prediction component. This inclusion of covariates acts as a "highway" to prevent the loss of covariate information.

Finally, we add a global residual connection layer within the model [29]. Through a linear layer, the original data $y^t_{1:L}$ is mapped, and then added to $\tilde{d}^t_{(L+1):(L+H)}$ to obtain the final prediction result $\tilde{y}^t_{(L+1):(L+H)}$. This linear layer essentially represents the core of the DLinear model. By incorporating a global residual connection, the WTMP model theoretically performs no worse than the DLinear model. In mathematical notation:

$$\tilde{y}^t_{(L+1):(L+H)} = y^t_{(L+1):(L+H)} + \text{LinearLayer}(y^t_{1:L}) \tag{13}$$

### 3.4. Evaluation

The precision of the predictive model's forecasts is gauged by a metric that assesses the proximity of its predictions to the true outcomes. Specifically, Mean Squared Error (MSE) is utilized as the training loss function, the lower values of MSE indicate better forecasting results.the quality of the model's fit is evaluated by the following criterion:

$$MSE\left(\left\{y^t_{(L+1:L+H)}\right\}_{t=1}^{N}, \left\{\tilde{y}^t_{(L+1:L+H)}\right\}_{t=1}^{N}\right) = \frac{1}{NH}\sum_{t=1}^{N}\left\|y^t_{(L+1:L+H)} - \tilde{y}^i_{(L+1:L+H)}\right\|_2^2. \tag{14}$$

An epoch is comprised of all possible look-back and forecast pairs that can be formed from the training data, allowing for potential overlap between mini-batches. This approach aligns with standard practices in previous studies on long-term forecasting.

Evaluation of the model is conducted on a test set, examining every possible look-back and forecast pair that can be derived from the test data. This method is commonly referred to as rolling validation or evaluation. Optionally, a similar evaluation can be applied to a validation set to fine-tune model parameters for optimal selection.

## 4. Experiments

In our research, we utilize Min–Max normalization to bring the dataset into a standard form. Subsequently, the data is partitioned into three distinct subsets: training, validation, and testing, with the proportions being 70%, 20%, and 10%, respectively. For the training phase, we implement a batch size of 64 and leverage the Adam optimization algorithm, which is set with a learning rate of 0.001. The training process is designed to halt after 200 epochs or if there is no reduction in the loss for 15 successive iterations. The reported metrics are derived from the mean of five independent experimental runs to ensure reliability and consistency.

### 4.1. Baselines and Setup

We choose SOTA Tranformer based models for time-series including Transformer, Informer, Pyraformer, Autoformer. Recent findings by DLinear have demonstrated that straightforward linear models can surpass the above models. Consequently, the DLinear model stands as a critical benchmark for comparison.Furthermore, we have decided to include an LSTM in our comparative evaluation [39], utilizing a one-layer LSTM with a 128-dimensional configuration.

In this study, we aim to accurately predict atmospheric pollutant concentrations in a specific region. Following widely adopted benchmarking standards, we selected 336 and 720 h as the look-back windows and {96, 168, 336} h as the forecast horizons. These time spans are chosen to capture mid-term trends and seasonal patterns in the time series, with data sampled on an hourly basis. The two- to four-week observation period not only covers short-term fluctuations but also accounts for potential cyclical factors. The forecast horizons are set to 4 to 14 days, providing an effective range for distinguishing the long-term prediction performance of different models while avoiding excessive errors that can arise from overly extended forecasts, which might obscure the true differences in model performance.

### 4.2. Results

The Mean Squared Error (MSE) and Mean Absolute Error (MAE) for all models at different forecast lengths are presented in Table 2. The final results are the average values taken from 5 independent runs, with the best experimental results in bold for the same forecast length.

**Table 2.** Comparison of Different Prediction Horizons of Atmospheric Pollutant Data.

| Models | 336/96 | 336/168 | 720/336 | 720/168 |
|---|---|---|---|---|
| Transformer | 0.698 0.487 | 0.674 0.567 | 0.691 0.475 | 0.655 0.548 |
| Pyraformer | 0.580 0.563 | 0.582 0.511 | 0.865 0.423 | 0.635 0.496 |
| Autoformer | 1.190 0.830 | 4.657 1.505 | 1.811 0.532 | 1.137 1.639 |
| Informer | 0.643 0.576 | 0.639 0.545 | 0.654 0.555 | 0.645 0.574 |
| LSTM | 0.611 0.554 | 0.720 0.652 | 0.812 0.651 | 0.812 0.598 |
| DLinear | 0.482 0.442 | 0.517 0.442 | 0.706 0.477 | 0.679 0.479 |
| WTMP | **0.341 0.377** | **0.387 0.412** | **0.545 0.415** | **0.526 0.417** |

Note: The entry "336/96" indicates that the original data sequence is 336 points long, with a prediction sequence of 96 points; subsequent entries follow the same format.

Furthermore, we have applied our trained models to real-world data to evaluate and compare their predictive capabilities, with the comparative curves illustrated in the Figure 5. Our findings indicate that initially, the LSTM model's forecasting accuracy is competitive with those based on Transformer architectures. Nonetheless, with an extended forecast sequence, there is a gradual increase in cumulative errors, which diminishes its predictive accuracy. The DLinear model also shows superiority over Transformer-based models for shorter forecast sequences but faces challenges in capturing the complex dynamics between pollutants as the forecasting horizon grows. Even so, it retains a level of performance that is still respectable. The WTMP model, enhancing the foundation laid by DLinear, demonstrates the most impressive practical performance, thanks to the MODWT decomposition that adeptly manages non-stationary data characteristics. By sidestepping the complexity of the self-attention mechanism, the MLP within the WTMP model excels particularly in long-term forecasting tasks. The experimental outcomes underscore its markedly superior forecasting performance in comparison to other models.

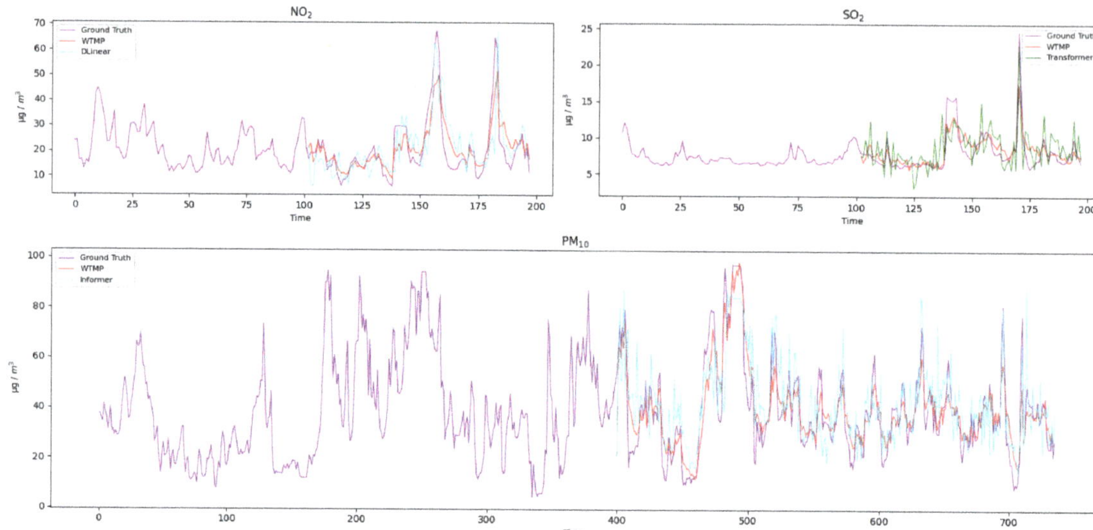


**Figure 5.** Trained model used for predicting different pollutant data and generating long-term forecast curves.

## 4.3. Ablation Study

We aimed to validate the efficacy of the MODWT algorithm, particularly when utilizing the Haar wavelet basis, in the context of time series forecasting for atmospheric pollutant concentrations. To this end, we devised an ablation study comparing various models with distinct configurations of wavelet transformations and other time series decomposition techniques.

### 4.3.1. Model A: Direct Feature Fusion

This model is a variation of the WTMP framework, where we have intentionally omitted the wavelet transformation process. Instead, the model directly integrates the features from residual blocks with covariate information after multiple passes, yielding the final output. This setup serves as a control to assess the impact of wavelet transformation on predictive performance.

### 4.3.2. Model B: Sliding Average Decomposition

Inspired by the effectiveness demonstrated by the sliding window average method in the DLinear model, we replaced the MODWT algorithm with a simple moving average technique in this model. The decomposed input sequence is merged by a straightforward addition, emphasizing the simplicity and efficiency of this approach. This model tests the hypothesis that a more traditional smoothing technique can be competitive with the MODWT algorithm.

### 4.3.3. Model C: MODWT with db4 Wavelet Basis

Here, we selected the 'db4' wavelet basis for the MODWT algorithm to explore the impact of wavelet choice on the forecasting accuracy. The 'db4' basis is known for its smoothness properties, which may be beneficial for certain types of time series data [40].

#### 4.3.4. Model D: MODWT with coif4 Wavelet Basis

In this model, we opted for the 'coif4' wavelet basis, which is designed to provide a more detailed analysis of the time series, potentially capturing higher-frequency variations that could be significant in pollution data.

#### 4.3.5. Model E: MODWT with fk4 Wavelet Basis

We chose the 'fk4' wavelet basis for this model, which is known for its balance between smoothness and detail, offering a comprehensive representation of the time series at various scales.

As before, we processed the validation dataset through the meticulously trained models. By scrutinizing the Mean Squared Error (MSE) and Mean Absolute Error (MAE) metrics, we discerned the comparative efficacy of each model. The Table 3 presents the empirical results of our experiment.

**Table 3.** Comparison of Different Prediction Horizons of Atmospheric Pollutant Data.

| Models | 336/96 | 336/168 | 720/336 | 720/168 |
|---|---|---|---|---|
| WTMP | 0.341 0.377 | 0.387 0.412 | 0.545 0.415 | 0.526 0.417 |
| Model A (Direct Feature Fusion) | 0.763 0.901 | 0.892 0.628 | 1.163 0.837 | 1.121 0.831 |
| Model B (Sliding Average Decomposition) | 0.631 0.510 | 0.692 0.561 | 0.796 0.640 | 0.764 0.603 |
| Model C (db4) | 0.410 0.386 | 0.423 0.465 | 0.596 0.458 | 0.586 0.429 |
| Model D (coif4) | 0.333 0.370 | 0.381 0.422 | 0.552 0.422 | 0.525 0.423 |
| Model E (fk4) | 0.461 0.390 | 0.399 0.409 | 0.610 0.430 | 0.596 0.421 |

From the experimental outcomes, we deduce the following conclusions: Model A, devoid of any decomposition technique, exhibits the weakest predictive performance, with a notably increased disparity from WTMP as the forecast horizon lengthens. Model B, employing a sliding average method, is simpler than wavelet transformations but potentially lacks the finesse to capture the intricate temporal features that MODWT can discern, thus demonstrating subpar predictive capabilities. Subsequently, we have Models C, D, and E that incorporate MODWT decomposition. Model D, leveraging the coif4 wavelet basis, is particularly adept at capturing high-frequency fluctuations and thus rivals the performance of Haar, occasionally outperforming it. Conversely, Models C and E, which employ the 'db4' and 'fk4' wavelet bases respectively, have demonstrated inferior performance in the experiment. They are not adept at managing the variables involved in this study.

### 4.4. More Analyses on Different Models

The LSTM model employs gating mechanisms, but the sequential processing of sequences can lead to error accumulation and vanishing gradients. In contrast, the Transformer model, which uses self-attention mechanisms, can more effectively capture global dependencies, thus compensating for the vanishing gradient problem. The improved Transformer architecture, Informer, utilizes the DMS (Direct Multi-Step) forecasting strategy to address error accumulation. However, the fundamental nature of the Transformer model is to model the semantic association between paired elements, thereby disrupting the temporal order of the elements. Although the introduction of positional encoding and the use of tokens to embed subsequences can help retain some sequential information, this is fatal for time series forecasting. The Dlinear model points out this shortcoming, retaining the sequential characteristics of time series and achieving results close to or surpassing those of Transformer models with just two simple linear layers. However, due to its simplicity, it abandons further processing of temporal information, making it difficult to model non-linear dependencies among the time-series sequence and time-independent covariates. The WTMP model designed based on this effectively addresses the aforementioned issues.

## 5. Conclusions

We propose a novel forecasting model known as WTMP, which integrates wavelet decomposition into an enhanced MLP model. Predictive experiments conducted using atmospheric pollutant data from the Rui'an Economic Development Zone have demonstrated the effectiveness of this method for long-sequence forecasting tasks of industrial park air pollutant concentrations. The key findings are as follows:

1. Prediction models based on the Transformer architecture underperform when compared to a simple MLP model. At least for the task of atmospheric pollutant forecasting, there is no need to employ the complex self-attention mechanism—central to Transformer models—to learn the periodicity and trends of long sequences.

2. The introduction of a wavelet decomposition module and residual connection blocks within the model has significantly improved the capture of non-stationarity and non-linear dependencies within the sequences. This enhancement has further elevated the predictive performance and effectively circumvented the issues of low fitting accuracy and poor generalization associated with linear models.

## 6. Future Work

Next, we will investigate the limitations of the WTMP model and verify its generalization ability by testing it on pollutant data from other regions. Where feasible, we will develop an atmospheric pollutant concentration model based on spatiotemporal data from multiple monitoring stations.

**Author Contributions:** Conceptualization, L.T., M.F. and X.W.; Methodology, L.T. and M.F.; Software, M.F.; Validation, L.T.; Formal Analysis, L.T.; Investigation, Q.W.; Resources, Q.W.; Data Curation, Q.W.; Writing—Original Draft Preparation, L.T.; Writing—Review and Editing, X.W.; Visualization, M.F.; Supervision, X.W.; Project Administration, X.W.; Funding Acquisition, X.W. and M.F. All authors have read and agreed to the published version of the manuscript.

**Funding:** This work was supported by the National Natural Science Foundation of China (62033012).

**Institutional Review Board Statement:** Not applicable.

**Informed Consent Statement:** Not applicable.

**Data Availability Statement:** The datasets used in this study will be available from the corresponding author upon reasonable request.

**Conflicts of Interest:** The authors declare no conflict of interest.

## References

1. Chanchan, G.; Fengying, Z.; Dekun, F.; Qingtao, W.; Min, L. Spatial characteristics of change trends of air pollutants in Chinese urban areas during 2016–2020: The impact of air pollution controls and the COVID-19 pandemic. *Atmos. Res.* **2023**, *283*, 106539.
2. Dw, D. An association between air pollution and mortality in six US cities. *N. Engl. J. Med.* **1993**, *329*, 1753–1759.
3. Heft-Neal, S.; Burney, J.; Bendavid, E.; Burke, M. Robust relationship between air quality and infant mortality in Africa. *Nature* **2018**, *559*, 254–258. [CrossRef]
4. Weijun, W.; Tianyu, M.; Lianru, W. Air pollutant concentration prediction based on a new hybrid model, feature selection, and secondary decomposition. *Air Qual. Atmos. Health* **2023**, *16*, 2019–2033.
5. Chen, L.; Han, B.; Wang, X.; Zhao, J.; Yang, W.; Yang, Z. Machine learning methods in weather and climate applications: A survey. *Appl. Sci.* **2023**, *13*, 12019. [CrossRef]
6. Box, G.E.; Jenkins, G.M.; Reinsel, G.C.; Ljung, G.M. *Time Series Analysis: Forecasting and Control*; John Wiley & Sons: Hoboken, NJ, USA, 2015.
7. Arystanbekova, N.K. Application of Gaussian plume models for air pollution simulation at instantaneous emissions. *Math. Comput. Simul.* **2004**, *67*, 451–458. [CrossRef]
8. Deb, C.; Zhang, F.; Yang, J.; Lee, S.E.; Shah, K.W. A review on time series forecasting techniques for building energy consumption. *Renew. Sustain. Energy Rev.* **2017**, *74*, 902–924. [CrossRef]
9. Tao, Q.; Liu, F.; Li, Y.; Sidorov, D.N. Air Pollution Forecasting Using a Deep Learning Model Based on 1D Convnets and Bidirectional GRU. *IEEE Access* **2019**, *7*, 76690–76698. [CrossRef]
10. David, S.; Valentin, F.; Jan, G.; Tim, J. DeepAR: Probabilistic forecasting with autoregressive recurrent networks. *Int. J. Forecast.* **2020**, *36*, 1181–1191.

11. Hoedt, P.J.; Kratzert, F.; Klotz, D.; Halmich, C.; Holzleitner, M.; Nearing, G.S.; Hochreiter, S.; Klambauer, G. Mc-lstm: Mass-conserving lstm. In Proceedings of the International Conference on Machine Learning, PMLR, Virtual Event, 18–24 July 2021; pp. 4275–4286.
12. Awasthi, P.; Das, A.; Sen, R.; Suresh, A.T. On the benefits of maximum likelihood estimation for regression and forecasting. *arXiv* **2021**, arXiv:2106.10370.
13. Shumway, R.H.; Stoffer, D.S.; Shumway, R.H.; Stoffer, D.S. ARIMA models. In *Time Series Analysis and Its Applications: With R Examples*; Springer: New York, NY, USA, 2017; pp. 75–163.
14. Dhiman, H.S.; Deb, D.; Guerrero, J.M. Hybrid machine intelligent SVR variants for wind forecasting and ramp events. *Renew. Sustain. Energy Rev.* **2019**, *108*, 369–379. [CrossRef]
15. Wu, Q.; Lin, H. A novel optimal-hybrid model for daily air quality index prediction considering air pollutant factors. *Sci. Total Environ.* **2019**, *683*, 808–821. [CrossRef] [PubMed]
16. Wu, C.; Li, Q.; Hou, J.; Karimian, H.; Chen, G. PM2.5 concentration prediction using convolutional neural networks. *Sci. Surv. Mapp* **2018**, *43*, 68–75.
17. Huang, C.J.; Kuo, P.H. A deep CNN-LSTM model for particulate matter (PM2.5) forecasting in smart cities. *Sensors* **2018**, *18*, 2220. [CrossRef] [PubMed]
18. Pak, U.; Kim, C.; Ryu, U.; Sok, K.; Pak, S. A hybrid model based on convolutional neural networks and long short-term memory for ozone concentration prediction. *Air Qual. Atmos. Health* **2018**, *11*, 883–895. [CrossRef]
19. Qi, Y.; Li, Q.; Karimian, H.; Liu, D. A hybrid model for spatiotemporal forecasting of PM2.5 based on graph convolutional neural network and long short-term memory. *Sci. Total Environ.* **2019**, *664*, 1–10. [CrossRef]
20. Vaswani, A.; Shazeer, N.; Parmar, N.; Uszkoreit, J.; Jones, L.; Gomez, A.N.; Kaiser, Ł.; Polosukhin, I. Attention is all you need. *Adv. Neural Inf. Process. Syst.* **2017**, *30*.
21. Zheng, S.; Lu, J.; Zhao, H.; Zhu, X.; Luo, Z.; Wang, Y.; Fu, Y.; Feng, J.; Xiang, T.; Torr, P.H.; et al. Rethinking semantic segmentation from a sequence-to-sequence perspective with transformers. In Proceedings of the IEEE/CVF Conference on Computer Vision and Pattern Recognition, Nashville, TN, USA, 20–25 June 2021; pp. 6881–6890.
22. Lee, J.; Toutanova, K. Pre-training of deep bidirectional transformers for language understanding. *arXiv* **2018**, arXiv:1810.04805.
23. Zhou, H.; Zhang, S.; Peng, J.; Zhang, S.; Li, J.; Xiong, H.; Zhang, W. Informer: Beyond efficient transformer for long sequence time-series forecasting. In Proceedings of the AAAI Conference on Artificial Intelligence, Virtual Event, 2–9 February 2021; Volume 35, pp. 11106–11115.
24. Wu, H.; Xu, J.; Wang, J.; Long, M. Autoformer: Decomposition transformers with auto-correlation for long-term series forecasting. *Adv. Neural Inf. Process. Syst.* **2021**, *34*, 22419–22430.
25. Zhou, T.; Ma, Z.; Wen, Q.; Wang, X.; Sun, L.; Jin, R. Fedformer: Frequency enhanced decomposed transformer for long-term series forecasting. In Proceedings of the International Conference on Machine Learning, PMLR, Baltimore, MD, USA, 17–23 July 2022; pp. 27268–27286.
26. Liu, S.; Yu, H.; Liao, C.; Li, J.; Lin, W.; Liu, A.X.; Dustdar, S. Pyraformer: Low-complexity pyramidal attention for long-range time series modeling and forecasting. In Proceedings of the International Conference on Learning Representations, Vienna, Austria, 4 May 2021.
27. Zeng, A.; Chen, M.; Zhang, L.; Xu, Q. Are transformers effective for time series forecasting? In Proceedings of the AAAI Conference on Artificial Intelligence, Washington, DC, USA, 7–14 February 2023; Volume 37, pp. 11121–11128. [CrossRef]
28. Lyu, H.; Sha, N.; Qin, S.; Yan, M.; Xie, Y.; Wang, R. Advances in neural information processing systems. *Adv. Neural Inf. Process. Syst.* **2019**, *32*.
29. Das, A.; Kong, W.; Leach, A.; Sen, R.; Yu, R. Long-term forecasting with tide: Time-series dense encoder. *arXiv* **2023**, arXiv:2304.08424.
30. Targ, S.; Almeida, D.; Lyman, K. Resnet in resnet: Generalizing residual architectures. *arXiv* **2016**, arXiv:1603.08029.
31. Rhif, M.; Ben Abbes, A.; Farah, I.R.; Martínez, B.; Sang, Y. Wavelet transform application for/in non-stationary time-series analysis: A review. *Appl. Sci.* **2019**, *9*, 1345. [CrossRef]
32. Yoo, Y.; Baek, J.G. A novel image feature for the remaining useful lifetime prediction of bearings based on continuous wavelet transform and convolutional neural network. *Appl. Sci.* **2018**, *8*, 1102. [CrossRef]
33. Sasal, L.; Chakraborty, T.; Hadid, A. W-transformers: A wavelet-based transformer framework for univariate time series forecasting. In Proceedings of the 2022 21st IEEE International Conference on Machine Learning and Applications (ICMLA), Nassau, Bahamas, 12–14 December 2022; pp. 671–676.
34. Heddam, S.; Merabet, K.; Difi, S.; Kim, S.; Ptak, M.; Sojka, M.; Zounemat-Kermani, M.; Kisi, O. River water temperature prediction using hybrid machine learning coupled signal decomposition: EWT versus MODWT. *Ecol. Inform.* **2023**, *78*, 102376. [CrossRef]
35. Yoon, J.; Jordon, J.; Schaar, M. Gain: Missing data imputation using generative adversarial nets. In Proceedings of the International Conference on Machine Learning, PMLR, Stockholm, Sweden, 10–15 July 2018; pp. 5689–5698.
36. Stekhoven, D.J.; Bühlmann, P. MissForest—Non-parametric missing value imputation for mixed-type data. *Bioinformatics* **2012**, *28*, 112–118. [CrossRef]
37. Bank, D.; Koenigstein, N.; Giryes, R. Autoencoders. In *Machine Learning for Data Science Handbook: Data Mining and Knowledge Discovery Handbook*; Springer: Berlin/Heidelberg, Germany, 2023; pp. 353–374.
38. Lara-Benítez, P.; Carranza-García, M.; Luna-Romera, J.M.; Riquelme, J.C. Temporal convolutional networks applied to energy-related time series forecasting. *Appl. Sci.* **2020**, *10*, 2322. [CrossRef]

39. Sherstinsky, A. Fundamentals of recurrent neural network (RNN) and long short-term memory (LSTM) network. *Phys. D Nonlinear Phenom.* **2020**, *404*, 132306. [CrossRef]
40. Safara, F.; Doraisamy, S.; Azman, A.; Jantan, A.; Ramaiah, A.R.A. Multi-level basis selection of wavelet packet decomposition tree for heart sound classification. *Comput. Biol. Med.* **2013**, *43*, 1407–1414. [CrossRef]

*atmosphere*

MDPI

*Article*

# Two-Stream Networks for COPERT Correction Model with Time-Frequency Features Fusion

**Zhenyi Xu [1,2,*], Ruibin Wang [1], Kai Pan [1,3], Jiaren Li [1,3] and Qilai Wu [1,3]**

[1] Institute of Artificial Intelligence, Hefei Comprehensive National Science Center, Hefei 230088, China; e20201061@stu.ahu.edu.cn (R.W.); wa21301055@stu.ahu.edu.cn (K.P.); wa22301176@stu.ahu.edu.cn (J.L.); wa23201062@stu.ahu.edu.cn (Q.W.)
[2] Key Laboratory of System Control and Information Processing, Ministry of Education, Shanghai 200240, China
[3] College of Artificial Intelligence, Anhui University, Hefei 230031, China
* Correspondence: xuzhenyi@mail.ustc.edu.cn

**Abstract:** Emission factors serve as a valuable tool for quantifying the release of pollutants from road vehicles and predicting emissions within a specific time or area. In order to overcome the limitation of the computer program to calculate emissions from the road transport (COPERT) model in directly obtaining precise emission factors from on-board diagnostic (OBD) data, we propose a novel two-stream network that combines time-series features and time-frequency features to enhance the accuracy of the COPERT model. Firstly, for the instantaneous emission factors of NOx from multiple driving segments provided by heavy-duty diesel vehicles in actual driving, we select the monitored attributes with a high correlation to the emission factor of NOx considering the data scale and employing Spearman rank correlation analysis to obtain the final dataset composed of them and emission factors. Subsequently, we construct an information matrix to capture the impact of past data on emission factors. Each attribute of the time series is then converted into a time-frequency matrix using the continuous wavelet transform. These individual time-frequency matrices are combined to create a multi-channel time-frequency matrix, which represents the historical information. Finally, the historical information matrix and the time-frequency matrix are inputted into a two-stream parallel model that consists of ResNet50 and a convolutional block attention module. This model effectively integrates time-series features and time-frequency features, thereby enhancing the representation of emission characteristics. The reliability and accuracy of the proposed method were validated through a comparative analysis with existing mainstream models.

**Keywords:** OBD; emission factors; COPERT; two-stream network; time-frequency features

## 1. Introduction

Heavy-duty diesel vehicles are one of the main tools for long-distance cargo transportation around the world, which have the advantages of flexible, convenient and low-cost transportation and play an important role in the national economy. However, due to their high fuel consumption, long mileage and high ownership, heavy-duty diesel vehicles have become an important source of nitrogen oxide emissions from motor vehicles. According to the "*China Mobile Source Environmental Management Annual Report* (2021) [1]", diesel vehicle emissions of carbon monoxide (CO), hydrocarbons (HC), nitrogen oxides (NOx) and particulate matter (PM) accounted for 9.9%, 18.7%, 57.3% and 77.8% of national vehicle emissions in 2020, respectively. Therefore, the NOx emission control is the difficulty and focus of the current pollution reduction in heavy-duty diesel vehicles, but the effectiveness of standard control measures for heavy-duty diesel vehicles needs to quantitatively be evaluated based on accurate diesel vehicle transient emission models. The formulation of the current emission limit standard is based on the emission factor model under the laboratory cycle conditions, and the vehicle emissions are calculated by establishing the correlation between the emission characteristics of the motor vehicle and various influencing factors of the motor vehicle. The emission factor model serves as the basis for the emission inventory accounting and the formulation of emission reduction policies.

The current vehicle emission models are different in terms of computational requirements and the experimental physical input data. Some emission factor models, such as the MOBILE model [2] developed by the U.S. Environmental Protection Agency (U.S. EPA), the EMFAC model [3] developed by the California Air Resources Board (CARB) and the COPERT model [4] developed by the European Commission, calculate the emissions based on the average speed of specific driving cycles. These models are usually used to estimate or predict macro-level traffic emissions in a specific area (regional area or city scale) for a specific time period of total pollution emissions (usually a quarter or a year), while ignoring the dynamic driving conditions. Another kind of emission model incorporate the instantaneous speed, acceleration/deceleration and other vehicle driving characteristic data to accurately reflect the emissions under different vehicle operating modes such as idle, steady-state cruise and various acceleration/deceleration levels, which are suitable for analyzing the emission of a single vehicle or a number of specific roads emissions calculation tasks. These models mainly include the IVE model [5] and the CMEM model [6] developed by the University of California, Riverside (UCR). The MOVES is a comprehensive model [7] developed by the EPA, which can estimate emissions based on average velocity as well as operating conditions. It can be appropriate for both micro, meso and macro emission estimation. Due to the lack of basic data, the domestic research on emission factor models started late, and the foreign default values were directly used to evaluate local vehicle pollution emissions, resulting in large estimation errors.

In the last few years, there have been significant advancements in the field of vehicle emission modeling, driven by the widespread adoption of technologies such as Portable Emission Measurement System (PEMS) [8] and OBD remote online monitoring [9]. These technologies have enabled researchers to obtain actual road emission factors, which can be used to refine and improve emission models used in developed countries. Quirama et al. [10] utilized PEMS to develop a methodology based on energy consumption analysis at a micro-trip level. This mode facilitated the assessment of real-world energy consumption and emissions for a fleet operating within a defined geographical region. Tsinghua University and other domestic institutions have developed the Beijing Vehicle Emission Factor Model (EMBEV) based on established emission models from foreign sources [11]. Wang et al. [12] employed a computational approach based on decision optimization to estimate road segment speeds using low-frequency GPS trajectory data. They combined this with a micro-emission model to estimate carbon dioxide emissions. Kumar et al. [13] used 6 years of air pollution data from 23 Indian cities, selected relevant features through correlation analysis, solved the data imbalance problem by using resampling techniques and, finally, identified key pollutants directly affecting the Air Quality Index (AQI) using a machine learning approach. Wang et al. [14] considered the influence of historical vehicle operating states and developed a microscopic emission model based on a BP neural network using short-term driving conditions. Xu et al. [15] proposed a deep learning-based approach for exhaust gas telemetry data using COPERT emission factors. This approach involved constructing a three-layer self-encoder network to extract features from heterogeneous data sources such as meteorological data, road network data, traffic flow data and urban functional areas, thereby correcting the COPERT emission model. Yang et al. [16] analyzed the actual $CO_2$ and NOx emissions of National IV buses using remote on-board diagnostic (OBD) data. They established a calculation method for $CO_2$ and NOx emissions based on OBD data. In light of the swift progress in artificial intelligence technology, there has been a notable emergence of innovative theoretical perspectives and sophisticated technical approaches. These advancements have brought about substantial changes in the landscape of research and analysis across traditional industries. Recently, there has been a surge in scholarly research focusing on the online monitoring of OBD vehicle networks. Xu et al. [17] proposed a transfer learning-based approach for predicting mobile source pollution in the context of OBD pollution monitoring, specifically focusing on the impact of multiple external factors. The study utilized diesel vehicles as a case study and successfully achieved knowledge transfer across different vehicle models. Molina [18] employed random forest to determine the most influential driver variables based on the best attributes of the training model using OBD II data. Rivera-Campoverde et al. [19] addressed the challenge of estimating vehicle pollutant emission levels in the absence of an accurate model and limited measurement campaigns. They proposed a novel method for pollutant emission estimation by utilizing vehicle driving variables such as vehicle gear, engine speed and gas pedal position as inputs to a neural network model. The results of their approach closely aligned with those obtained from the IVE model and real driving emissions (RDE) test results. Wang et al. [20] proposed a coherent methodology that utilizes the OBD system to collect operational data from heavy-duty diesel vehicles. Subsequently, an artificial neural network is constructed to develop an emission prediction model. This approach enables real-time monitoring of the vehicle's emission status, providing valuable insights for environmental protection and vehicle management purposes. Chen et al. [21] presented a methodology for gathering vehicle parameters, including speed, RPM, throttle position, engine load, etc., through the OBD interface. Subsequently, they employed the AdaBoost algorithm to classify driving behaviors, achieving an impressive accuracy rate of 99.8% across various driving scenarios.

However, the current emission model uses artificially designed parameters such as vehicle speed and acceleration to characterize the relationship between vehicle driving conditions and pollution emissions, where only the time domain information of vehicle driving conditions data came from multiple sensors is used to construct the emission model, without considering the frequency domain representation information of the vehicle engine states. The current emission models are mainly based on the traditional regression learning method, in which it is difficult to ensure the accuracy of the model representation and comprehensively describe the mapping relationship between the vehicle driving state and emissions under different working conditions.

The representative models are shown in Table 1 (where VSP represents the instantaneous output power per unit mass of a motor vehicle). The data of these models come from bench tests and on-board tests and it is difficult to obtain accurate emission factors directly using these models on OBD data. Under the circumstance that it is difficult to fully verify the results of most emission models in China, the process application should be an important indicator of emission model selection. When calculating emissions, it is not necessary to choose the latest model. Different models are suitable for different regional situations and development stages Although the emission model based on operating conditions can better reflect the relationship between actual driving conditions and emissions, vehicle pollution control is still in the early stage in most cities in China, and there is a lack of statistical data on vehicle ownership and in-vehicle driving conditions. When the model input parameters are difficult to obtain and the accuracy cannot be guaranteed, it will not help even if there is a model that is fully applicable to the local area. The purpose of this work is to revise the emission model, and the primary consideration is the availability of model parameters for model selection, and the ease of building an a priori emission model. Considering the congruence between the emission standards and vehicular control technologies employed in China and Europe, coupled with the accessibility of the requisite parameters for the European-developed COPERT model, it is not surprising that COPERT has found extensive application in research related to pollution emissions from on-road mobile sources within China [22–25]. In light of these factors, this study has elected to utilize COPERT as the emission model of choice, subjecting it to amend for the purpose of calculating NOx emission factors within the OBD dataset.

Vehicle driving conditions play a pivotal role in determining emissions. When modeling on-road mobile source emissions, the focus is typically on establishing the relationship between the operational status of the mobile source and the level of pollutant emissions. The accuracy and validity of the model rely on the precise representation of the on-road mobile source's operational status. Therefore, in order to modify the COPERT model to achieve the application of COPERT model on the OBD dataset, it is necessary to accurately represent the on-road mobile source operation state. However, the mobile source driving state is complicated and variable due to the influence of driving behavior, external environment and other factors, and it is not enough to accurately represent the mobile source driving state by only relying on the monitoring data at a certain moment. Therefore, considering the influence of historical information on the calculation of emission factors at the current moment is the key to our modification of the COPERT model.

**Table 1.** Typical traffic emission models.

| Characterization Parameters | Model | Application Scale | Calculation Method | Data Source | Parameter Numbers |
|---|---|---|---|---|---|
| Average speed | MOBILE [2] | Macro, Meso | Statistical Regression | Bench Test | 27 |
| | COPERT [4] | Macro, Meso | Statistical Regression | Bench Test | 15 |
| | EMFAC [3] | Macro, Meso | Statistical Regression | Bench Test | 16 |
| VSP | IVE [5] | Macro, Meso | Statistical Regression | Bench Test and On-road Emission Test | 19 |
| Engine power requirement | CMEM [6] | Macro, Meso, Microscopic | Physical Modeling | On-road Emission Test | 47 |
| Average speed, VSP, Traffic volume | MOVES [7] | Macro, Meso, Microscopic | Statistical Regression | Bench Test and On-road Emission Test | - |

To address the aforementioned challenges in emission model construction, we propose a two-stream modification model based on the fusion of time-series and time-frequency features with historical information (HI_TTFTS). First, the actual emission factors of the OBD, using Spearman rank correlation analysis, are used to extract the correlation factors from the data volume for the measured data of the OBD, which are presented as not strictly continuous. Then, by utilizing the NOx emission factors alongside the corresponding attributes at the same time point, the historical information matrix is constructed. Subsequently, the continuous wavelet transform (CWT) is applied to obtain a multi-channel time-frequency matrix representation of the historical information. Finally, the two-stream model, consisting of ResNet and the convolutional block attention module (CBAM), is employed to incorporate both the time series matrix and the time-frequency matrix for model refinement. To assess the effectiveness of the proposed model, a comprehensive set of experiments was conducted. The results obtained from these experiments demonstrate the significant potential of our approach in accurately estimating NOx emission factors using online monitoring data from OBD vehicle networks. This innovative approach opens up new avenues for future research in this field.

The remaining parts of this paper are arranged as follows. We clarify the COPERT model and relevant deep learning models in Section 2. In Section 3, we present the details of the proposed time-frequency two-stream network model. The experiments and result analysis are presented in Section 4. Finally, we conclude the paper and discuss the future possible extending work in Section 5.

## 2. Related Works

### *2.1. COPERT*

The development of the COPERT model was supported by funding from the European Environment Agency (EEA) to provide motor vehicle emissions data for the European National Atmospheric Emissions Inventory (NAEI). The model uses average speed to characterize vehicle travel characteristics and is based on a large number of test data, which are compatible with different national standards and parameter variables. Since China adopts European vehicle emission test protocols and regulatory systems similar to China, the COPERT model has good applicability in calculating Chinese motor vehicle exhaust emissions [26].

The emission factors for different pollutants under different standards, all as a function of velocity, are calculated as follows: [27].

$$EF_{COPERT} = (a + c \cdot V_r + e \cdot V_r^2)/(1 + b \cdot V_r + d \cdot V_r^2) \tag{1}$$

where $EF_{COPERT}$ (g/km) is the emission factor of NOx; $a, b, c, d, e, f$ are the constant factors and $V_r$ (km/h) is the average speed on the road section. The values of the constant coefficients associated with the calculation of NOx emission factors under different standards are given in Table 2, and the paper is calculated for the Euro 5 standard.

**Table 2.** Coefficients of COPERT model [27].

| Coefficients | Euro 1 | Euro 2 | Euro 3 | Euro 4 | Euro 5 |
|---|---|---|---|---|---|
| $a$ | $5.25 \times 10^{-1}$ | $2.84 \times 10^{-1}$ | $9.24 \times 10^{-2}$ | $1.06 \times 10^{-1}$ | $1.89 \times 10^{-1}$ |
| $b$ | / | $-2.34 \times 10^{-2}$ | $-1.22 \times 10^{-2}$ | / | 1.57 |
| $c$ | $-1.00 \times 10^{-2}$ | $-8.69 \times 10^{-3}$ | $-1.49 \times 10^{-3}$ | $-1.58 \times 10^{-3}$ | $8.15 \times 10^{-2}$ |
| $d$ | / | $4.43 \times 10^{-4}$ | $3.97 \times 10^{-5}$ | / | $2.73 \times 10^{-2}$ |
| $e$ | $9.36 \times 10^{-5}$ | $1.14 \times 10^{-4}$ | $6.53 \times 10^{-6}$ | $7.10 \times 10^{-6}$ | $-2.49 \times 10^{-4}$ |
| $f$ | / | / | / | / | $-2.68 \times 10^{-1}$ |

### 2.2. Emission Factor Obtaining of NOx Based on OBD

In the actual monitoring of OBD, the specific values of emission factors cannot be obtained directly through monitoring, and it is necessary to calculate the reliable actual emission factors with the help of its related data, which is calculated as follows: [28].

$$EF\_OBD = \frac{0.454}{0.746} \cdot \frac{\sum_{t_1}^{t_2} 0.001587 \cdot \mathrm{NOx} \cdot Q_{exh}}{\sum_{t_1}^{t_2} \frac{Q_{FR}}{3600} \cdot \rho_f} \cdot BSFC \tag{2}$$

where $EF\_OBD$ (g/kWh) is the emission factor of NOx and $BSFC$ is the specific engine fuel consumption, which can be obtained based on bench or PEMS measurements, or by referring to the US EPA report (EPA-420-R-02-005) [29]. NOx (ppm) is the NOx volume concentration downstream of SCR read by the OBD; $Q_{FR}$ (L/h) is the engine fuel volume flow rate read by the OBD; $t_1$ and $t_2$ are the start and end of a driving time, respectively, $\rho_f$ (kg/L) is the density of the fuel used in the engine and $Q_{exh}$ (kg/h) is the instantaneous exhaust mass flow rate, which is calculated as follows.

$$Q_{exh} = \frac{3.6 \cdot Q_{FR} \cdot \left(1 + \frac{x + \frac{y}{4} - \frac{z}{2}}{12x + y + 16z} \cdot 22.4/20.9\% \cdot \rho_{air}\right)}{1 - \frac{O_2 \cdot \rho_{air}}{20.9 \rho_{exh}}} \tag{3}$$

where $x$, $y$ and $z$ are the molar ratio of carbon to carbon (C/C), the molar ratio of hydrogen to carbon (H/C) and the molar ratio of oxygen to carbon (O/C) in the fuel, respectively. According to GB17691-2018 [30], diesel is $CH_{1.86}O_{0.006}$; $O_2$ (%) which is the volumetric concentration of $O_2$ in the exhaust gas monitored by the OBD; $\rho_{air}$ is the density of ambient air at 0 °C and 101.3 kPa. The value is 1.293 kg/m$^3$; $\rho_{exh}$ for the exhaust density in kg/m$^3$, according to GB17691-2018 [30]. The exhaust density of heavy-duty vehicles burning diesel is 1.2943 kg/m$^3$.

To ensure consistency between the COPERT model and the OBD model in calculating emission factor units for heavy-duty diesel vehicles, a conversion factor of 1.27 [28] is applied to the OBD-calculated NOx emission factor to convert it from grams per kilowatt-hour (g/kWh) to grams per kilometer (g/km).

## 2.3. ResNet

The CNN model structure has a notable influence on the representational ability of the network. In recent years, there has been a proliferation of CNN network architectures, including notable ones such as AlexNet [31], GoogLeNet [32], VGGNet [33] and ResNet [34]. These architectures have been proposed successively to address various challenges. ResNet, in particular, introduces the concept of residual learning to mitigate the degradation problem encountered in deep networks, where increasing the number of layers leads to a decline in performance. By incorporating residual learning, ResNet enables deeper networks while maintaining accuracy and controlling computational speed. The key building block of ResNet is the residual unit, which is depicted in Figure 1.

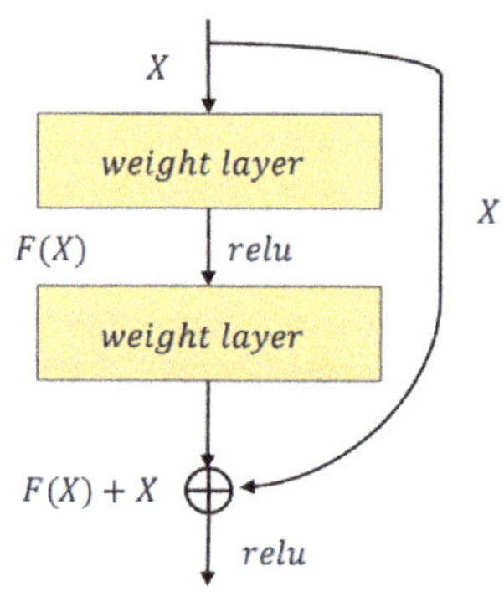


**Figure 1.** Structure of the residual module.

Currently, ResNet is one of the most extensively employed networks for feature extraction. The ResNet family includes popular models such as ResNet18, ResNet34, ResNet50, ResNet101 and ResNet152. In this study, we have selected ResNet50 as the primary backbone network. The detailed architecture of ResNet50 is presented in Table 3.

**Table 3.** Architecture of ResNet50.

| Stage 0 | $7 \times 7$, 64, strider 2 |
|---|---|
| Stage 1 | $3 \times 3$ max pool, stride 2 $\begin{bmatrix} 1 \times 1, 64 \\ 3 \times 3, 64 \\ 1 \times 1, 256 \end{bmatrix} \times 3$ |
| Stage 2 | $\begin{bmatrix} 1 \times 1, 128 \\ 3 \times 3, 128 \\ 1 \times 1, 512 \end{bmatrix} \times 4$ |
| Stage 3 | $\begin{bmatrix} 1 \times 1, 256 \\ 3 \times 3, 256 \\ 1 \times 1, 1024 \end{bmatrix} \times 6$ |
| Stage 4 | $\begin{bmatrix} 1 \times 1, 512 \\ 3 \times 3, 512 \\ 1 \times 1, 2048 \end{bmatrix} \times 3$ |

## 2.4. Attention in CNN

In the cognitive process of things, attention helps the brain to automatically ignore low-value and low-likelihood information so as to think deeply about certain issues without being distracted by other things. With reference to this property, data scientists have carried out a series of studies on it in the field of deep learning [35,36]. In recent years, there have been numerous studies to improve the performance of CNNs by introducing attention mechanisms, which can be classified as a mixture of spatial attention, channel attention already spatial attention and channel attention according to their scopes.

For CNNs, each layer of the CNN outputs a $C \times W \times H$ feature map, where C represents the number of channels, and W and H represent their width and height. Spatial attention is assigned different weights to each point of the $W \times H$ scale matrix for all channels, and the assigned weight represents the contribution of that point to the current task. STN [37] and DCN [38], as its representatives, have shown good performance. Unlike spatial attention, the channel attention mechanism applies a weight to each feature map on each channel, and this weight represents the relevance of each channel to the current task. The larger the value, the more important the channel is in the current task. Among the representative models are SENet [39] and ECANet [40].

Since both the spatial attention mechanism and channel attention mechanism can improve the performance of CNN networks to some extent, some scholars have considered combining the two, giving a greater boost to CNNs. The authors of [41] proposed a convolutional block attention module (CBAM), given an intermediate feature map, so that the network inferred the attention weights along the space and channel sequentially, and then multiplied with the original feature map to make adaptive adjustments to the features, and the overall structure of CBAM is shown in Figure 2.

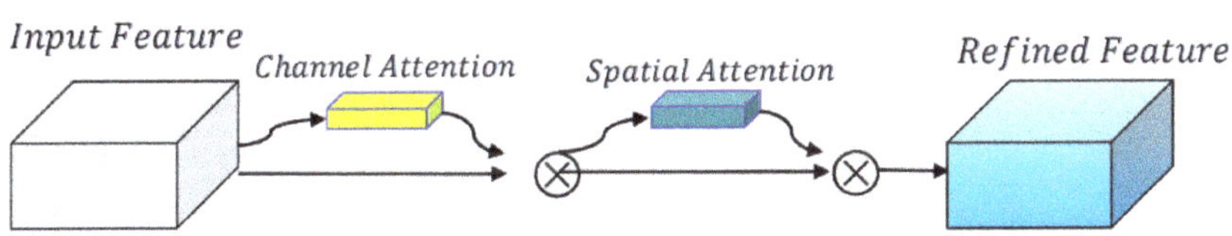


**Figure 2.** Structure of the CBAM (Convolutional Block Attention Module).

It decomposes the original one-dimensional time-series signal into a two-dimensional time-frequency matrix by applying scaling and translation operations to the wavelet basis function.

### 2.5. *Continuous Wavelet Transform*

The continuous wavelet transform (CWT) is a versatile method for time-frequency analysis. It decomposes the original one-dimensional time-series signal into a two-dimensional time-frequency matrix by applying scaling and translation operations to the wavelet basis function [42]. A two-dimensional time-frequency matrix can be obtained by performing a continuous wavelet transform on a one-dimensional time-series signal. This matrix provides a comprehensive representation of the frequency information of the signal as it evolves over time. Equation (4) [43] illustrates the specific details of the CWT conversion for any function $f(t)$ in the $L^2(R)$ space.

$$CWT_f(a,\tau) = \langle f(t), \psi_{a,\tau}(t) \rangle = \frac{1}{\sqrt{a}} \int f(t)\psi(\frac{t-\tau}{a})dt \tag{4}$$

where $\psi_{a,\tau}(t) = \frac{1}{\sqrt{a}}\psi(\frac{t-\tau}{a})$, $a$ and $\tau$ are parameters representing stretching and translation, respectively.The function $\psi(t)$ is a wavelet basis function derived from the mother wavelet through operations of stretching and translation.

## 3. Methodology

In this section, we focus on the specific introduction of the model. Refer to Figure 3, where we present a two-stream network model based on the fusion of time-series and time-frequency features of historical information to modify the COPERT model. Firstly, in view of the fact that the OBD data actually monitored are not strictly continuous in temporal terms, we divide the whole dataset into driving segments to obtain multiple continuous driving segments; Then, based on the amount of data in each driving segment, we use Spearman rank correlation analysis to select the correlation attributes with the emission factors of NOx obtained from the OBD data. Finally, the selected relevant attributes are used to construct the historical information matrix, and the corresponding historical information matrix is converted into a time-frequency matrix using CWT, and the historical information matrix and the time-frequency matrix are used as inputs in two parallel structures combined by ResNet50 and CBAM to complete the correction of the COPERT model and achieve the emission factor of NOx using the COPERT model on the OBD dataset.

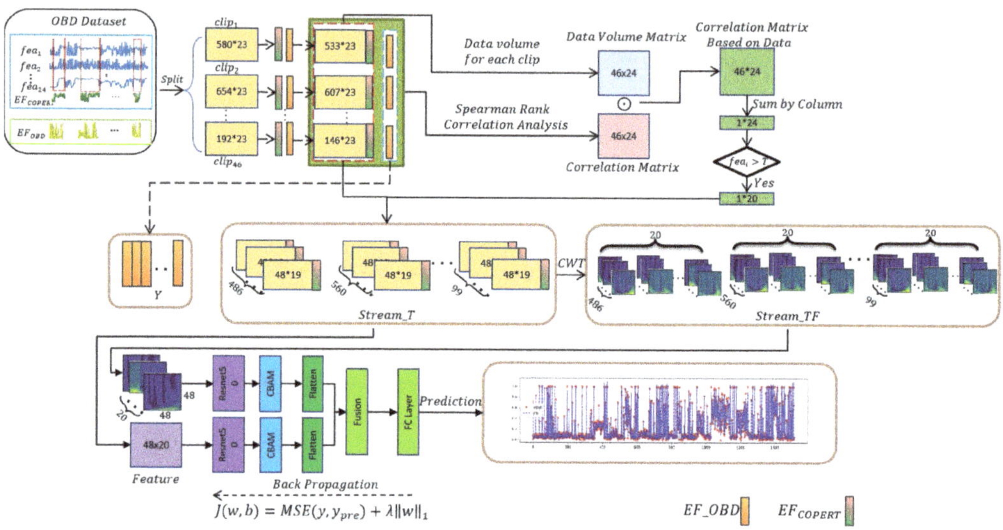


**Figure 3.** The architecture of the HI_TTFTS network.

### 3.1. Data Description

In this paper, the experiment uses the actual travel OBD data of a diesel vehicle collected in Hefei in 2020, the engine attributes and sampling standard of this diesel vehicle are shown in Table 4.

**Table 4.** Engine section parameters and sampling standards.

| Fuel Type | Diesel Fuel |
|---|---|
| Displacement | 11.596 L |
| Maximum Horsepower | 375 PS |
| Maximum Torque | 1800 N· m |
| Rated Rotation Speed | 2100 rpm |
| Engine Type | WP12.375E51 |
| Brand | WEICHAI |
| Sampling Interval | 5 s |
| Time Period | 8 June 2020–29 November 2020 |
| Number of Sampling | 27,562 |

In the data preprocessing phase, the collected raw data underwent several steps. Initially, irrelevant attributes were removed, followed by anomalous records with values of

0 for vehicle speed, NOx concentration and instantaneous engine fuel consumption rate. Table 5 provides an overview of the attributes and their corresponding symbols in each dataset after the completion of the preprocessing stage.

**Table 5.** Attribute Comparison Table.

| Name | Label | Name | Label |
|---|---|---|---|
| Engine speed | $E_{speed}$ | Actual output torque percentage | AOTP |
| Engine water temperature | EWT | Engine fuel temperature | EFT |
| Post-treatment downstream NOx | NOx | Post-treatment downstream oxygen | O2 |
| Atmospheric pressure | AP | Environmental temperature | ET |
| Post-treatment exhaust gas mass flow rate | $PT_{EGM}$ | Urea tank level | UTL |
| Urea tank temperature | UTT | Vehicle speed | $V_{speed}$ |
| Gas peddal opening | GPO | Single Driving Miles | SDM |
| Engine fuel consumption rate (instantaneous) | EFCR | Average Engine fuel consumption rate | $EFCR_{avg}$ |
| Engine fuel consumption for single driving | $EFC_{SD}$ | Total engine fuel consumption | $EFC_{total}$ |
| Battery voltage | BV | Fuel tank level | FTL |
| Cumulative engine runtime | CER | Longitude | LNG |
| Latitude | LAT | / | / |

## 3.2. Data Processing

In the actual correction phase of the COPERT model, we need data that are as continuous as possible in time series with complete labeling of each record. However, the OBD data collected for the actual road driving cannot be collected directly due to their inherent non-strict continuity in time series and the $EF_{COPERT}$ to be corrected and the $EF\_OBD$ as a label. We need to divide the pre-processed dataset into continuous driving segments and obtain the emission factors for NOx. The data processing methods in this paper are Driving Segment Division and Get Emission factor of NOx.

### 3.2.1. Driving Segment Division

Considering that the dataset consists of real-world driving data from a diesel vehicle collected across several days, it contains consecutive driving segments, resulting in a presentation of the data that is not strictly continuous. During preprocessing, irrelevant records were removed, causing a division in the continuous driving segments and further division in the already non-strictly continuous records. However, it is important to note that the records without delay preceding and following the removed erroneous records are still regarded as continuous. Therefore, It is essential to determine an optimal upper limit for the continuous time interval when partitioning the dataset into driving segments. This ensures that the impact of removing invalid data during preprocessing is minimized.

In this study, we have defined a maximum time interval of 180 s. Any records with intervals exceeding this threshold are considered as separate driving segments. After the initial screening, each piece of the drive contains a varying number of records. To ascertain that each selected driving segment is of sufficient length for calculating emission factors based on historical information, we have set the minimum number of records per segment to 180, which is equivalent to a duration of 15 min. The data volume for each segment is presented in Figure 4.

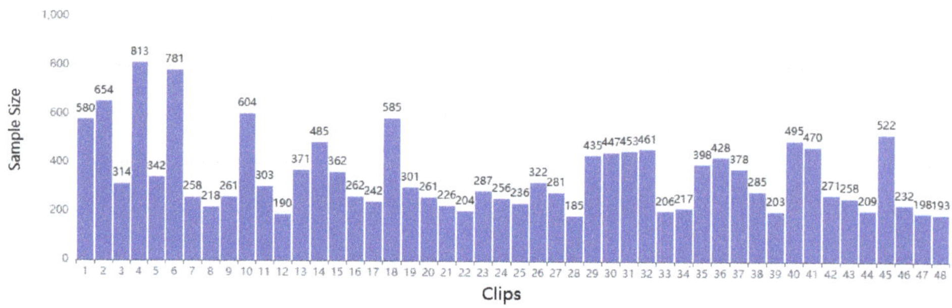

**Figure 4.** Valid Driving Clip.

#### 3.2.2. Obtaining NOx Emission Factors

In the process of obtaining emission factors, the emission factor for the *k*-th record is determined by considering the collective data from the first record to the *k*th record. The specific criteria for selecting *k* will be explained in more detail in Section 4. As we encounter the challenge of computing the NOx emission factor for the initial *k*-1 records within each driving segment during emission factor calculation, we initiate the process for each driving segment from the *k*th record onwards and align the computed emission factors $EF\_OBD$ and $EF_{COPERT}$ with the corresponding timestamps. Furthermore, our dataset lacks the measurement of the engine fuel volume flow rate $Q_{FR}$, thus we derive $Q_{FR}$ using the following formula.

$$Q_{FR} = 0.01 \cdot EFCR \cdot V_{Speed} \tag{5}$$

By using Equations (1)–(3), $EF_{COPERT}$ and $EF\_OBD$ were computed. The results are shown in Figure 5 (taking *k* = 48 as an example). The figure illustrates that, in the majority of data records, the peaks of $EF_{COPERT}$ are lower than those of $EF\_OBD$, while the troughs are higher. Directly observing common characteristics of the two trends is challenging. This indicates a substantial disparity in the NOx emission factors calculated by the COPERT and OBD models. Simply deriving emission factors from the COPERT model and adjusting them to fit the OBD data using specific mathematical formulas is difficult.

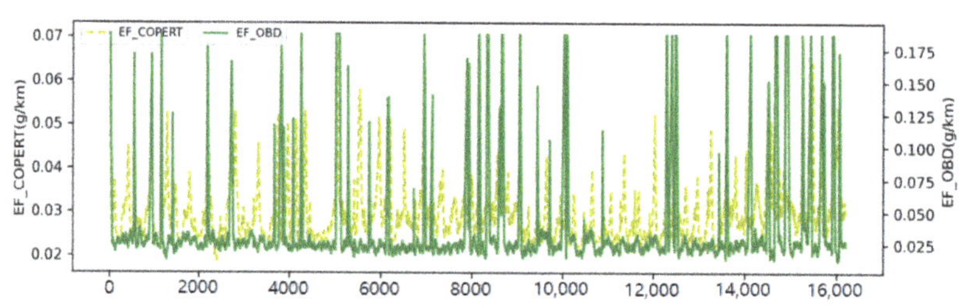

**Figure 5.** Emission factors of NOx.

### 3.3. Screening of Correlation Factors

The impact of each attribute derived from OBD data on the calculation of emission factors varies. Attributes with low correlations have minimal influence during the training of the model. This not only hampers the enhancement of model performance but also increases the number of model parameters. Thus, it is essential to choose factors that demonstrate strong correlations with $EF\_OBD$ before initiating model training.

In our work, Spearman correlation analysis is used by us to select the attributes that have high correlation with $EF\_OBD$. Using Spearman correlation analysis for a continuous dataset, we can easily derive correlations among attributes using the $t$-value test; however, our data are multiple continuous driving segments and there are differences in data volume. Therefore, we propose a correlation factor screening based on the amount of data for each model.

The data correlation analysis for multiple driving segments is shown in Figure 3. We computed the direct Spearman correlation coefficients between $EF\_OBD$ and other factors (including $EF_{COPERT}$) in all driving segments separately and chose $t$ = 1.645 for hypothesis testing to determine whether $EF\_OBD$ is strongly correlated with other attributes in that form segment. Based on this, we extracted all other driving segment attributes that were correlated with $EF\_OBD$ and constructed a correlation matrix $C$. Each row in $C$ represents the correlation of each attribute in the driving segment with $EF_{COPERT}$ and $EF\_OBD$, where 1 represents a strong correlation and 0 represents a weak correlation, and each column in $C$ represented a specific attribute type. To accommodate variations in data volume across different driving segments, we generated a data volume matrix $D$. Each row in $D$ corresponded to a specific driving segment, and the values in each column represented the data volume of that driving segment. We then computed the Hadamard product of matrices $C$ and $D$ (with $C$ being the element-wise matrix of $D$). The values in the resulting matrix represent the correlation of each driving segment attribute with respect to $EF\_OBD$ based on the current amount of data. Finally, the summation of each column in this matrix yielded the correlation data volume matrix $R_C$, which is visually depicted as illustrated below.

$$\begin{aligned} R_C &= \sum_{i=0}^{num(fea)} C \odot D \\ &= \left[nums_{f1}, nums_{f2}, \cdots, nums_{fn}, nums_{EF_{COPERT}}\right] \end{aligned} \tag{6}$$

$$C = \begin{bmatrix} S_{1,1} & S_{1,2} & \cdots & S_{1,copert} \\ S_{2,1} & S_{2,2} & \cdots & S_{2,copert} \\ \vdots & \vdots & \vdots & \vdots \\ S_{48,1} & S_{48,2} & \cdots & S_{48,copert} \end{bmatrix} \tag{7}$$

$$D = \begin{bmatrix} nums_1 & nums_1 & \cdots & nums_1 \\ nums_2 & nums_2 & \cdots & nums_2 \\ \vdots & \vdots & \cdots & \vdots \\ nums_{48} & nums_{48} & \cdots & nums_{48} \end{bmatrix} \tag{8}$$

Within the equation, $nums_{fi}$ denotes the data quantity linked to $EF\_OBD$ for the $i$-th attribute, $numsEF_{COPERT}$ represents the data quantity associated with $EF_{COPERT}$ versus $EF_OBD$, and $\odot$ signifies the Hadamard product. The data quantity for each attribute in $R_C$ is compared against the predefined threshold of correlation data $P_C$. An attribute is considered strong correlation if $P_C \geq numsf_i$. The final correlation attributes obtained are then combined with $EF_OBD$ to obtain the final dataset.

### 3.4. Two-Stream Model Based on Historical Information

As shown above, we have constructed an input matrix based on the NOx emission factor about $k$-1 segment history information under multiple external factors. Briefly, each column of the input matrix represents one-dimensional time-series information of a certain attribute, which indicates that the whole matrix is from the time domain and is a time-series matrix containing multidimensional time-series information. To convert the one-dimensional temporal signal into a two-dimensional time-frequency matrix for each column of the input matrix, the Continuous Wavelet Transform (CWT) can be utilized. This transformed time-frequency matrix can represent both the time and frequency domain information of the information collected from heavy-duty diesel vehicles. It offers a clear representation of the correlation between signal frequency and temporal transformation.

### 3.4.1. Historical Information Matrix Construction

After the above series of operations, we get the driving segments that combine emission factors and keep only the attributes strongly related to $EF\_OBD$. For each driving segment, with $k$ as a step and $EF\_OBD$ as the label, In order to construct a historical information matrix, we take the driving segment with $k$ = 3 and length m as a representative example. Within each matrix, the first two records represent the historical information, while the third record represents the current information. The corresponding $EF\_OBD$ value for the current information is used as the label. This procedure yields a total of $m - 2$ matrices, as shown in Figure 6.

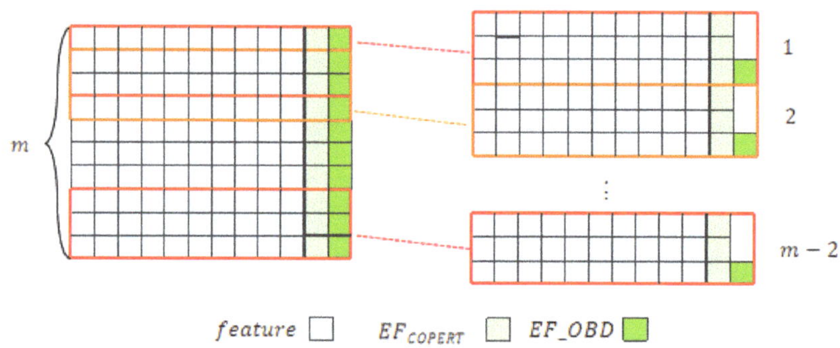


**Figure 6.** History Information Matrix Construction Process.

### 3.4.2. CWT

For each column in the historical information matrix, the application of the continuous wavelet transform allows for the conversion of the one-dimensional time-series signal into a two-dimensional time-frequency matrix. This transformation facilitates the visualization of the emission attribute signals from the mobile source, revealing their characteristics in both the time and frequency domains. It provides a clear depiction of the relationship between signal frequency transformation and time. CWT is performed on a historical information matrix ($k$ = 48, $P_c$ = Q2), and the resulting time-frequency matrix is visualized as shown in Figure 7.

For CWT, the choice of wavelet basis functions has a large impact on the effect of wavelet transform. To evaluate the conversion effect of CWT, for a variety of wavelet basis functions, we consider the converted time-frequency matrix as a gray-scale map, and based on the amount of data, we construct a quality index ($QI_{avg}$) index using information entropy (IE) and Laplacian gradient (LG) to evaluate its converted time-frequency map, where IE is used to measure the uncertainty of the information source and LG is used to evaluate the clarity of the gray-scale map. The formulaic description of the quality index ($QI_{avg}$) is as follows:

$$\begin{aligned} QI_{avg} &= \frac{QI_{sum}}{count(feature)+count(train,val)} \\ &= \frac{\sum_{train,val}\sum_{feaature}(QI)}{count(feature)+count(train,val)} \end{aligned} \tag{9}$$

$$QI = \frac{IE + LB}{2} \tag{10}$$

$$IE = -\sum_{i=0}^{255} p(i)\log_2 p(i) \tag{11}$$

$$LB = \sum_x \sum_y |G(x,y)| \tag{12}$$

In the provided equation, $count(feature)$ represents the number of features, while $count(train, val)$ denotes the cardinality of the combined training and validation datasets. The expression $p(i) = count(p = i)/(m * n)$ is utilized to quantify the relative frequency of grayscale values with pixel value i in an image of dimensions mn, where the grayscale range spans from 0 to 255. In this context, $count(p = i)$ signifies the occurrence count of pixel value $i$ within the image. The symbol $G(x, y)$ represents the convolution operation involving the Laplacian operator at the pixel coordinate $(x, y)$, and Laplacian operator $L = \begin{bmatrix} 0 & 1 & 0 \\ 1 & -4 & 1 \\ 0 & 1 & 0 \end{bmatrix}$.

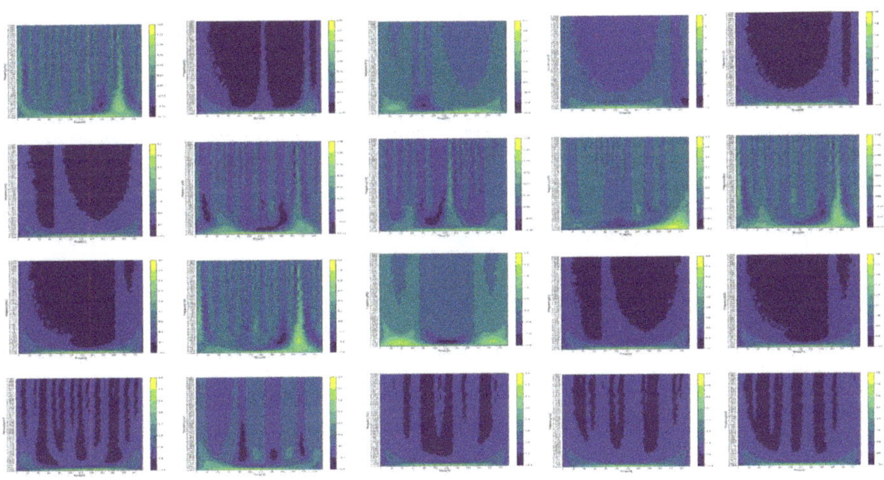

**Figure 7.** Visualization of the time-frequency matrices.

By using the $QI_{avg}$ indicator, the quality of each wavelet basis function is shown Figure 8 ($k$ = 48), where the mexh wavelet basis function has the largest $QI_{avg}$ value and the best conversion. Therefore, the *mexh* function will be used for the continuous wavelet transform in the subsequent experiments of this paper. After CWT, the time-frequency matrix obtained for each relevant attribute and $EF_{COPERT}$ is superimposed into a multi-channel matrix as the final time-frequency stream input.

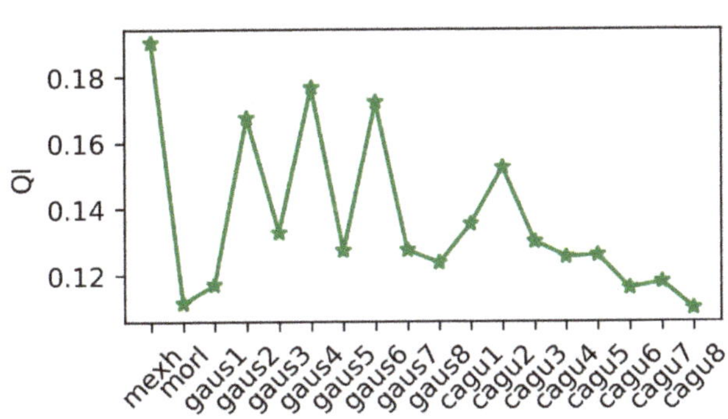


**Figure 8.** $QI_{avg}$ of each wavelet basis function.

3.4.3. Two Stream

In this paper, we use a combination of ResNet50 and CBAM to extract historical features. The CBAM module is added after the stage four output.

The two stream models use a combination of two parallel Resnet50 and CBAM, and after averaging the pooling layers, the output size of the two stream models is $BathSize \times 2048 \times 1 \times 1$, which is reshaped to $BathSize \times 2048$, and then the reshaped features are fused. Feature fusion is a mid-level fusion between data fusion and decision fusion [44]. It can eliminate redundant information due to correlation between different feature sets. The features extracted from the time-series stream and the time-frequency stream are fused and the strategy is defined as follows.

$$d^b = concat(T, TF) = [t_1^b, t_2^b, \cdots, t_{2048}^b, tf_1^b, tf_2^b, \cdots, tf_{2048}^b] \tag{13}$$

where $b$ represents the bath sequence number, $T$ is the temporal flow characteristics and $TF$ is the time-frequency flow characteristics.

After passing through the feature fusion layer, the model can obtain a more comprehensive and accurate evaluation structure. Finally, the fused feature vectors are fed into the FC layer to achieve one regression prediction. The loss function is as follows, where the first term MSE is used to ensure the prediction performance in training and the second term $L1$ norm is used to prevent overfitting.

$$J(w, b) = MSE(y, \hat{y}) + \lambda \|w\|_1 \tag{14}$$

where $MSE(y, \hat{y}) = \frac{1}{n} \sum_{i=1}^{n} (wx_i + b - y_i)^2$, $w$ is the weight, $b$ is the bias, $y$ is the label and $\hat{y}$ is the model prediction.

## 4. Experiments

In this study, we employ Min–Max normalization to standardize the dataset. The dataset is then divided into training, validation and test sets in a ratio of 7:2:1. During training, a batch size of 64 is used, and the Adam optimizer is applied with a learning rate of 0.0001. The maximum number of iterations is set to 200. Evaluation metrics, including MAE, RMSE and MAPE, are used to assess the model's performance. Training is stopped if the loss fails to decrease for 15 consecutive iterations. The values of each metric are the average values of five repeated experiments.

### 4.1. Historical Information Step Setting

For our method, the length $k$ of historical information has an important impact on the performance of the model. A too short length $k$ will make the historical information not be considered completely enough, resulting in a lower performance of the trained model. A too long length $k$ will cause the constructed historical information matrix to carry too much information and increase the unnecessary computation. Therefore, the determination of the value of the historical information step $k$ in the model is particularly important, however, the existing theoretical knowledge does not support us to make a direct determination of the value of $k$. Therefore, we design to use different values of $k$ to experiment on the two-stream model to select the optimal $k$.

We experimented in the case of for the deletion of any attributes. Historical information is determined at intervals of 1 min (12 records) to obtain the historical information set $K_L = \{12, 24, 36, 48, 60, 72, 84\}$, where the numbers represent the amount of data that make up a matrix. The experimental results are shown in Table 6.

Figure 9 is a visual depiction of the various metrics in Table 6. From the figure, it is easy to see that at $k$ = 48, our model has the smallest values in the three metrics MAE, MAPE and RMSE. Therefore, for our dataset, the historical information step $k$ is finally set to 48.

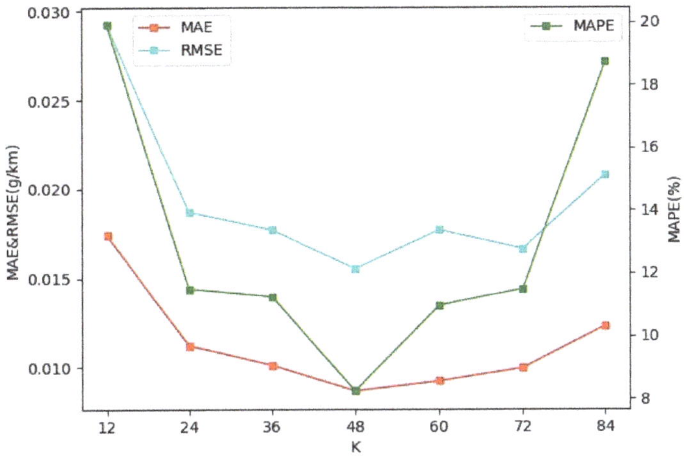


**Figure 9.** Model performance at different $k$ values.

**Table 6.** Performance of the Model for Various Values of $k$.

| Metrics / K | MAE | MAPE | RMSE |
|---|---|---|---|
| 12 | 0.0173 | 19.90% | 0.0292 |
| 24 | 0.0112 | 11.49% | 0.0187 |
| 36 | 0.0101 | 13.27% | 0.0177 |
| 48 | **0.0086** | **8.23%** | **0.0159** |
| 60 | 0.0092 | 10.96% | 0.0181 |
| 72 | 0.0099 | 11.44% | 0.0166 |
| 84 | 0.0123 | 18.72% | 0.0208 |

### 4.2. Setting of Correlation Data Volume

The amount of relevant data, $P_C$, is also an important parameter for the performance of the model. In this paper, the candidate set of $P_C$ is set to consist of quartiles of uncensored attributes and total data volume $L_{P_C}$ = [0, Q1, Q2, Q3], which is experimented with $k$ = 48, and the relevance of each attribute at $k$ = 48 is shown in Table 7.

As shown in Table 7, if setting the amount of relevant data is set to 25% (Q1) of the total data, only the attribute AP needs to be removed from all attributes; if setting the amount of relevant data is set to 50% (Q2) of the total data, the four attributes AP, $PT_{EGM}$, UTL and UTT are removed from all attributes. If it is set to 75% (Q3) of the total data, the attributes $E_{Speed}$, AOTP, EWT, NOx, AP, ET, $PT_{EGM}$, UTL, UTT, GPO, EFCR, $EFCR_{avg}$, BV, FTL and CER are removed from all attributes.

The experimental results are shown in Table 8, where 0 means no irrelevant factor deletion is performed. From Table 8, we can see that only one attribute of AP is deleted in Q1, so the model performance is similar to that of the model without irrelevant factor deletion, and the best performance of the model is at Q2, which has the smallest index, and the rest of the relevant data volume has a significant gap compared with it. Therefore, we finally choose the relevant data volume $P_C$ as Q2.

**Table 7.** The correlation of each feature under different $P_C$.

| Feature \ $P_C$ | $Q_1$ | $Q_2$ | $Q_3$ | Feature \ $P_c$ | $Q_1$ | $Q_2$ | $Q_3$ | Feature \ $P_c$ | $Q_1$ | $Q_2$ | $Q_3$ |
|---|---|---|---|---|---|---|---|---|---|---|---|
| $E_{speed}$ | ✓ | ✓ | | AOTP | ✓ | ✓ | | EWT | ✓ | ✓ | |
| EFT | ✓ | ✓ | ✓ | $NO_x$ | ✓ | ✓ | | O2 | ✓ | ✓ | ✓ |
| AP | | | | ET | ✓ | ✓ | | $PT_{EGM}$ | ✓ | | |
| UTL | ✓ | | | UTT | ✓ | | | $V_{speed}$ | ✓ | ✓ | ✓ |
| GPO | ✓ | ✓ | | SDM | ✓ | ✓ | ✓ | EFCR | ✓ | ✓ | |
| $EFCR_{avg}$ | ✓ | ✓ | | $EFC_{SD}$ | ✓ | ✓ | ✓ | $EFC_{total}$ | ✓ | ✓ | ✓ |
| BV | ✓ | ✓ | | FTL | ✓ | ✓ | | CER | ✓ | ✓ | |
| LNG | ✓ | ✓ | ✓ | LAT | ✓ | ✓ | ✓ | $EF_{COPERT}$ | ✓ | ✓ | ✓ |

**Table 8.** Model performance at different $P_C$.

| $P_C$ \ Metrics | MAE | MAPE | RMSE |
|---|---|---|---|
| 0 | 0.0086 | 8.23% | 0.0159 |
| $Q_1$ | 0.0089 | 9.03% | 0.0153 |
| $Q_2$ | **0.0077** | **7.38%** | **0.0137** |
| $Q_3$ | 0.0086 | 8.82% | 0.0156 |

### *4.3. Ablation Experiments*

In order to verify the validity of the module models in this paper, we perform relevant ablation experiments with the already selected parameters $k = 48$, $P_C = Q2$, where the main experimental models are as follows.

(a) ResNet50: single-stream resnet50 with the historical information matrix as input.
(b) ResNet50 (CBAM): Resnet50 of a single stream combined with CBAM, with the historical information matrix as input.
(c) HI_TTFTS (without CBAM): two-stream model composed of two resnet50, the two-stream inputs are the historical information matrix and the time-frequency matrix, respectively.
(d) HI_TTFTS (left): two-stream model composed of two resnet50, left combined with CBAM, the left input is the historical information matrix and the right input is the time-frequency matrix.
(e) HI_TTFTS (right): two-stream model composed of two resnet50, right combines CBAM, the left input is the historical information matrix, and the right input is the time-frequency matrix.
(f) HI_TTFTS (all): two-stream model composed of two resnet50, respectively, with CBAM, the left input is the historical information matrix and the right input is the time-frequency matrix.

The results of the ablation experiments are shown in Table 9, and the RMSE, MAE and MAPE metrics obtained for each model experiment show that the HI_TTFTS (CBAM) model obtains a richer feature representation using two streams of time-frequency and time-sequence streams in parallel, while the key information for making modifications to the COPERT model is obtained from the large amount of information by adding the CBAM module to both streams, and its model performance is better than the other five models. Specifically, by comparing the ResNet50 (CBAM) and Resnet50 models, it can be seen that CBAM brings a positive effect on the correction of emission factors for the ResNet model, which can improve its performance; and by comparing the ResNet50 and Two stream (without cbam) models, it can be seen that the positive effect of the two-stream model on the model performance improvement. Finally, by comparing HI_TTFTS (without

CBAM), HI_TTFTS (left) and HI_TTFTS (right), the performance of the model can be improved to some extent by adding the CBAM module to any stream in the two streams, while the multi-channel time-frequency matrix obtained by CWT through transforming the historical information matrix brings the number of parametric larger and more informative feature representation, it is easier to obtain an effective feature map representation by imposing an attention mechanism on the informative time-frequency stream, which makes the HI_TTFTS (right) model perform slightly better than the HI_TTFTS (left) model.

**Table 9.** Ablation model performance comparison.

| Model \ Metrics | MAE | MAPE | RMSE |
|---|---|---|---|
| ResNet50 | 0.0156 | 17.36% | 0.0256 |
| ResNet50 (CBAM) | 0.0098 | 9.37% | 0.0186 |
| HI_TTFTS (Without CBAM) | 0.0138 | 16.69% | 0.0225 |
| HI_TTFTS (Left) | 0.0123 | 13.70% | 0.0198 |
| HI_TTFTS (Right) | 0.0112 | 13.23% | 0.0181 |
| HI_TTFTS (all) | **0.0077** | **7.38%** | **0.0137** |

### *4.4. Comparative Experiment*

To verify the excellent performance of our proposed model for COPERT correction, we selected the following mainstream models for comparison with our proposed dual-flow model based on historical information.

(a) Support Vector Regression (SVR): only consider the influence of each attribute on the emission factor at the current moment, the input data is only the combination of each attribute and $EF_{COPERT}$ at the current moment.
(b) Artificial Neural Network (ANN): only consider the influence of each attribute on the emission factor at the current moment, the input data is only the combination of each attribute and $EF_{COPERT}$ at the current moment.
(c) Convolutional Neural Network (CNN): a general CNN model with time-frequency matrix as input.
(d) TF: ResNet50 model with time-frequency matrix as input.
(e) TSFF: Each column of the historical information matrix is pieced together into a 6*8 matrix [44] and multi-channel superposition is performed with dual-stream inputs of the historical information matrix and the constructed 2D construction matrix, respectively, to a parallel structure combining ResNet50 and CBAM.

With mae, mape and rmse as evaluation indicators, the experimental results are shown in Table 10, and HI_TTFTS (ours) has the best indicators. Specifically, SVR has the worst effect and ANN is the second best, which is due to the fact that both SVR model and ANN model do not take the historical information matrix as input, and only consider the influence of each attribute on NOx emission at the current moment, and the correction effect is poor, while ANN can obtain better performance by training with sufficient amount of data; In the case of considering the influence of historical information, the CNN is limited by the network structure, and the model performance is poor, while the two-stream method TSFF, which constructs a two-dimensional matrix by splicing to obtain richer information, is second, and the time-frequency conversion is more informative than the method of splicing to construct a two-dimensional matrix, so the effect is better. Our proposed method combines the CBAM module on the basis of time-frequency stream ResNet50, and obtains the best performance by supplementing the time-frequency information with the new time-series history matrix.

Figure 10 shows the prediction results of each model corrected for COPERT, and we have chosen points in the range of 400–500 in the test set to show the details, from which we can see that the HI_TTFTS proposed in this paper fits better with *EF_OBD*.

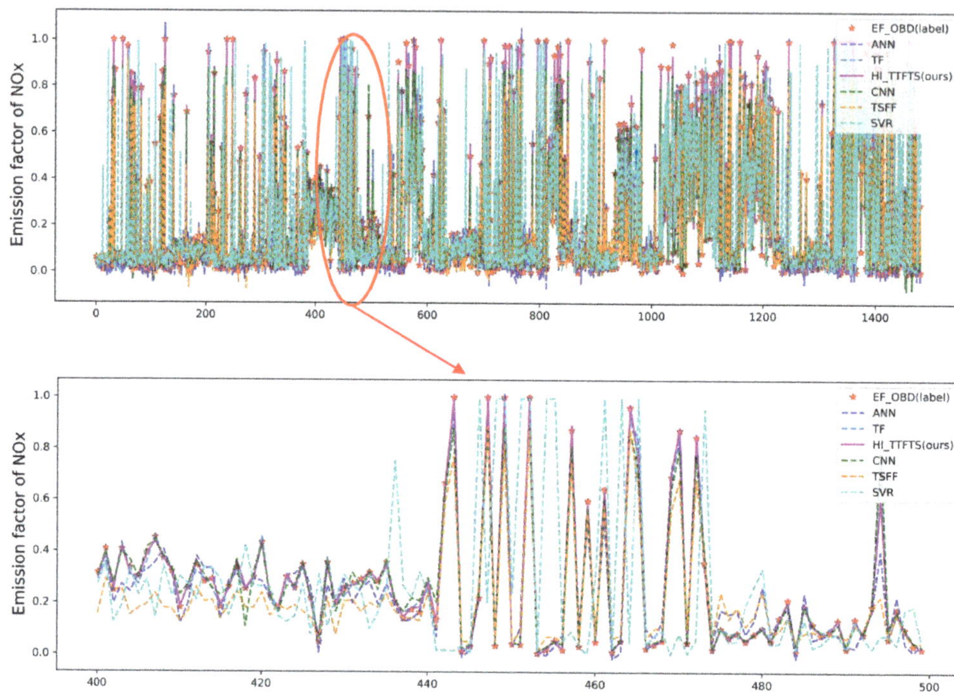


**Figure 10.** Comparing the predictions of various models.

**Table 10.** Performance of mainstream model vs. HI_TTFTS for COPERT modification.

| Model \ Metrics | MAE | MAPE | RMSE |
|---|---|---|---|
| SVR | 0.0864 | 87.18% | 0.1524 |
| ANN | 0.0229 | 138.43% | 0.0275 |
| CNN | 0.01963 | 60.6% | 0.0292 |
| TF | 0.0135 | 16.76% | 0.0216 |
| TSFF | 0.0163 | 20.13% | 0.0274 |
| HI_TTFTS | **0.0077** | **7.38%** | **0.0137** |

## 5. Conclusions

In order to perform an accurate NOx emission factor calculation on the OBD online monitoring dataset using the COPERT model, we propose a two-stream amendment model based on the fusion of time-series and time-frequency features with historical information (HI_TTFTS). The model uses Spearman's rank correlation analysis based on data volume, which solves the situation that the actual dataset presents multiple consecutive segments in the time series, which makes the correlation analysis complicated. Our proposed model considers the influence of historical information on current moment emissions, and the historical information matrix and the corresponding time-frequency matrix are constructed as inputs to the parallel ResNet50 and CBAM combined network. This approach achieves the fusion of temporal features and two-dimensional time-frequency features, which improves the COPERT model correction effect. By comparing our model with mainstream models, we have verified its more stable performance and reliable prediction accuracy.

In this study, we used data from a diesel vehicle for about six months for the experiment. Due to the limitation of the difficulty of dataset collection, we could not obtain more data. Therefore, our conclusions may be limited by the size of the data and the model of the car being taken, and we hope that we can further improve the quality of the data in the later studies to support us to get more generalizable results. In addition, the spatial and temporal features are not considered separately in this work, but are treated as one kind of feature, and in future research, the temporal and spatial features can be considered separately to carry out more in-depth research.

**Author Contributions:** Conceptualization, Z.X. and R.W.; methodology, Z.X. and R.W.; software, K.P.; validation, J.L.; formal analysis, K.P. and Q.W.; investigation, K.P.; resources, Q.W.; data curation, K.P.; writing–original draft preparation, Z.X.; writing–review and editing, J.L.; visualization, K.P.; supervision, Z.X.; project administration, Z.X.; funding acquisition, Z.X. All authors have read and agreed to the published version of the manuscript.

**Funding:** This work was supported in part by the National Natural Science Foundation of China (62103124, 62033012), Open Project Program of Key Laboratory of Ministry of Education of System Control and Information Processing (grant No. SCIP20230109).

**Institutional Review Board Statement:** Not applicable.

**Informed Consent Statement:** Not applicable.

**Data Availability Statement:** The datasets used in this study will be available from the corresponding author upon reasonable request.

**Conflicts of Interest:** The authors declare no conflict of interest.

## References

1. Ministry of Ecology and Environment of the People's Republic of China. *China Mobile Source Environmental Management Annual Report*; Ministry of Ecology and Environment of the People's Republic of China: Beijing, China, 2021.
2. US Environmental Protection Agency. *User's Guide to MOBILE6.1 and MOBILE6.2: Mobile Source Emission Factor Model (EPA Report EPA-420-R-03-010)*; Office of Transportation and Air Quality: Washington, DC, USA, 2003.
3. California Air Resources Board. *EMFAC2021 User's Guide*; California Environmental Protection Agency: Sacramento, CA, USA, 2021.
4. Ntziachristos, L.; Gkatzoflias, D.; Kouridis, C.; Samaras, Z. COPERT: A European road transport emission inventory model. In *Information Technologies in Environmental Engineering: Proceedings of the 4th International ICSC Symposium Thessaloniki, Greece, 28–29 May 2009*; Springer: Heidelberg/Berlin, Germany, 2009; pp. 491–504.
5. Lents, J.; Davis, N. *IVE Model User's Guide, Model and Data Files*; Technical Report Submitted to the US Environmental Protection Agency; US Environmental Protection Agency: Washington, DC, USA, 2009.
6. Scora, G.; Barth, M. *Comprehensive Modal Emissions Model (CMEM), Version 3.01. User Guide*; Centre for Environmental Research and Technology, University of California: Riverside, CA, USA, 2006; Volume 1070, p. 1580.
7. Vallamsundar, S.; Lin, J. Overview of US EPA new generation emission model: MOVES. *Int. J. Transp. Urban Dev.* **2011**, *1*, 39.
8. Söderena, P.; Laurikko, J.; Weber, C.; Tilli, A.; Kuikka, K.; Kousa, A.; Väkevä, O.; Venho, A.; Haaparanta, S.; Nuottimäki, J. Monitoring Euro 6 diesel passenger cars NOx emissions for one year in various ambient conditions with PEMS and NOx sensors. *Sci. Total Environ.* **2020**, *746*, 140971. [CrossRef] [PubMed]
9. Oluwaseyi, M.M.; Sunday, A.M. Specifications and analysis of digitized diagnostics of automobiles: A case study of on board diagnostic (OBD II). *Int. J. Eng. Res. Technol.* **2020**, *9*, 91–105. [CrossRef]
10. Quirama, L.F.; Giraldo, M.; Huertas, J.I.; Jaller, M. Driving cycles that reproduce driving patterns, energy consumptions and tailpipe emissions. *Transp. Res. Part Transp. Environ.* **2020**, *82*, 102294. [CrossRef]
11. Yang, D.; Zhang, S.; Niu, T.; Wang, Y.; Xu, H.; Zhang, K.M.; Wu, Y. High-resolution mapping of vehicle emissions of atmospheric pollutants based on large-scale, real-world traffic datasets. *Atmos. Chem. Phys.* **2019**, *19*, 8831–8843. [CrossRef]
12. Wang, S.; Li, Z.; Tan, J.; Guo, W.; Li, L. A method for estimating carbon dioxide emissions based on low frequency GPS trajectories. In Proceedings of the 2017 Chinese Automation Congress (CAC), Jinan, China, 20–22 October 2017; pp. 1960–1964.
13. Kumar, K.; Pande, B.P. Air pollution prediction with machine learning: A case study of Indian cities. *Int. J. Environ. Sci. Technol.* **2023**, *20*, 5333–5348. [CrossRef]
14. Wang, Z.; Guan, W.; Bi, J.; Ma, J. Microscopic emission model of motor vehicle based on short-time real driving cycle. *Chin. J. Environ. Eng.* **2016**, *10*, 5803–5807.

15. Xu, Z.; Kang, Y.; Cao, Y.; Li, Z. Deep amended COPERT model for regional vehicle emission prediction. *Sci. China Inf. Sci.* **2021**, *64*, 1–3. [CrossRef]
16. Yang, L.; Zhang, S.; Wu, Y.; Chen, Q.; Niu, T.; Huang, X.; Zhang, S.; Zhang, L.; Zhou, Y.; Hao, J. Evaluating real-world $CO_2$ and NOx emissions for public transit buses using a remote wireless on-board diagnostic(OBD) approach. *Environ. Pollut.* **2016**, *218*, 453–462. [CrossRef]
17. Xu, Z.; Wang, R.; Kang, Y.; Zhang, Y.; Xia, X.; Wang, R. A Deep Transfer NOx Emission Inversion Model of Diesel Vehicles with Multisource External Influence. *J. Adv. Transp.* **2021**, *2021*, 4892855. [CrossRef]
18. Molina Campoverde, J.J. Driving mode estimation model based in machine learning through PID's signals analysis obtained from OBD II. *Int. Conf. Appl. Technol.* **2020**, 80–91.
19. Rivera-Campoverde, N.D.; Muñoz-Sanz, J.L.; Arenas-Ramirez, B.d.V. Estimation of pollutant emissions in real driving conditions based on data from OBD and machine learning. *Sensors* **2021**, *21*, 6344. [CrossRef] [PubMed]
20. Wang, J.; Wang, L.; Ji, Z.; Qi, S.; Xie, Z.; Yang, Z.; Zhang, X. Research on Actual Road Emission Prediction Model of Heavy-Duty Diesel Vehicles Based on OBD Remote Method and Artificial Neural Network. *J. Phys. Conf. Ser.* **2021**, *2005*, 23–25. [CrossRef]
21. Chen, S.H.; Pan, J.S.; Lu, K. Driving behavior analysis based on vehicle OBD information and adaboost algorithms. In Proceedings of the International Multiconference of Engineers and Computer Scientists, Hong Kong, China, 18–20 March 2015; Volume 1, pp. 18–20.
22. Liu, Y.H.; Liao, W.Y.; Li, L.; Huang, Y.T.; Xu, W.J. Vehicle emission trends in China's Guangdong Province from 1994 to 2014. *Sci. Total Environ.* **2017**, *586*, 512–521. [CrossRef] [PubMed]
23. Sun, S.; Jiang, W.; Gao, W. Vehicle emission trends and spatial distribution in Shandong province, China, from 2000 to 2014. *Atmos. Environ.* **2016**, *147*, 190–199. [CrossRef]
24. D'Angiola, A.; Dawidowski, L.E.; Gómez, D.R.; Osses, M. On-road traffic emissions in a megacity. *Atmos. Environ.* **2010**, *44*, 483–493. [CrossRef]
25. Wang, R.; Xia, X.; Xu, Z. Heavy-Duty Emission Prediction Model Using Wavelet Features and ResNet. In Proceedings of the Asian Simulation Conference, Changsha, China, 9–11 December 2022; pp. 525–537.
26. Cai, H.; Xie, S. Estimation of vehicular emission inventories in China from 1980 to 2005. *Atmos. Environ.* **2007**, *41*, 8963–8979. [CrossRef]
27. Li, R.; Yang, F.; Liu, Z.; Shang, P.; Wang, H. Effect of taxis on emissions and fuel consumption in a city based on license plate recognition data: A case study in Nanning, China. *J. Clean. Prod.* **2019**, *215*, 913–925. [CrossRef]
28. Nelson English, Loxton, and Andrews Pty. Ltd. *A Review of Dynamometer Correlations, In-service Strategies and Engine Deterioration*; National Environment Protection Council: Adelaide, Australia, 2000.
29. US EPA (US Environmental Protection Agency). *Analysis of BSFCs and Calculation of Heavy-Duty Engine Emission Conversion Factors. EPA-420-R-02-005*; US EPA: Washington DC, USA, 2002.
30. *GB 17691-2018*; Limits and Measurement Methods for Emissions from Diesel Fueled Heavy-Duty Vehicles (CHINA VI). Ministry of Ecology and Environment of People's Republic of China: Beijing, China, 2018.
31. Krizhevsky, A.; Sutskever, I.; Hinton, G.E. Imagenet classification with deep convolutional neural networks. *Adv. Neural Inf. Process. Syst.* **2012**, *60*, 84–90. [CrossRef]
32. Szegedy, C.; Liu, W.; Jia, Y.; Sermanet, P.; Reed, S.; Anguelov, D.; Erhan, D.; Vanhoucke, V.; Rabinovich, A. Going deeper with convolutions. In Proceedings of the IEEE Conference on Computer Vision and Pattern Recognition, Boston, MA, USA, 7–12 June 2015; pp. 1–9.
33. Yue-Hei Ng, J.; Hausknecht, M.; Vijayanarasimhan, S.; Vinyals, O.; Monga, R.; Toderici, G. Beyond short snippets: Deep networks for video classification. In Proceedings of the IEEE Conference on Computer Vision and Pattern Recognition, Boston, MA, USA, 7–12 June 2015; pp. 4694–4702.
34. He, K.; Zhang, X.; Ren, S.; Sun, J. Deep residual learning for image recognition. In Proceedings of the IEEE Conference on Computer Vision and Pattern Recognition, Boston, MA, USA, 7–12 June 2015; pp. 770–778.
35. Guo, M.; Xu, T.; Liu, J.; Liu, Z.; Jiang, P.; Mu, T.; Zhang, S.; Martin, R.; Cheng, M.; Hu, S. Attention mechanisms in computer vision: A survey. *Comput. Vis. Media* **2022**, *8*, 331–368. [CrossRef]
36. Hu, D. An introductory survey on attention mechanisms in NLP problems. In Proceedings of the SAI Intelligent Systems Conference, London, UK, 5–6 September 2019; Volume 2, pp. 432–448.
37. Jaderberg, M.; Simonyan, K.; Zisserman, A. Spatial transformer networks. *Adv. Neural Inf. Process. Syst.* **2015**, *2*, 2017–2025.
38. Almahairi, A.; Ballas, N.; Cooijmans, T.; Zheng, Y.; Larochelle, H.; Courville, A. Dynamic capacity networks. In Proceedings of the International Conference on Machine Learning, New York, NY, USA, 19–24 June 2016; pp. 2549–2558.
39. Hu, J.; Shen, L.; Sun, G. Squeeze-and-excitation networks. In Proceedings of the IEEE Conference on Computer Vision and Pattern Recognition, Salt Lake City, UT, USA, 18–29 June 2018; pp. 7132–7141.
40. Wang, Q.; Wu, B.; Zhu, P.; Li, P.; Zuo, W.; Hu, Q. ECA-Net: Efficient channel attention for deep convolutional neural networks. In Proceedings of the IEEE/CVF Conference on Computer Vision and Pattern Recognition, Seattle, WA, USA, 13–19 June 2018; pp. 11534–11542.
41. Woo, S.; Park, J.; Lee, J.Y.; Kweon, I.S. Cbam: Convolutional block attention module. In Proceedings of the European Conference on Computer Vision (ECCV), Munich, Germany, 8–14 September 2018; pp. 3–19.

42. Yoo, Y.; Baek, J. A novel image feature for the remaining useful lifetime prediction of bearings based on continuous wavelet transform and convolutional neural network. *Appl. Sci.* **2018**, *8*, 1102. [CrossRef]
43. Miao, R.; Gao, Y.; Ge, L.; Jiang, Z.; Zhang, J. Online defect recognition of narrow overlap weld based on two-stage recognition model combining continuous wavelet transform and convolutional neural network. *Comput. Ind.* **2019**, *112*, 103115. [CrossRef]
44. Xue, F.; Zhang, W.; Xue, F.; Li, D.; Xie, S.; Fleischer, J. A novel intelligent fault diagnosis method of rolling bearing based on two-stream feature fusion convolutional neural network. *Measurement* **2021**, *176*, 109226. [CrossRef]

*atmosphere*

MDPI

*Article*

# A Hybrid Autoformer Network for Air Pollution Forecasting Based on External Factor Optimization

**Kai Pan [1,2], Jiang Lu [3], Jiaren Li [1,2] and Zhenyi Xu [2,*]**

1 AHU-IAI AI Joint Laboratory, Anhui University, Hefei 230031, China
2 Institute of Artificial Intelligence, Hefei Comprehensive National Science Center, Hefei 230088, China
3 Hefei Branch of China Telecom Co., Ltd., Hefei 230031, China
* Correspondence: xuzhenyi@mail.ustc.edu.cn

**Abstract:** Exposure to air pollution will pose a serious threat to human health. Accurate air pollution forecasting can help people to reduce exposure risks and promote environmental pollution control, and it is also an extremely important part of smart city management. However, the current deep-learning-based models for air pollution forecasting usually focus on prediction accuracy improvement without considering the model interpretability. These models usually fail to explain the complex relationships between prediction targets and external factors (e.g., ozone concentration ($O_3$), wind speed, temperature variation, etc.) The relationships between variables in air pollution time series prediction problems are very complex, with intricate relationships between different types of variables, often with nonlinear multivariate dependencies. To address these problems mentioned above, we proposed a hybrid autoformer network with a genetic algorithm optimization to predict air pollution temporal variation as well as establish interpretable relationships between pollutants and external variables. Furthermore, an elite variable voting operator was designed to better filter out more important external factors such as elite variables, so as to perform a more refined search for elite variables. Moreover, we designed an archive storage operator to reduce the effect of neural network model initialization on the search for external variables. Finally, we conducted comprehensive experiments on the Ma'anshan air pollution dataset to verify the proposed model, where the prediction accuracy was improved by 2–8%, and the selection of model influencing factors was more interpretable.

**Keywords:** air pollution forecasting; external factor optimization; genetic algorithm, autoformer; temporal variation; multivariate dependencies; relationships analysis; model interpretability

## 1. Introduction

In recent years, with the development of urbanization and industrialization, air quality issues have been on the agenda [1], and air pollution has been placed on an increasingly important position in policy formulation and implementation. Air pollution in cities is mainly caused by industrial emissions and transportation, which can produce pollutants such as $NO_2$, $O_3$, and $SO_2$ [2]. According to the 2021 China Ecological Environment Status Bulletin, only 64.3% of China's 339 cities at the prefecture level and above met the ambient air quality standards in 2021. In the Yangtze River Delta region, the average percentage of days with air quality exceeding standards was 13.3%, with $O_3$ and $PM_{2.5}$ as the primary pollutants accounting for 55.4% and 30.7% of the total exceedance days, respectively. The synergistic treatment of multiple pollutants has become the focus of air pollution prevention and control in China [3].

Air pollution prediction refers to the extraction of information and characteristics from historical air pollution data to predict the future trend of air pollution [4]. Multivariate time prediction means that, for the predicted time series, there may be very many external factors that affect the predicted target, such as the value of some pollutants related to the historical value of the target pollutant, as well as the closely related external factors including temperature, humidity, wind direction etc. [5]. Many cities have established monitoring stations in various locations to detect ozone ($O_3$), nitric oxide(NO), $PM_{2.5}$, and other detection data. The main source of air pollution from industrial emissions, human activities, transportation, natural causes (wild fires), and other factors, such as weather factors, wind speed, temperature, and humidity, will also affect the settlement of pollutants and, thus, affect their monitoring values. One contaminant may also be a precursor to another, and Figure 1 shows the correlation relationships of different pollutants. Thus, air pollution is affected by many complex factors, and these factors can interact with each other, thus making the prediction of air pollution a difficult problem. If it were possible to predict the place with high pollution probability one or two days in advance, more efficient actions could be taken to alleviate the potential regional pollution [6].

**Figure 1.** The daily trend of $O_3$ with $PM_{2.5}$ and precursor $NO_x$ on a certain day at a monitoring station:(**a**) $O_3$-$PM_{2.5}$; (**b**) $O_3$-$NO_x$.

Traditional air pollution prediction research are based on statistical methods, such as the Autoregressive Model (AR), the Moving Average Model (MA), and the Auto-Regression and Moving Average Model (ARMA), as well as the Autoregressive Integrated Moving Average (ARIMA) [7]. Although these methods can model the time series well, they all need the time series to have large smoothness, which requires high requirements on the dataset. However, in air pollution monitoring, there will often be problems such as missing data due to sensor failure, so pre-processing of the data is usually required in the application [8]. However, most statistical methods focus only on air pollution values in their predictions, without considering the changes in pollutant concentrations themselves by other factors, such as weather factors and the effects of other pollutants on them. With the development of artificial intelligence approaches and big data, many research projects have utilized machine learning and deep learning techniques for air pollution prediction. In traditional machine learning methods, Fan et al. [9] used a heuristic algorithm combined with SVM to predict daily diffuse solar radiation in air-polluted regions. S.Gocheva-Ilieva et al. [10] proposed a novel framework for stacked regression based on machine learning to predict the daily average concentrations of particulate matter ($PM_{10}$), where four base models were built and evaluated. Johansson, C et al. [11] applied different machine learning (ML) algorithms—including Random Forest (RF), Extreme Gradient Boosting (XGB), and Long Short-Term Memory (LSTM)—to improve deterministic predictions of $PM_{10}$, $NO_x$, and $O_3$ for 1, 2, and 3 days at different locations in Greater Stockholm, Sweden.

In the problem of air pollution prediction, the use of sample features in traditional machine learning methods mainly requires expert knowledge in air pollution, which is usually time-consuming and laborious. In particular, different regions have different environmental conditions and different characteristics of air pollution changes, and different structural characteristics of atmospheric flow due to topography and population density [12] make it more difficult to extract relevant features. Moreover, air pollutants have very complex chemical reactions; for example, $NO_x$ are important precursor pollutants in

$O_3$ formation and produce complex photochemical reactions, which makes it difficult to construct complex nonlinear feature mappings through traditional machine learning.

Deep learning has made many promising advances in the field of air pollution prediction and analysis. Shikhovtsev A.Yu et al. [13] used a deep neural network based on GMDH to estimate and predict the characteristics of turbulence intensity in the stratosphere. M.Catalano et al. [14] used Autoregressive Integrated Moving Average with Explanatory Variable(ARIMAX) and Artificial Neural Network (ANN) models to compare the result of urban transportation networks for air pollution peak predictions. The results showed that neural networks were predicting peaks at a superior rate to the ARIMAX model. However, the ANN does not reflect the temporal characteristics of air pollution variation well, where the air pollution variation is highly time-dependent and is closely related to the recent air pollution observation, as well as the previous period of variation. Recurrent Neural Networks (RNNs) can use the output of the previous moment as the input of the next moment to achieve feature extraction and the learning of time series. B.T. Ong et al. [15] proposed a Deep Recurrent Neural Network (DRNN) with a novel pre-training method to predict $PM_{2.5}$ in Japan. However, RNNs suffer from gradient disappearance and gradient explosion when dealing with long time sequence problems. M. Krishan et al. [16] used the Long Short-Term Memory (LSTM) approach to predict $O_3$, $PM_{2.5}$, $NO_x$, and CO concentrations at a location in the NCT of Delhi. Li et al. [17] proposed a hybrid CNN-LSTM model by combining a Convolutional Neural Network (CNN) with a Long Short-Term Memory (LSTM) neural network for forecasting the next 24 h of $PM_{2.5}$ concentration in Beijing. In recent years, Transformer has made incredible progress in sequential problem processing by using a multi-headed self-attentive mechanism to obtain point-in-time correlations. Chen et al. [18] combined a CNN with Transformer to predict $O_3$ concentrations and achieved good results in both short- and long-term predictions. Wenfeng Zheng et al. [19–24] used various deep learning methods to achieve better results on both haze time scale and spatio-temporal prediction.

Several contributions have been made in combining genetic algorithms with neural network models to explore hyperparameters; for example, Rana Muhammad Adnan et al. used ALO to optimize the number of hidden layer neurons and the learning rate of an LSTM [25]. The ANFIS-GBO model used two operators to optimize the learning parameters to improve the prediction accuracy of the ANFIS [26]. In addition, the PSOGWO and PSOGSA algorithms have also been used to optimize the control parameters of the ELM model [27]. The SVR-SAMOA model integrates the Simulated Annealing (SA) algorithm with the Mayor Optimization Algorithm (MOA) to determine the optimal hyperparameters for Vector Regression (SVR) [28], and the ANN-EMPA combines mutation and crossover operators with the ANN to produce robust hybrid prediction model [29]. The CNN-INFO is highly efficient in optimizing complex phenomena with unacknowledged search areas [30].

However, all these machine-learning-based methods are hard to interpret and require manual feature engineering based on a priori knowledge, which is prone to prediction errors. For the models built by deep learning training, although the prediction accuracy is high, their trained weights are of little use to us, because they have little physical meaning for real-world problems. For the air pollution prediction problem, the external variables have a very strong correlation with the prediction target, as shown in Figure 1, where $PM_{2.5}$, $NO_x$, and $O_3$ are significantly negatively correlated, and, as the PM2.5 concentration increases, the chemical reaction is suppressed, thus reducing the rate of $O_3$ production. In addition, the non-homogeneous chemical reactions occurring on the surface of the particles due to the increase of $PM_{2.5}$ concentration also affect the concentration of $O_3$, while $NO_x$ is a precursor of $O_3$ and undergoes photochemical reactions to produce ozone. These correlations can help the authorities in predicting atmospheric pollution while helping them to develop effective policies to mitigate pollution, but this information is difficult to obtain in deep learning, and a model is needed that can explore the impact of current external factors on atmospheric pollution while also predicting them.

All these challenges inspire us to rethink the air pollution prediction problems based on deep learning models with model interpretability. Specifically, a Hybrid Autoformer Network with a Genetic Algorithm Model (GA-Autoformer) was proposed to predict air pollution temporal variation, as well as explore the relationship between external variables and target pollution. The main contributions of the proposed method are summarized as follows:

(1) A Hybrid Autoformer Network with a Genetic Algorithm Model was proposed to predict the air pollution variation, where the genetic algorithm was used to optimize the external variable weighting problem for different variables that have different effects on the target pollution.
(2) The Elite Variable Operator was proposed to vote at fixed intervals of generations to find out the variables that have a greater impact on the target prediction to be selected as elite variables, which are explored to perform a more refined search.
(3) The proposed Archive Storage Operator, using genetic algorithms, led to deviations in the final results due to the influence of the initialization of individual models, where the individuals with better value may be less effective due to initialization and vice versa. The archive mechanism was used to store the individuals with good results and to filter them to get the really good ones.
(4) We conducted comprehensive experiments on the Ma'anshan air pollution dataset to verify the proposed model, where the prediction accuracy was greatly improved, and the selection of model influencing factors was more interpretable.

The rest parts of this paper is arranged as follows. We describe our study area and data set in detail in Section 2. Sections 3 and 4 review the related work and detail our model. The experiments and result analysis are presented in Section 5. Finally, we discuss and conclude the paper in Sections 6 and 7.

## 2. Study Area and Data Requirement

### 2.1. Study Area

Ma'anshan is located in East China (Figure 2), east of Anhui Province, in the lower reaches of the Yangtze River, bordering Nanjing, located between 31°46′42″–31°17′26″ north latitude and 118°21′38″–118°52′44″ east longitude. Furthermore, it is an important node city in the South Anhui International Tourism and Culture Demonstration Zone. The overall terrain of the city is relatively flat, slightly higher in the north and lower in the south. It has a north subtropical monsoon climate with four distinct seasons. Ma'anshan Port is one of the top ten ports on the Yangtze River, and Zhengpu Port is the only 10,000-ton deep-water port in the Jiangbei region of Anhui Province, where an Anhui river–sea intermodal transport hub is being built.

As an important industrial city in the Yangtze River Delta region, Ma'anshan straddles the Yangtze River and has a large impact on the surrounding air quality. Our study area covers residential, industrial, and rurial areas of Ma'anshan. The air quality data collected from the above areas can reflect the main air quality conditions of representative areas in Ma'anshan.

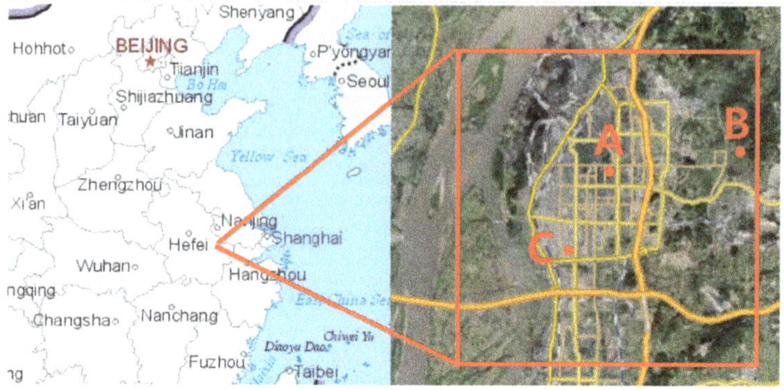


**Figure 2.** The map on the right identifies the location of the urban area of Ma'anshan, and the red dots represent the location of the monitoring stations where the data are concentrated. Location A is located in Ma'anshan Hudong Road Fourth Primary School (Resident area), Location B is located in Poutang National Resort Park (Rurial area), and C is located in Maanshan Economic and Technological Development Zone (industrial area), which correspond to Location A, Location B, and Location C used in the experiment, respectively.

### 2.2. Data Requirement

We used the air pollutant data from the air pollution quality testing station in Ma'anshan, Anhui Province, and our predicted targets were ozone concentration ($O_3$), particulate matter with a particle size below 2.5 microns($PM_{2.5}$), and the air quality index (AQI). The final model input includes three kinds of information:

(1) Air pollution gas detection content, including carbon monoxide (CO), nitrogen dioxide ($NO_2$), and sulfur dioxide ($SO_2$).
(2) Air pollution index, including particulate matter with a particle size below 10 microns (PM10) and total suspended particulates (TSPs).
(3) Environmental factors of the target AQI station, including wind direction, wind speed, precipitation, vapor pressure, humidity, visibility, atmospheric pressure, and temperature.

The specific data set partitioning and description will be detailed in Section 5.1.

## 3. Preliminary

### 3.1. Transformer for Time Series Forecasting

Recently, the Transformer model has achieved very excellent results in natural language processing. The self-attention mechanism enables the network to capture long sequences of features while avoiding the circular structure that exists in RNNs and LSTMs by using the parallel design that allows for a significant reduction in prediction time [31].

The Transformer model is a multi-layer model, including a multi-headed self-attentive layer, followed by a feed-forward layer, plus residual connections and layer normalization, as shown in Figure 3.

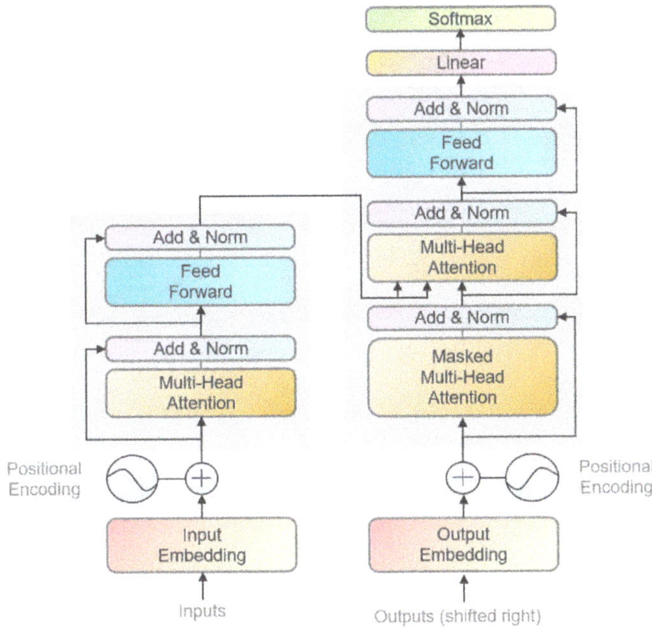


**Figure 3.** Architecture of Transformer.

Due to its great potential in sequence modeling, the Transformer soon attracted the attention of researchers. Wu et al. [32] first applied the Transformer to the influenza outbreak prediction problem. Subsequently, researchers carried out optimization of their model, and Li et al. [33] proposed a convolutional self-attention by producing queries and keys with causal convolution and sparse bias to reduce the computational complexity from $O(L^2)$ to $O(LlogL)$. Zhou et al. [34] proposed the ProbSparse Self-Attention mechanism, called Informer, and designed a generative style decoder to produce long-term forecasting. Xu et al. [35] decomposed the time series into trend parts and seasonal parts and used an auto-correlation mechanism.

### 3.2. Genetic Algorithm

The genetic algorithm was proposed by Professor Holland in 1975 [36]. It is a computational model that simulates Darwin's biological evolutionary process of genetic selection and natural elimination, and, so far, genetic algorithms have been considered as the basis of intelligent optimization algorithms [37]. Genetic algorithms have been widely used in many fields, such as function optimization, path planning, production scheduling, neural-network-structured searches, and other problems [38].

The basic idea of the genetic algorithm is to borrow the law of biological evolution through reproduction–competition to achieve superiority and inferiority, so that the problem is approached to the optimal solution step-by-step. In the process of problem solving, we modeled the data as similar to genes in an organism, and approximate the optimal solution through genetic, selection, crossover, and mutation operations of genes.

The general process of the genetic algorithm is as follows in Figure 4.

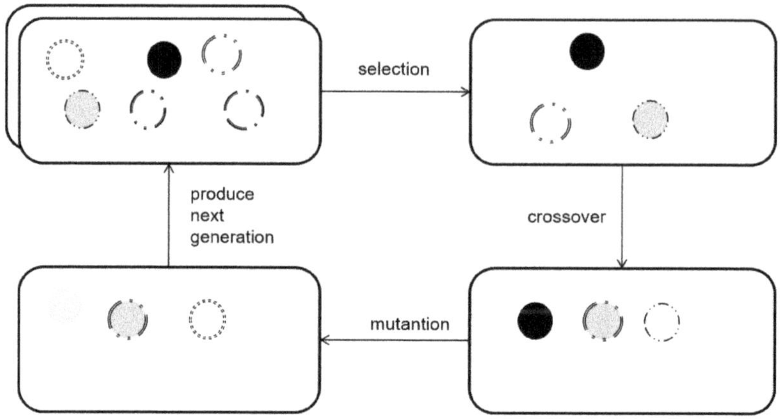


**Figure 4.** Framework of the genetic algorithm.

(1) Selection

Individuals who are suitable as parents are selected from the population according to certain criteria, and the offspring are reproduced by mating. There are various selection methods, such as the fitness proportion method, roulette method, essence preservation method, etc.

(2) Crossover

Crossover is the operation of swapping two chromosomes into groups (recombination). There are various methods of crossover, such as single-point crossover, multi-point crossover, partial mapping crossover (PMX), sequential crossover (OX), heuristic crossover, etc.

(3) Mutation

Mutation is a change in a gene with a certain probability. Mutation has the function of local search, while crossover has the function of global search compared to variation. The crossover and mutation operations help to maintain the diversity of the population and avoid falling into local optima in the early stage of the search.

From the above three basic operations, selection embodies the competitive evolutionary idea of superiority and inferiority, wherein superior individuals are obtained by crossover and mutation operations.Through several iterations, the individuals approach the optimal solution.

In recent years, genetic algorithms have also played a role in air pollution control. The Artificial Neural Network with Back Propagation (BP) with a middle layer and sigmoid activation function and its hybrid with a Genetic Algorithm (BP-GA) were used to predict $PM_{10}$ levels by M. Asghari [39]. G. Nunnari et al. [40] combined the use of wavelets and genetic algorithms to search for the best wavelet parameters to predict the daily average of $SO_2$.

### 3.3. Problem Formulation

In this section, we provide the air pollution prediction problem definition. With a fixed sliding window, the inputs for external factors (other pollutants, meteorological factors, etc.) are $X_T = (X_1, X_2, ......., X_{T-1}, X_T)$, where $X_i = (X_i^1, X_i^2, ......., X_i^{L-1}, X_i^L)$ represents a particular external factor belonging to $X_T$, and the sliding window size is $L$. Furthermore, we set the atmospheric pollutants history value as $y_T = (y_1, y_2, ......., y_{T-1})$.

$$\tilde{y_T} = f(X, y) \tag{1}$$

Our goal is to build a model that uses external factors $X_T$ and historical pollution values $y$ to obtain predicted air pollution $\tilde{y_T} = f(X, y)$, where $f$ is the nonlinear mapping model to be learned.

## 4. Methodology

In this section, we will give a detailed description of the Hybrid Autoformer Network with a Genetic Algorithm model (GA-autoformer). A genetic algorithm was used to explore the influence of external factors on the prediction target, determine the degree of influence of each factor on the target, and then put the approximate optimal results into the neural network model. The back-bone neural network model adopted in this paper is Autoformer, which can obtain the best particles and the best prediction accuracy through $n$ iterations. The overall structure of the model is as shown in Figure 5. We will explain the process of the whole structure in detail in Sections 4.1–4.7. The pseudo code for the whole algorithm is shown in Algorithm 1.

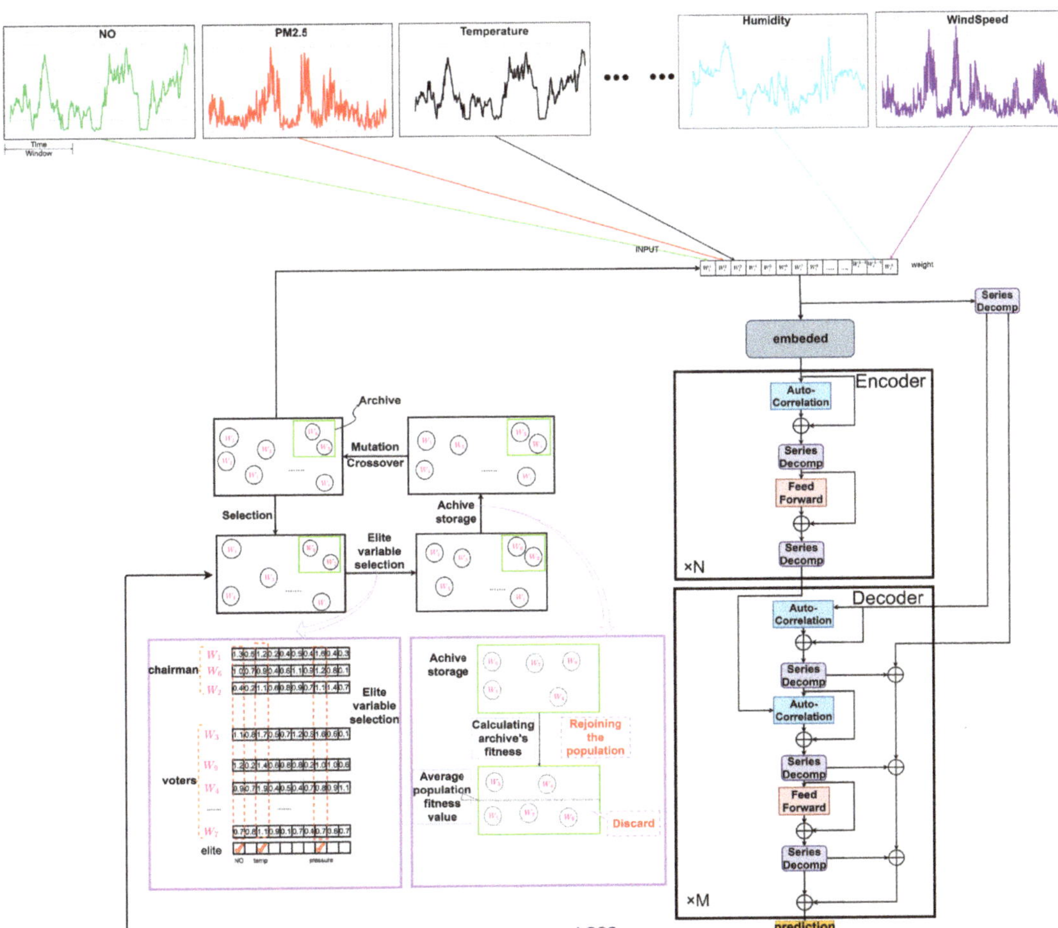


**Figure 5.** Architecture of GA-autoformer. The whole neural architecture consists of two parts. The right side is the neural network architecture, and the autoformer model is used as back-bone network in this paper. The left side is the genetic algorithm, and the lower left side is two operators that we proposed: Elite Voting Operator and Archive Storage Operator.

**Algorithm 1** Frame process of the GA-autoformer model

**Require:** $D$: Multivariate time series dataset on air pollution used in this iteration,
    $t$: Number of iterations,
    $p$: Population number,
    $k$: Interval iteration of executive operator,
    $M$: Neural network model
**Ensure:** $W_b$: Best weight
  Randomly generate W = $[W_1, W_2, W_3, \ldots, W_{p-1}, W_p]$
  $archive\_weight = \emptyset$
  $archive\_fitness = \emptyset$
  $elite = \emptyset$
  **for** $i = 1$ to $t$ **do**
    $Fitness_W \leftarrow M(D, W)$
    $Fitness_a \leftarrow M(D, archive) \cup Fitness_a$
    $W$ = tournament_selection($W$,$Fitness_W$)

    **if** $i\%k$==0&&$i$!=0 **then**
      $elite \leftarrow$ elite_voting($W'$,$Fitness_W$)
      $archive \leftarrow$ archive_storage($archive\_weight$, $archive\_fitness$,$W'$,$Fitness_a$,$Fitness_W$)
    **end if**
    $W' \leftarrow$ crossover($Fitness_W$,$W'$,$elite$)
    $W' \leftarrow$ mutation($Fitness_W$,$W'$,$elite$)
    $W \leftarrow$ generate_population($Fitness_W$,$W'$)
  **end for**
  $W_b \leftarrow$ select_best($archive$)
  return $W_b$

### 4.1. Generate Random Individuals

Firstly, $n$ individuals are randomly generated as the population, and each individual represents a candidate solution, namely, the weight of external factors.

$$W_i = [W_i^1, W_i^2, W_i^3, \ldots, W_i^{L-1}, W_i^L] \quad (2)$$

where $L$ is the number of external variables $i \in (1, 2, 3, \ldots p)$, and $p$ is the number of populations.

### 4.2. Calculation of Fitness Values

The weight value of each individual is cross multiplied with the multivariate variable value in the current air pollution data set $D$, and the obtained result is input into the neural network as a new data set $D'$ to obtain the corresponding prediction accuracy, so the prediction accuracy can be used as the fitness value of the current individual.

### 4.3. Selection

In order not to lose information, we put the best particle into the new population.Then, we used the Tournament Selection Algorithm to randomly select $n$ individuals from the rest of the population and let these $n$ individuals compete. Then we put the best individuals into the new population. We keep iterating until the population of the new population meets the requirements. This new population is the produced offspring and has a higher fitness value than before. Furthermore, $n$ is usually set as 2.

### 4.4. Elite Variable Voting Operator/Archive Storage Operator

Elite Variable Voting and Archive Storage are performed every $k$ generations. The Elite Variable Voting Operator can find elite variables from excellent individuals and search more finely in the subsequent mutation and crossover. The Archive Storage operator can reduce the instability caused by the randomness of neural model initialization by sending the weight into the neural network model. The specific implementation details are expanded in Sections 4.8 and 4.9.

### 4.5. Crossover

Although the average fitness value is improved in the process of selection, it cannot produce new individuals. Crossover mimics the method of biological clock hybridization to produce new varieties, transposes some parts of chromosomes, and uses the random pairing method to determine the parents of individuals.

We adopted the following strategy on the crossover of elite variables: elite variables tend to have high weight values, and if they are crossed with some non-elite, it may lead to the loss of information preserved by elite variables, thus leading to population non-convergence, but moderate elite crossover with non-elite variables may also lead to an increase in population diversity, for which we proposed that elite variables be crossed with non-elite variables in the early stage and only with elite variables in the later stage, thus maintaining population convergence. We treated the first 50% of the iteration as the early stage, where elite variables can cross with non-elite variables, and, in the later stage, elite variables could only cross with elite variables. We adopted shuffle crossover as our crossover method [41]. The flow chart of the crossover is shown in Figure 6.

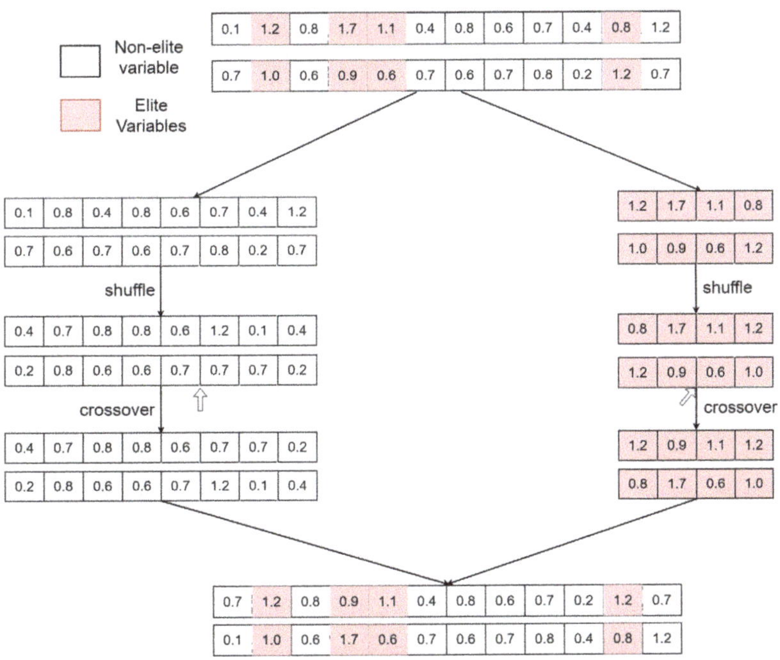


**Figure 6.** Description of crossover. We separated the elite and non-elite variables for shuffle crossover and then combined them.

### 4.6. Mutation

Crossover and selection can ensure that excellent genes are left in each evolution, but this may lead to the local optimization of the whole population. When we cross generate a new chromosome, we can randomly select several genes on the chromosome, and randomly modify the value of genes [42].

$$f(x) = \frac{1}{\sqrt{2\pi}\sigma} \exp\left(-\frac{(x-\mu)^2}{2\sigma^2}\right) \tag{3}$$

$$W_i^k = W_i^k + \gamma * W_i^k * N(0,1) \tag{4}$$

Furthermore, we performed a Gaussian variation on $W_i^k$, wherein the Gaussian distribution random perturbation term $W_i^k * N(0,1)$ was added to the original state $W_i^k$. Equation (3) is the general form of the Gaussian distribution, where $N(0,1)$ is the standard normal distribution, $\mu$ is 0, and $\sigma$ is 1. In order to reflect the difference between elite and non-elite variables, different weighting coefficients $\gamma$ were used as in Equation 4. When non-elite variables are selected for variation, $\gamma$ = 1. Furthermore, when elite variables are selected for variation,$\gamma$ = 1.2. We can find that the elite variables are more important to find the global optimum. It can make the population jump out of the local optimum and improve the convergence speed.

### 4.7. Iteration

Through $n$ iterations of 1–6, the highest fitness value of the individual in the archive in the last generation was calculated, and the individual was taken as the optimal weight as the return value.

### 4.8. Elite Voting Operator

In the time series prediction of air pollution, certain variables will have more influence on time series forecasting; we call them elite variables. The elite variables vary in different problems. This operator can automatically find the elite variables and optimize them more finely. The pseudo code for the whole algorithm is in Algorithm 2. The following is the step of the Elite Variable Voting Operator.

$$\text{dist}(W_a, W_b) = \sqrt{\sum_{i=1}^{L} (x_a^i - y_b^i)^2} \tag{5}$$

Elite variable voting was conducted every k generations.

(1) Candidates are selected from the top 30% of the population based on population fitness, and these individuals are the best candidates in the population; they represent the evolutionary direction of the population.
(2) Among the candidates, we want to have some particles that can lead the candidates to the optimization direction more effectively; we call these particles "chairman". We appoint two particles with highest fitness value among the candidates as "chairman". In order to maintain diversity, the candidates that differ most from the "chairman" are added in "chairman". We use the Euclidean distance as Equation (5) to measure the distance between two particles.
(3) The elite variables chosen by vote are able to give special treatment in the process of mutation and crossover: there is a higher probability of becoming larger in the process of mutation and crossover.

**Algorithm 2** Elite Voting Operator

**Require:** $W$:Population,
  $Fitness_W$: Population fitness value
**Ensure:** *elite*: Elite Variables
  *chairman* ← ∅
  *voters* ← ∅
  *chairman* ← Take the top 2 fitness values in *voters* as *chairman*
  *chairman* ← *chairman*+Take the particle in $W$ that is most different from the particle in *chairman* *voters* ← *W- chairman*
  *elite* ← voting(*chairman*,*voters*)
  return *elite*

### 4.9. Archive Storage Operator

During the process of neural network training,the result can be bad due to different initialization weights, which can inter the judgment about the effect of weights of different external variables. Therefore, we introduced an archive storage mechanism to put potentially optimal solutions into the archive. The pseudocode for the whole algorithm is in Algorithm 3, and the process is as follows:

**Algorithm 3** Archive Storage Operator

```
Require: archive_weight,
        archive_fitness,
        W: population,
        Fitness_W: Population fitness value,
        Fitness_a: archive fitness value
Ensure: W: population
  d ← size(archive)
  \overline{Fitness_W} = Fitness_W / size(Fitness_W)
  for i = 1 to d do
     if Fitness_a^i > \overline{Fitness_W} then
        W ← rejoing_population(W_i, W)
     else
        Discard(W_i, archive_weight,archive_fitness)
     end if
  end for
  return W
```

(1) Every $k$ generations, the best particle in all $k$ generations is copied into the archive.
(2) In the next $k$ iterations, individuals in archive will not be involved in the process of the genetic algorithm, but only in the calculation of the fitness.
(3) Every $k$ generations, we perform an examination. If the fitness value of the individual is greater than the average fitness value of the current population, this individual will replace the individual with the lowest fitness in the population; otherwise, it will be discarded.

### 4.10. Prediction and Optimization

For the populations obtained in Sections 4.1–4.7, each particle represents a candidate solution denoted as $W_i = (W_i^1, W_i^2, .......W_i^L)$, which represents the weight of each external factor. For the external factor input $X_T = (X_1, X_2, .......X_{T-1}, X_T)$, by multiplying each weight with its input counterpart, $X_t^{'} = (W_i^1 * X_t^1, W_i^2 * X_t^2......W_i^{L-1} * X_t^{L-1}, W_i^L * X_t^L)$. $X_t^{'}$ is fed into the Transformer network along with $y$.

Unlike traditional forecasting methods that decompose into seasonal parts and trend parts, we gradually decomposed trend and periodic parts from hidden variables in the learning process. This was based on the idea of sliding average, as shown in (6), to achieve progressive decomposition.

$$\begin{aligned} X_{t_{trend}}^{'} &= AvgPool(Padding(X_t^{'})) \\ X_{t_{seasonal}}^{'} &= X_t^{'} - X_{t_{trend}}^{'} \end{aligned} \tag{6}$$

After $X_t^{'}$ decomposition into seasonal terms $X_{t_{seasonal}}^{'}$ and trend terms $X_{t_{trend}}^{'}$, the similarity of the different seasonal terms is further aggregated at the encoder pair periodicity using an autocorrelation mechanism. The autocorrelation coefficients can be obtained using the fast Fourier transform, and, finally, the similar subsequence information is aggregated in (7) and (8), where the $\tau_k = \arg\operatorname{Topk}_{\tau\in\{1,\cdots,L\}}(\mathcal{R}_{\mathcal{Q},\mathcal{K}}(\tau))$. Here, the multi-headed form of query, key value, is still used so that the self-attentive mechanism can be replaced seamlessly.

Furthermore, the most probable cycle length is $k = \lfloor c \times \log L \rfloor$ to avoid fusion of irrelevant or even opposite subsequences.

$$[H]\widehat{\mathcal{R}}_{\mathcal{Q},\mathcal{K}}(\tau_1), \cdots, \widehat{\mathcal{R}}_{\mathcal{Q},\mathcal{K}}(\tau_k) = \mathrm{SoftMax}(\mathcal{R}_{\mathcal{Q},\mathcal{K}}(\tau_1), \cdots, \mathcal{R}_{\mathcal{Q},\mathcal{K}}(\tau_k)) \tag{7}$$

$$\mathrm{AutoCorrelation}\ (\mathcal{Q}, \mathcal{K}, \mathcal{V}) = \sum_{i=1}^{k} \mathrm{Roll}(\mathcal{V}, \tau_k)\widehat{\mathcal{R}}_{\mathcal{Q},\mathcal{K}}(\tau_k) \tag{8}$$

In the decoder, the trend and seasonal terms are predicted separately. For the seasonal term, the feature information obtained by using encoder is aggregated into predicted seasonal values. For the trend term, the information is gradually extracted from the predicted hidden variables using the cumulative method. Finally, the predicted values of the trend and periodicity terms are summed to obtain the predicted values.

By calculating the empirical loss between the trained predicted pollutant value $\tilde{y_T}$ and the real pollutant value $y_T$, we train the entire model. Our loss function is Root Mean Square Error(RMSE); the loss is not only propagated back from the decoder's outputs across the entire transformer model, it is also involved in the selection of the new generation of the population as the fitness value.

## 5. Experiments and Results

In this section, we give the parameter settings and experimental results, compare them with some current baselines, and attempt to translate the results into interpretable conclusions. We also try to prove and explain the role of each operator through a series of ablation experiments.

### 5.1. Dataset Descriptions

The details of the datasets are shown in Table 1. We cut the dataset into a training set and test set by 70% and 30%, respectively, in chronological order. In particular, when we made a prediction for one of the targets, the other two targets were entered into the model as external variables. To measure the importance of external factors, we normalized the dataset.

**Table 1.** Details of the datasets.

| Dataset | | Location A Location B Location C |
|---|---|---|
| Time_interval | | 1 h |
| Time span | | 1 January 2020–6 October 2020 |
| Prediction target | | $O_3$, $PM_{2.5}$, AQI |
| External factors | Relevant pollutants | CO, $NO_2$, $SO_2$ |
| | Air pollution index | $PM_{10}$, TSP |
| | Weather factors | wind direction, wind speed, precipitation, vapor pressure, humidity, visibility, atmospheric pressure, temperature. |

The units of CO (carbon monoxide) is $mg/m^3$, the units of $O_3$, $NO_2$,$PM_{2.5}$ and $PM_{10}$ are $\mu g/m^3$, and TSP is total suspended particulate matter (mg/L). The unit of wind speed is m/s, which is the average wind speed in 10 min. Furthermore, wind direction is measured by an anemometer, which is projected to the [0, 360°] interval. Precipitation (mm) refers to the amount of precipitation per hour, visibility (m) is the 10-min average visibility, humidity (%) is the relative humidity, pressure (hPa) is the atmosphere pressure measured by the monitoring point, and temperature (°C) is measured in degrees Celsius. In addition, we performed statistical analysis of the data in Table 2 to avoid the presence of extreme data.

**Table 2.** Descriptive statistics of the datasets.

| Variables | Measurement |
|---|---|
| $O_3$ | [2, 300] |
| $PM_{2.5}$ | [30, 320] |
| AQI | [30, 320] |
| CO | [0, 5] |
| $NO_2$ | [0, 150] |
| $SO_2$ | [5, 210] |
| $PM_{10}$ | [0, 330] |
| TSP | [0, 330] |
| WindDirection | [0, 360] |
| WindSpeed | [0, 15] |
| Precipitation | [0, 82.1] |
| vapor pressure | [0, 54.9] |
| humidity | [0, 99] |
| visibility | [0, 31,565] |
| atmospheric pressure | [0, 1019.8] |
| temperature | [−0.8, 37.1] |

The available time period was from 1 January 2020 to 6 October 2020, and Figure 7 shows the distribution of the amount of data by seasons of the available time period.

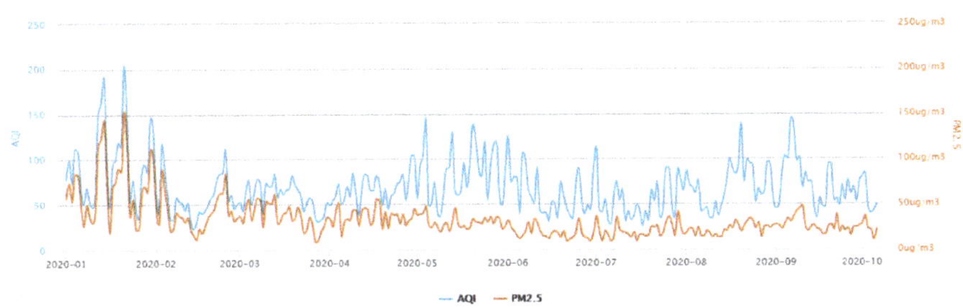


(**a**)

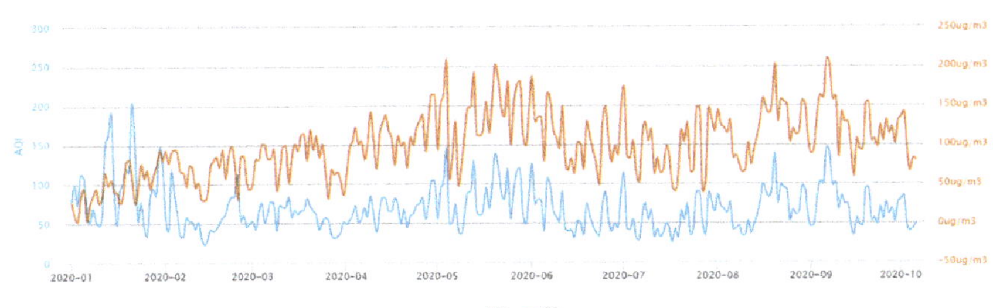


(**b**)

**Figure 7.** Distribution of the AQI, $PM_{2.5}$, and $O_3$ data by seasons. (**a**) AQI-$PM_{2.5}$. (**b**) AQI-$O_3$.

### 5.2. Parameter Setting

The length of the input sequence in the autoformer was 96, the length of the predicted sequence was 24, the number of headers was 8, the value of dropout was 0.05, the batch_size was 32, and the learning rate was 0.0001. In the genetic algorithm, the size of population was 20, the number of iterations was 50, the probability of crossover was 0.8, the probability of variation was 0.1, and the size of the archive was 5. We set k to be 10.

### 5.3. Evaluation Metrics

We chose the Root Mean Square Error (RMSE), Mean Absolute Error (MAE) and Mean Absolute Percentage Error (MAPE) as the criteria for evaluating the prediction performance.

$$RMSE = \sqrt{\frac{1}{n}\sum_{i=1}^{n}(\hat{y}_i - y_i)^2} \tag{9}$$

$$MAE = \frac{1}{n}\sum_{i=1}^{n}|\hat{y}_i - y_i| \tag{10}$$

$$MAPE = \frac{100\%}{n}\sum_{i=1}^{n}\left|\frac{\hat{y}_i - y_i}{y_i}\right| \tag{11}$$

where $n$ is the length of the time series prediction, $y_i$ is the target value of the model prediction, and $y_i$ is the actual target value.

### 5.4. Baselines

To verify the performance of our proposed model, we compared GA-autoformer with the following baseline models.

(1) RNN: RNN is a classical time series prediction model that is capable of extracting time series features. Unlike feed-forward neural networks, which use the output of the previous neuron as input to the next neuron, RNN involves a structure that gives the network the ability to remember information about trends and cycles [15].

(2) LSTM: LSTM belongs to the class of RNNs and also belongs to the recurrent network model. LSTM solves the problem that RNNs cannot extract long-term time dependence and uses multiple gate mechanisms to alleviate the gradient explosion and gradient disappearance problems that exist in RNNs [16].

(3) EA-LSTM: EA-LSTM is based on the attention LSTM and uses the genetic-algorithm-based competitive random search (CRS) instead of gradient-based approach to explore the attention layer weights; thus, it better assigns the weights of features within the time window [43].

(4) Transformer: Recently, the Transformer model has made a big breakthrough in time series prediction. Unlike the RNN and LSTM, Transformer is not a cyclic sequence model. Its prediction efficiency and its ability to predict long-term time series are greatly improved [32].

(5) Informer: The authors designed an efficient transformer-based long-time prediction model, named Informer, by proposing a ProbSparse self-attentive mechanism, which utilizes self-attentive distillation to highlight dominant attention by halving the cascade layer input with a generative decoder for the one-time prediction of long-time sequence sequences. A new solution to the long-time sequence prediction problem is provided [34].

(6) Autoformer: The authors used a deep decomposition architecture. The authors designed sequence decomposition units to embed deep models, implement progressively predictive, auto-correlation mechanisms, discard point-wise connected self-attention

mechanisms, and implement series-wise connected autocorrelation mechanisms to break information utilization bottlenecks [35].

*5.5. Analysis of Prediction Result*

We trained the GA-autoformer for 50 iterations. Table 3 shows the prediction accuracy comparison with different baselines, and the prediction line graphs are put in Figure 8. From Table 3, we can see that our model had a higher prediction accuracy compared to other baseline models, and in comparison with several baseline models, 23 out of 27 comparisons achieved the first, and the remaining four were in the second. For the LSTM and RNN, which all belong to recurrent network structure, they all had a big gap with our model. The EA-LSTM uses the genetic algorithm to optimize the attention layer, combined with the LSTM, but there was still a gap relative to Transformer. Most of the advantages over the Transformer and its other variants were also achieved, which shows that external variables do affect the time series prediction, and this model successfully found the approximate optimal solution of external variables by using a genetic algorithm to jump out of the local optimum. Furthermore, it can be seen from the Figure 9 that the Archive Storage Operator can effectively reduce the impact of the initialization of the neural network on the prediction model.

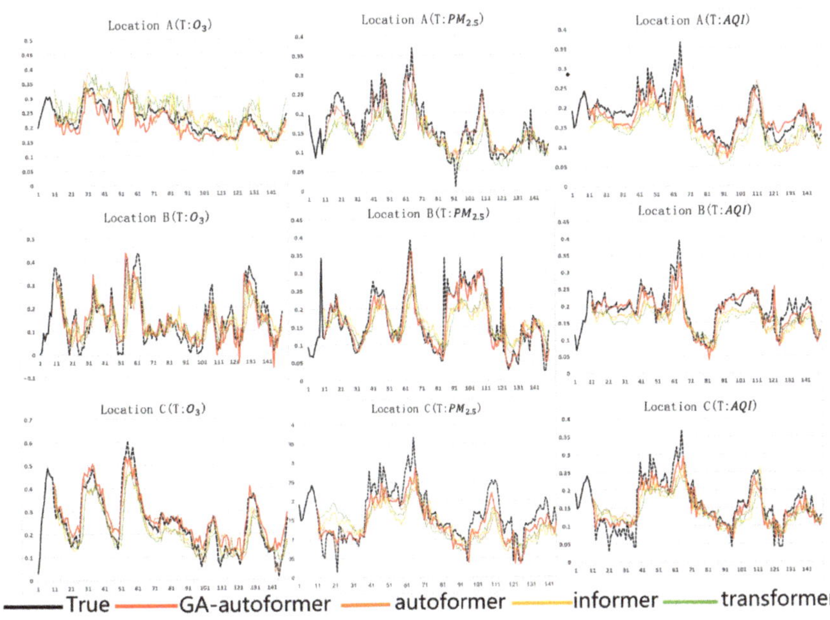


**Figure 8.** Actual normalization value and average predicted value for the test experimental run for each monitor station with different target pollutants.The y-axis represents the predicted values and ground truth, and the x-axis is the predicted time-step.

**Table 3.** Model prediction performance on air pollution test set.

| **Model** | | | | | **Location A** | | | | |
|---|---|---|---|---|---|---|---|---|---|
| **Target** | | $O_3$ | | | $PM_{2.5}$ | | | AQI | |
| | **MAE** | **RMSE** | **MAPE** | **MAE** | **RMSE** | **MAPE** | **MAE** | **RMSE** | **MAPE** |
| RNN | 0.634 | 1.341 | 1.469 | 0.033 | 0.052 | 0.423 | 0.253 | 0.481 | 1.151 |
| LSTM | 0.612 | 1.313 | 1.419 | 0.034 | 0.047 | 0.346 | 0.251 | 0.477 | 1.072 |
| EA-LSTM | 0.573 | 1.226 | 1.297 | 0.029 | 0.041 | 0.335 | 0.249 | 0.473 | 1.081 |
| transformer | 0.514 | 1.175 | 1.069 | 0.030 | 0.037 | 0.311 | 0.241 | 0.468 | 1.011 |
| informer | 0.481 | 1.056 | 0.816 | 0.029 | 0.284 | 0.285 | 0.239 | 0.461 | 1.036 |
| autoformer | 0.479 | 0.990 | 0.778 | **0.027** | 0.294 | 0.275 | 0.235 | 0.462 | 1.021 |
| GA-autoformer | **0.460** | **0.981** | **0.761** | 0.028 | **0.027** | **0.242** | **0.231** | **0.453** | **0.861** |
| | | | | | **Location B** | | | | |
| **Target** | | $O_3$ | | | $PM_{2.5}$ | | | AQI | |
| | **MAE** | **RMSE** | **MAPE** | **MAE** | **RMSE** | **MAPE** | **MAE** | **RMSE** | **MAPE** |
| RNN | 0.439 | 0.542 | 1.235 | 0.227 | 0.347 | 2.090 | 0.045 | 0.023 | 0.301 |
| LSTM | 0.422 | 0.513 | 1.238 | 0.156 | 0.246 | 1.654 | 0.041 | 0.212 | 0.284 |
| EA-LSTM | 0.342 | 0.459 | 1.239 | 0.107 | 0.106 | 0.715 | 0.034 | 0.213 | 0.265 |
| transformer | 0.333 | 0.454 | 1.234 | 0.031 | 0.044 | 0.225 | 0.036 | 0.189 | 0.235 |
| informer | 0.329 | 0.441 | 1.250 | 0.038 | 0.046 | 0.216 | 0.031 | 0.176 | 0.229 |
| autoformer | **0.305** | 0.432 | **1.205** | 0.028 | 0.041 | **0.208** | 0.034 | 0.172 | 0.217 |
| GA-autoformer | 0.315 | **0.406** | 1.215 | **0.026** | **0.039** | 0.215 | **0.029** | **0.170** | **0.207** |
| | | | | | **Location C** | | | | |
| **Target** | | $O_3$ | | | $PM_{2.5}$ | | | AQI | |
| | **MAE** | **RMSE** | **MAPE** | **MAE** | **RMSE** | **MAPE** | **MAE** | **RMSE** | **MAPE** |
| RNN | 0.081 | 0.246 | 0.510 | 0.294 | 0.332 | 1.513 | 0.169 | 0.290 | 0.360 |
| LSTM | 0.061 | 0.240 | 0.412 | 0.262 | 0.316 | 1.479 | 0.091 | 0.159 | 0.311 |
| EA-LSTM | 0.057 | 0.233 | 0.346 | 0.251 | 0.281 | 1.431 | 0.041 | 0.106 | 0.281 |
| transformer | 0.056 | 0.231 | 0.348 | 0.241 | 0.279 | 1.419 | 0.033 | 0.063 | 0.245 |
| informer | 0.054 | 0.229 | 0.339 | 0.236 | 0.269 | 1.410 | 0.036 | 0.062 | 0.240 |
| autoformer | 0.052 | 0.228 | 0.336 | 0.231 | **0.264** | 1.369 | 0.034 | 0.061 | 0.239 |
| GA-autoformer | **0.051** | **0.223** | **0.328** | **0.209** | 0.269 | **1.356** | **0.031** | **0.056** | **0.238** |

The RMSE, MAE, and MAPE are the normalization metrics.

Furthermore, we analyzed the effects of $\gamma$, archive size, and k on the experiment. Here, we used the dataset Location A, and the prediction target was $O_3$. From Figure 10, we can see that the best results were obtained when lambda was set to 1.2, archive size was set to 5, and k was set to 10.

And the results of the training process are visualized in Figure 11. It can be clearly seen in Figure 11 that, as the number of iterations increased, the population gradually converged and evolved in the right direction. Regarding the interpretability of the experimental results, we will elaborate in Section 5.6.

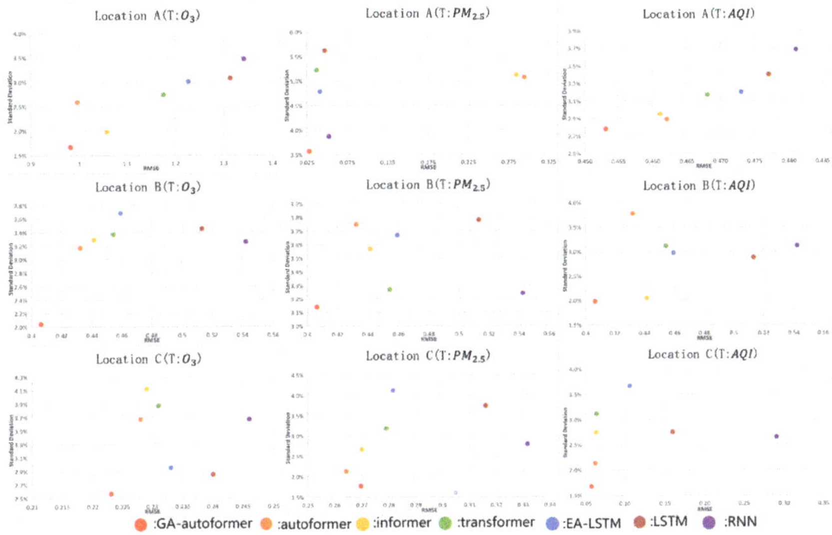


**Figure 9.** Scatter plot of RMSE vs. standard deviation.

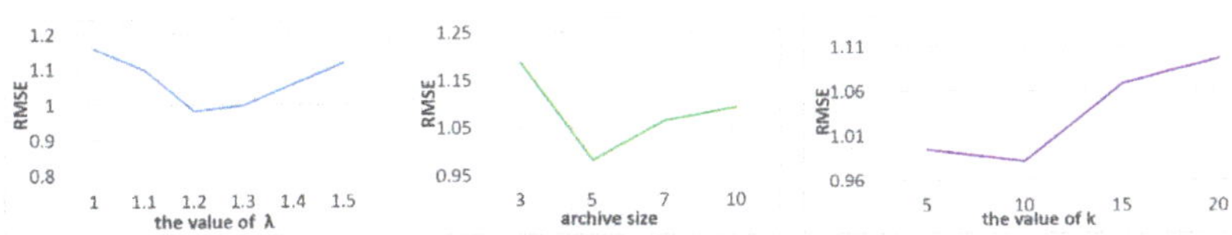


**Figure 10.** Sensitivities analysis of $\gamma$, archive size, and k.

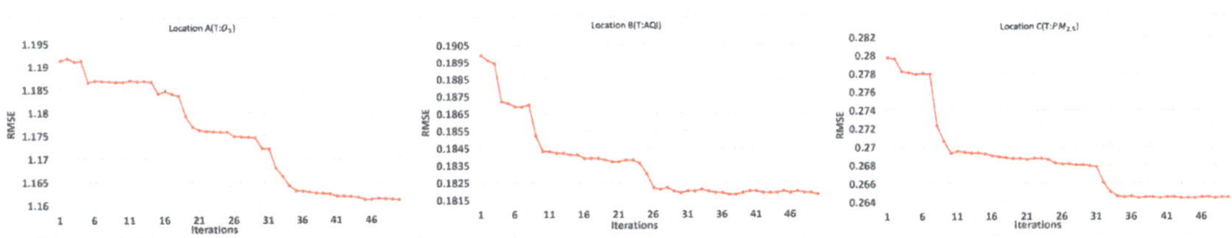


**Figure 11.** The training process of genetic algorithm.

## 5.6. Model Interpretability

Figure 12 shows the external factor optimization weights from the last generation of the population, as well as the elite variables that were taken at the last time in multiple experiments. Furthermore, the color is more red, which indicates that the factor was more important, while the yellow color indicates the less important factor to the target predicted pollution.

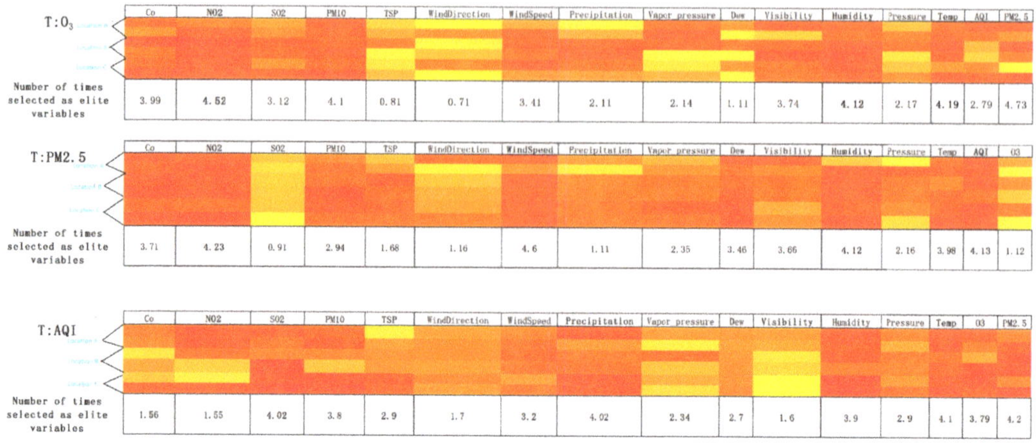


**Figure 12.** We took the weights of the external variables for six random individuals in the final population; the color is more red, which indicates that the variable was more important.

When selecting $O_3$ as the predicted target pollution, it can be seen that the optimization individuals found all had higher values on $NO_2$, temperature, etc. The control of $O_3$ pollution mainly involves the control of its precursors, which are mainly nitrogen oxides and carbon monoxide. Nitrogen oxides react with surrounding atmospheric ozone and subsequently form nitric acid [44]. The increase in temperature represents a flanking increase in solar radiation, which leads to an increase in ozone levels, but high temperatures also lead to increased vertical convective activity in the atmosphere, which facilitates local ozone and precursor diffusion dilution. High humidity facilitates $O_3$ pollution removal. On the other hand, water vapor in the atmosphere affects solar ultraviolet radiation and, thus, slows down photochemical reactions, and the humidity has a large negative correlation with $O_3$. Furthermore, $O_3$ pollution was mainly negatively correlated with wind speed, mainly because wind speed enhances the horizontal diffusion of ozone and contributes to ozone dilution. Those factors' weights were relatively large in the experiments and were all selected as elite variables several times, which is in accordance with our a prior studies on $O_3$ [45].

In terms of meteorological factors, $PM_{2.5}$ was positively correlated with air temperature and relative humidity and negatively correlated with wind speed. When the wind speed is low and the humidity is high, the intensity of inversion temperature increases, which is unfavorable to the diffusion of $PM_{2.5}$ and other pollutants in the vertical and horizontal directions and aggravates the accumulation of particulate matter pollution, thus making its mass concentration remain high, and when the temperature and relative humidity are both at high levels in autumn and winter, fog is easily produced. The suspended fog droplets easily adsorb and capture gaseous pollutants and particulate matter pollutants, which is favorable to the formation of secondary particles. The hourly concentration of $NO_2$ had a good positive correlation with the hourly concentration of $PM_{2.5}$, thus indicating that the contribution of traffic pollution emissions to $PM_{2.5}$ was larger. Traffic exhaust emissions are transformed into secondary particles after a period of chemical reaction, which affects the concentration level of $PM_{2.5}$. The results can also be clearly seen in the heat map [46].

When selecting the AQI as the predicted target, it is easy to know that there is a large relationship with $SO_2$, $O_3$, $PM_{2.5}$, and $PM_{10}$. As an indicator to measure air pollution, the AQI is closely related to the content of each pollutant. Not only that, wind speed, temperature, and humidity can usually affect the diffusion rate of atmospheric pollutants, and they had strong correlations with atmospheric pollutants, thus leading to strong correlations between the AQI and these factors. It can be seen that various variables were selected as elite variables several times [47].

In addition, the weights derived from the data sets at different locations differed when predicting the same target. In the industrial area (Location C), nitrogen oxide emissions were much larger than in the residential area (Location A) and had a greater weighting compared to the residential area where nitrogen oxide had a greater impact on pollutants. The main sources of pollutants in residential areas are domestic cookers and winter heating, which mostly consume coal and produce carbon monoxide and sulfide, and their corresponding weights were also higher.

From the above analysis, it can be seen that, in the prediction of different targets, our models successfully identified the relationship between external variables and the predicted target pollution, which is consistent with the knowledge of the relevant research, thus proving that the evolutionary direction of the final population of the genetic algorithm is correct. Through the exploration of different external variables, we can analyze and identify the sources of pollutants and help the government develop effective pollution mitigation policies.

### *5.7. Ablation Experiment*

To verify the effects of different genetic algorithm operators, we compared the base transformer with three variant models combined with genetic algorithm optimization, including autoformer (base model), GA-autoformer (our proposed model), autoformer using only the unchanged genetic algorithm (denoted as autoformer-GA(u)), the model using only the Elite Variable Operator(autoformer-GA(elite)) and the Archive Storage Operator (autoformer-GA(archive)), respectively, which are presented in Table 4.

From Table 4, we can find that the Transformer used only the traditional genetic algorithm (autoformer-GA(u)), which had some improvement comparing with the base Transformer. However, the improvement was not significant, and the effects were not as good as our proposed model(autoformer-GA). Furthermore, the two models using only one operator (autoformer-GA(elite) and autoformer-GA(archive)) were not as effective as the proposed model.

**Table 4.** Ablation experiment.

| **Model** | | | | | **Location A** | | | | |
|---|---|---|---|---|---|---|---|---|---|
| **Target** | | **$O_3$** | | | **$PM_{2.5}$** | | | **AQI** | |
| | **MAE** | **RMSE** | **MAPE** | **MAE** | **RMSE** | **MAPE** | **MAE** | **RMSE** | **MAPE** |
| autoformer | 0.479 | 0.995 | 0.778 | **0.027** | 0.029 | 0.275 | 0.235 | 0.462 | 1.021 |
| autoformer-GA | 0.478 | 0.991 | 0.782 | 0.036 | 0.029 | 0.269 | 0.234 | 0.461 | 0.979 |
| autoformer-GA(elite) | 0.476 | 0.986 | 0.771 | 0.029 | 0.029 | 0.252 | 0.232 | 0.457 | 0.957 |
| autoformer-GA(archive) | 0.469 | 0.984 | 0.769 | 0.028 | 0.028 | 0.261 | 0.234 | 0.451 | 0.892 |
| our-model | **0.460** | **0.981** | **0.761** | 0.028 | **0.027** | **0.242** | **0.231** | **0.453** | **0.861** |

**Table 4.** *Cont.*

| | Location B | | | | | | | | |
|---|---|---|---|---|---|---|---|---|---|
| **Target** | **$O_3$** | | | **$PM_{2.5}$** | | | **AQI** | | |
| | **MAE** | **RMSE** | **MAPE** | **MAE** | **RMSE** | **MAPE** | **MAE** | **RMSE** | **MAPE** |
| autoformer | **0.305** | 0.432 | **1.205** | 0.028 | 0.041 | 0.208 | 0.034 | 0.172 | 0.217 |
| autoformer-GA | 0.316 | 0.453 | 1.233 | 0.028 | **0.037** | 0.218 | 0.034 | 0.173 | 0.208 |
| autoformer-GA(elite) | 0.311 | 0.443 | 1.234 | 0.029 | 0.038 | **0.195** | 0.033 | 0.171 | 0.205 |
| autoformer-GA(archive) | 0.309 | 0.441 | 1.234 | 0.027 | 0.037 | 0.204 | 0.032 | 0.173 | 0.202 |
| our-model | 0.315 | **0.406** | 1.215 | **0.026** | 0.039 | 0.215 | **0.029** | **0.170** | **0.207** |
| | **Location C** | | | | | | | | |
| **Target** | **$O_3$** | | | **$PM_{2.5}$** | | | **AQI** | | |
| | **MAE** | **RMSE** | **MAPE** | **MAE** | **RMSE** | **MAPE** | **MAE** | **RMSE** | **MAPE** |
| autoformer | 0.052 | 0.228 | 0.336 | 0.231 | **0.264** | 1.369 | 0.034 | 0.061 | 0.240 |
| autoformer-GA | 0.053 | 0.226 | 0.331 | 0.221 | 0.261 | 1.359 | 0.032 | 0.058 | 0.239 |
| autoformer-GA(elite) | 0.052 | 0.229 | 0.337 | 0.218 | 0.262 | 1.362 | 0.033 | 0.059 | 0.237 |
| autoformer-GA(archive) | 0.051 | 0.224 | 0.329 | 0.212 | 0.261 | 1.358 | 0.034 | 0.059 | 0.237 |
| our-model | **0.051** | **0.223** | **0.328** | **0.209** | 0.267 | **1.356** | **0.031** | **0.056** | **0.238** |

The RMSE, MAE, and MAPE are the normalization metrics.

For the Elite Variable Operator, as can be seen in the heat map Figure 12, the variables selected as elite variables for many times were larger than the other variables, thus indicating that elite variable operators can find variables that have a greater impact on prediction accuracy and give special treatment to make the search more refined when crossover and mutation occur.

For the Archive Storage Operator, it can be seen in Figure 13 that the model using the Archive Storage Operator not only improved the overall prediction, but also the variance was greatly reduced. This is because the fluctuations caused by the randomized weights of the neural network model could potentially affect the accuracy of the air pollution prediction. Therefore, we stored the potentially good particles and evaluated them several times to filter out the individuals which had better fitness and put them back into the population.

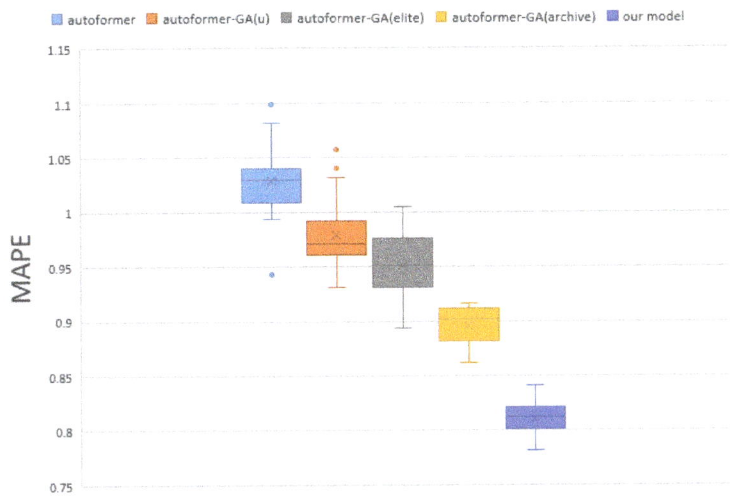


**Figure 13.** The box plot represents the distribution of the final prediction results of different models. The experimental result is the average fitness value of the population for the last iteration using Location A, with ozone as the prediction target.

## 6. Discussion

For the air pollution time series prediction problem, Johansson, C et al. [11] used various machine learning methods (e.g., Random Forest (RF), Extreme Gradient Boosting (XGB), and Long Short-Term Memory (LSTM)) for multiple pollutants ($PM_{10}$, $NO_x$, and $O_3$) at multiple locations with multi-temporal predictions. It can be seen that, in the time-series prediction problem, different pollutants and different locations had different environmental conditions and air pollution change characteristics, which require certain a priori knowledge. Prediction models using metaheuristics combined with neural networks are also evolving, such as ANN-EMPA [29], which combines crossover and mutation operators with ANN to enhance the prediction models. However, most of these models generally use metaheuristics to optimize the hyperparameters and the structure of the neural network model, but the results obtained from the optimization are not interpretable and do not explain why the optimized hyperparameters can better improve the pollution prediction model.

To address the above problem, we proposed a Hybrid Autoformer Network with a Genetic Algorithm model to predict air pollution temporal variation, as well as explore the relationship between external variables and target pollution. Unlike the above model, our model combines genetic algorithm with autoformer, wherein autoformer has the ability of long-time series prediction, and the genetic algorithm can explore the influence of external variables on the predicted target, which makes our model interpretable.

From Table 3 and Figure 8, we can find that the prediction accuracy of the GA-autoformer was higher than other baseline models. As shown in Figure 9, the standard deviation of the GA-autoformer was also lower than the rest of the models, which indicates that our Archive Storage Operator was able to preserve excellent particles in the iterations. Meanwhile, Figure 12 shows the effects exhibited by different external variables for different prediction targets in different prediction locations, such as industrial areas, residential areas, and suburban areas, which demonstrates the robustness and interpretability of our proposed model.

## 7. Conclusions

In this study, we proposed a Hybrid Autoformer Network with a Genetic Algorithm model to predict air pollution. A genetic algorithm was used to explore the influence of external variables on pollutants, thus making the model explanatory. In addition, we proposed two operators to better explore external variables: the Elite Variable Voting Operator was used to screen out more important external factors as elite variables and search them more finely; the Archive Storage Operator to store outstanding individuals to alleviate model fluctuations caused by random weights in deep learning. Finally, we conducted a comprehensive experiment on the Ma'anshan air pollution dataset to validate the proposed model. In comparison with the current state-of-the-art models, the prediction accuracy was improved by 2–8%, and the selection of model influencing factors was more interpretable.

However, the air pollution prediction model proposed in this work has some shortcomings, such as that the model cannot handle streaming monitoring data and cannot support online learning. Moreover, the external factors were only explored for a single monitoring site. In future work, we will continue to explore the relationship between external variables in air pollution time series prediction across multiple monitoring sites and continue to explore intelligent optimization algorithms, such as genetic algorithms combined with transfer learning models, and apply them to cross-city air pollution forecasting.

**Author Contributions:** Conceptualization, Z.X. and K.P.; methodology, Z.X. and K.P.; software, K.P.; validation, J.L. (Jiang Lu) and J.L. (Jiaren Li); formal analysis, K.P. and J.L. (Jiang Lu); investigation, K.P.; resources, J.L. (Jiang Lu) and J.L. (Jiaren Li); data curation, K.P.; writing—original draft preparation, K.P.; writing—review and editing, Z.X.; visualization, K.P. and J.L. (Jiang Lu); supervision, Z.X.; project administration, Z.X.; funding acquisition, Z.X. All authors have read and agreed to the published version of the manuscript.

**Funding:** This work was supported in part by the National Natural Science Foundation of China (62103124, 62033012), as well as the Major Special Science and Technology Project of Anhui, China (202003a07020009).

**Institutional Review Board Statement:** Not applicable.

**Informed Consent Statement:** Not applicable.

**Data Availability Statement:** No new data were created or analyzed in this study. Data availability 864 is not applicable as no datasets were generated or analyzed during the current study.

**Conflicts of Interest:** The authors declare no conflict of interest.

## References


1. Organization, W.H. Air Pollution. Available online: https://www.who.int/health-topics/air-pollution. (accessed on 1 November 2022).
2. Steinfeld, J.I. Atmospheric chemistry and physics: From air pollution to climate change. *Environ. Sci. Policy Sustain. Dev.* **1998**, *40*, 26. [CrossRef]
3. 2021 China Ecological Environment Status Bulletin. Available online: https://www.mee.gov.cn/hjzl/sthjzk/zghjzkgb/202205/P020220608338202870777.pdf (accessed on 1 November 2022).
4. Fan, J.; Li, Q.; Hou, J.; Feng, X.; Karimian, H.; Lin, S. A spatiotemporal prediction framework for air pollution based on deep RNN. *ISPRS Ann. Photogramm. Remote. Sens. Spat. Inf. Sci.* **2017**, *4*, 15. [CrossRef]
5. Lengyel, A.; Héberger, K.; Paksy, L.; Bánhidi, O.; Rajkó, R. Prediction of ozone concentration in ambient air using multivariate methods. *Chemosphere* **2004**, *57*, 889–896. [CrossRef]
6. Xu, Z.; Cao, Y.; Kang, Y. Deep spatiotemporal residual early-late fusion network for city region vehicle emission pollution prediction. *Neurocomputing* **2019**, *355*, 183–199. [CrossRef]
7. Badicu, A.; Suciu, G.; Balanescu, M.; Dobrea, M.; Birdici, A.; Orza, O.; Pasat, A. PMs concentration forecasting using ARIMA algorithm. In Proceedings of the 2020 IEEE 91st Vehicular Technology Conference (VTC2020-Spring), Antwerp, Belgium, 25–28 May 2020; pp. 1–5.
8. Kumar, U.; Jain, V. ARIMA forecasting of ambient air pollutants (O3, NO, NO2 and CO). *Stoch. Environ. Res. Risk Assess.* **2010**, *24*, 751–760. [CrossRef]

9. Fan, J.; Wu, L.; Ma, X.; Zhou, H.; Zhang, F. Hybrid support vector machines with heuristic algorithms for prediction of daily diffuse solar radiation in air-polluted regions. *Renew. Energy* **2020**, *145*, 2034–2045. [CrossRef]
10. Gocheva-Ilieva, S.; Ivanov, A.; Stoimenova-Minova, M. Prediction of Daily Mean PM10 Concentrations Using Random Forest, CART Ensemble and Bagging Stacked by MARS. *Sustainability* **2022**, *14*, 798. [CrossRef]
11. Johansson, C.; Zhang, Z.; Engardt, M.; Stafoggia, M.; Ma, X. Improving 3-day deterministic air pollution forecasts using machine learning algorithms. *Atmos. Chem. Phys. Discuss.* **2023**, 1–52. [CrossRef]
12. Shikhovtsev, A.Y.; Kovadlo, P.G.; Lezhenin, A.A.; Korobov, O.A.; Kiselev, A.V.; Russkikh, I.V.; Kolobov, D.Y.; Shikhovtsev, M.Y. Influence of Atmospheric Flow Structure on Optical Turbulence Characteristics. *Appl. Sci.* **2023**, *13*, 1282. [CrossRef]
13. Shikhovtsev, A.Y.; Kovadlo, P.G.; Kiselev, A.V.; Eselevich, M.V.; Lukin, V.P. Application of Neural Networks to Estimation and Prediction of Seeing at the Large Solar Telescope Site. *Publ. Astron. Soc. Pac.* **2023**, *135*, 014503. [CrossRef]
14. Catalano, M.; Galatioto, F.; Bell, M.; Namdeo, A.; Bergantino, A.S. Improving the prediction of air pollution peak episodes generated by urban transport networks. *Environ. Sci. Policy* **2016**, *60*, 69–83. [CrossRef]
15. Ong, B.T.; Sugiura, K.; Zettsu, K. Dynamically pre-trained deep recurrent neural networks using environmental monitoring data for predicting PM2.5. *Neural Comput. Appl.* **2016**, *27*, 1553–1566. [CrossRef]
16. Krishan, M.; Jha, S.; Das, J.; Singh, A.; Goyal, M.K.; Sekar, C. Air quality modelling using long short-term memory (LSTM) over NCT-Delhi, India. *Air Qual. Atmos. Health* **2019**, *12*, 899–908. [CrossRef]
17. Li, T.; Hua, M.; Wu, X. A hybrid CNN-LSTM model for forecasting particulate matter (PM2. 5). *IEEE Access* **2020**, *8*, 26933–26940. [CrossRef]
18. Chen, Y.; Chen, X.; Xu, A.; Sun, Q.; Peng, X. A hybrid CNN-Transformer model for ozone concentration prediction. *Air Qual. Atmos. Health* **2022**, *15*, 1533–1546. [CrossRef]
19. Yin, L.; Wang, L.; Huang, W.; Liu, S.; Yang, B.; Zheng, W. Spatiotemporal Analysis of Haze in Beijing Based on the Multi-Convolution Model. *Atmosphere* **2021**, *12*, 1408. [CrossRef]
20. Tian, J.; Liu, Y.; Zheng, W.; Yin, L. Smog prediction based on the deep belief—BP neural network model (DBN-BP). *Urban Clim.* **2022**, *41*, 101078. .: 10.1016/j.uclim.2021.101078. [CrossRef]
21. Yin, L.; Wang, L.; Huang, W.; Tian, J.; Liu, S.; Yang, B.; Zheng, W. Haze Grading Using the Convolutional Neural Networks. *Atmosphere* **2022**, *13*, 522. [CrossRef]
22. Zhang, Z.; Tian, J.; Huang, W.; Yin, L.; Zheng, W.; Liu, S. A Haze Prediction Method Based on One-Dimensional Convolutional Neural Network. *Atmosphere* **2021**, *12*, 1327. [CrossRef]
23. Liu, Y.; Tian, J.; Zheng, W.; Yin, L. Spatial and temporal distribution characteristics of haze and pollution particles in China based on spatial statistics. *Urban Climate* **2022**, *41*, 101031. .: 10.1016/j.uclim.2021.101031. [CrossRef]
24. Wu, X.; Liu, Z.; Yin, L.; Zheng, W.; Song, L.; Tian, J.; Yang, B.; Liu, S. A Haze Prediction Model in Chengdu Based on LSTM. *Atmosphere* **2021**, *12*, 1479. [CrossRef]
25. Yuan, X.; Chen, C.; Lei, X.; Yuan, Y.; Adnan, R.M. Monthly runoff forecasting based on LSTM–ALO model. *Stoch. Environ. Res. Risk Assess.* **2018**, *32*, 2199–2212. [CrossRef]
26. Adnan, R.M.; Mostafa, R.R.; Elbeltagi, A.; Yaseen, Z.M.; Shahid, S.; Kisi, O. Development of new machine learning model for streamflow prediction: Case studies in Pakistan. *Stoch. Environ. Res. Risk Assess.* **2021**, *36*, 999–1033. [CrossRef]
27. Adnan, R.M.; Mostafa, R.R.; Kisi, O.; Yaseen, Z.M.; Shahid, S.; Zounemat-Kermani, M. Improving streamflow prediction using a new hybrid ELM model combined with hybrid particle swarm optimization and grey wolf optimization. *Knowl.-Based Syst.* **2021**, *230*, 107379. [CrossRef]
28. Adnan, R.M.; Kisi, O.; Mostafa, R.R.; Ahmed, A.N.; El-Shafie, A. The potential of a novel support vector machine trained with modified mayfly optimization algorithm for streamflow prediction. *Hydrol. Sci. J.* **2022**, *67*, 161–174. [CrossRef]
29. Ikram, R.M.A.; Ewees, A.A.; Parmar, K.S.; Yaseen, Z.M.; Shahid, S.; Kisi, O. The viability of extended marine predators algorithm-based artificial neural networks for streamflow prediction. *Appl. Soft Comput.* **2022**, *131*, 109739. [CrossRef]
30. Adnan, R.M.; Dai, H.L.; Mostafa, R.R.; Islam, A.R.M.T.; Kisi, O.; Elbeltagi, A.; Zounemat-Kermani, M. Application of novel binary optimized machine learning models for monthly streamflow prediction. *Appl. Water Sci.* **2023**, *13*, 110. [CrossRef]
31. Wen, Q.; Zhou, T.; Zhang, C.; Chen, W.; Ma, Z.; Yan, J.; Sun, L. Transformers in Time Series: A Survey. *arXiv* **2022**, arXiv:2202.07125.
32. Wu, N.; Green, B.; Ben, X.; O'Banion, S. Deep transformer models for time series forecasting: The influenza prevalence case. *arXiv* **2020**, arXiv:2001.08317.
33. Li, S.; Jin, X.; Xuan, Y.; Zhou, X.; Chen, W.; Wang, Y.X.; Yan, X. Enhancing the locality and breaking the memory bottleneck of transformer on time series forecasting. In Proceedings of the Advances in Neural Information Processing Systems 32 (NeurIPS 2019), Processing Systems 2019, NeurIPS 2019, Vancouver, BC, Canada, 8–14 December 2019.
34. Zhou, H.; Zhang, S.; Peng, J.; Zhang, S.; Li, J.; Xiong, H.; Zhang, W. Informer: Beyond efficient transformer for long sequence time-series forecasting. In Proceedings of the AAAI, Vurtual, 2–9 February 2021.
35. Xu, J.; Wang, J.; Long, M.; Wu, H. Autoformer: Decomposition transformers with auto-correlation for long-term series forecasting. *Adv. Neural Inf. Process. Syst.* **2021**, *34*, 20.
36. Sampson, J.R. Adaptation in natural and artificial systems (John H. Holland). *SIAM Rev.* **1976**, *18*, 529. [CrossRef]
37. Whitley, D. A genetic algorithm tutorial. *Stat. Comput.* **1994**, *4*, 65–85. [CrossRef]
38. Kumar, M.; Husain, M.; Upreti, N.; Gupta, D. Genetic algorithm: Review and application. *Int. J. Inf. Technol. Knowl. Manag.* **2010**, *2*, 451–454. [CrossRef]

39. Asghari, M.; Nematzadeh, H. Predicting air pollution in Tehran: Genetic algorithm and back propagation neural network. *J. Data Min.* **2016**, *4*, 49–54.
40. Nunnari, G. Modelling air pollution time-series by using wavelet functions and genetic algorithms. *Soft Comput.* **2004**, *8*, 173–178. [CrossRef]
41. Caruana, R.A.; Eshelman, L.J.; Schaffer, J.D. Representation and hidden bias II: Eliminating defining length bias in genetic search via shuffle crossover. In Proceedings of the 11th International Joint Conference on Artificial intelligence, Detroit, MI, USA, 20–25 August 1989; Volume 1, pp. 750–755.
42. Higashi, N.; Iba, H. Particle swarm optimization with Gaussian mutation. In Proceedings of the 2003 IEEE Swarm Intelligence Symposium. SIS'03 (Cat. No. 03EX706) IEEE, Indianapolis, IN, USA, 26 April 2003; pp. 72–79.
43. Li, Y.; Zhu, Z.; Kong, D.; Han, H.; Zhao, Y. EA-LSTM: Evolutionary attention-based LSTM for time series prediction. *Knowl.-Based Syst.* **2019**, *181*, 104785. [CrossRef]
44. Obolkin, V.; Molozhnikova, E.; Shikhovtsev, M.; Netsvetaeva, O.; Khodzher, T. Sulfur and Nitrogen Oxides in the Atmosphere of Lake Baikal: Sources, Automatic Monitoring, and Environmental Risks. *Atmosphere* **2021**, *12*, 1348. [CrossRef]
45. Liu, H.; Liu, J.; Liu, Y.; Yi, K.; Yang, H.; Xiang, S.; Ma, J.; Tao, S. Spatiotemporal variability and driving factors of ground-level summertime ozone pollution over eastern China. *Atmos. Environ.* **2021**, *265*, 118686. [CrossRef]
46. Wang, Y.; Liu, C.; Wang, Q.; Qin, Q.; Ren, H.; Cao, J. Impacts of natural and socioeconomic factors on PM2. 5 from 2014 to 2017. *J. Environ. Manag.* **2021**, *284*, 112071. [CrossRef]
47. Miao, L.; Liu, C.; Yang, X.; Kwan, M.P.; Zhang, K. Spatiotemporal heterogeneity analysis of air quality in the Yangtze River Delta, China. *Sustain. Cities Soc.* **2022**, *78*, 103603. [CrossRef]

atmosphere

MDPI

*Review*

# Deep Learning in Airborne Particulate Matter Sensing and Surface Plasmon Resonance for Environmental Monitoring

**Balendra V. S. Chauhan [1,2,*], Sneha Verma [2,3], B. M. Azizur Rahman [4] and Kevin P. Wyche [1,5,*]**

1 Centre for Earth Observation Science, School of Applied Sciences, University of Brighton, Brighton BN2 4GJ, UK
2 Tech GPT Ltd., Newcastle upon Tyne NE6 2SR, UK; sneha.verma@newcastle.ac.uk
3 School of Engineering, Newcastle University, Newcastle upon Tyne NE1 7RU, UK
4 School of Science and Technology, City University of London, London EC1V 0HB, UK; b.m.a.rahman@city.ac.uk
5 School of Science and Technology, Nottingham Trent University, Nottingham NG1 4BU, UK
* Correspondence: b.chauhan@brighton.ac.uk (B.V.S.C.); kevinpwyche@gmail.com (K.P.W.)

**Abstract:** This review explores advanced sensing technologies and deep learning (DL) methodologies for monitoring airborne particulate matter (PM), which is critical for environmental health assessments. It begins with discussing the significance of PM monitoring and introduces surface plasmon resonance (SPR) as a promising technique in environmental applications, alongside the role of DL neural networks in enhancing these technologies. This review analyzes advancements in airborne PM sensing technologies and the integration of DL methodologies for environmental monitoring. This review emphasizes the importance of PM monitoring for public health, environmental policy, and scientific research. Traditional PM sensing methods, including their principles, advantages, and limitations, are discussed, covering gravimetric techniques, continuous monitoring, optical and electrical methods, and microscopy. The integration of DL with PM sensing offers potential for enhancing monitoring accuracy, efficiency, and data interpretation. DL techniques, such as convolutional neural networks (CNNs), autoencoders, recurrent neural networks (RNNs), and their variants, are examined for applications like PM estimation from satellite data, air quality prediction, and sensor calibration. This review highlights the data acquisition and quality challenges in developing effective DL models for air quality monitoring. Techniques for handling large and noisy datasets are explored, emphasizing the importance of data quality for model performance, generalizability, and interpretability. The emergence of low-cost sensor technologies and hybrid systems for PM monitoring is discussed, acknowledging their promise while recognizing the need for addressing data quality, standardization, and integration issues. This review identifies areas for future research, including the development of robust DL models, advanced data fusion techniques, applications of deep reinforcement learning, and considerations of ethical implications.

**Keywords:** surface plasmon resonance (SPR); particulate matter (PM); sensors; machine learning (ML); deep learning (DL); artificial intelligence (AI); environmental monitoring; air quality

## 1. Introduction

Airborne PM is a complex mixture of solid particles and liquid droplets suspended in the air, varying in size, composition, and origin. Monitoring PM has become increasingly crucial due to its profound impacts on human health, environmental safety, and climate change. The significance of PM monitoring spans several key areas, including public health protection, environmental preservation, and policy development. From a public health perspective, PM monitoring is vital due to the severe health impacts associated with exposure. The World Health Organization (WHO) estimates that exposure to fine PM ($PM_{2.5}$) contributes to approximately 4.2 million premature deaths globally each year, primarily due to cardiovascular and respiratory diseases, and lung cancer [1]. PM, especially fine ($PM_{2.5}$) and ultrafine particles, can penetrate deep into the respiratory system and even enter the bloodstream, leading to a wide range of health problems [2].

Long-term exposure to PM has been associated with increased mortality rates, reduced life expectancy, and a higher incidence of chronic diseases such as asthma, chronic obstructive pulmonary disease (COPD), and cardiovascular disorders. Short-term exposure to high levels of PM can exacerbate existing health conditions and lead to acute health effects, including increased hospital admissions for respiratory and cardiovascular issues [3]. Children, the elderly, and individuals with pre-existing health conditions are particularly vulnerable to the adverse effects of PM exposure. Recent studies have also suggested potential links between PM exposure and neurological disorders, including cognitive decline and an increased risk of dementia [4]. Given the wide-ranging health impacts, accurate monitoring of PM levels is essential for assessing public health risks and implementing timely interventions to protect vulnerable populations.

From an environmental perspective, PM monitoring is crucial for understanding and mitigating its impacts on ecosystems and the climate. Different components of PM can have varying effects on the climate. While some particles, such as black carbon, contribute to global warming by absorbing solar radiation, others, like sulfates, have a cooling effect by reflecting sunlight back into space [5]. Understanding the composition and distribution of PM through comprehensive monitoring is essential for accurate climate modeling and predictions.

PM deposition can harm plant life by blocking stomata, reducing photosynthesis, and altering the soil chemistry. This can lead to reduced agricultural productivity and changes in ecosystem composition [6]. Furthermore, PM can be deposited in water bodies, affecting aquatic ecosystems and potentially contaminating drinking water sources [7]. Fine particles are also the primary cause of reduced visibility in many urban and natural areas, including national parks and wilderness areas [8]. From a regulatory and policy standpoint, effective monitoring of PM is essential for developing and enforcing air quality standards and regulations.

Many countries and international organizations have established air quality guidelines and standards based on PM concentrations. For instance, the United States Environmental Protection Agency (EPA) has set National Ambient Air Quality Standards (NAAQS) for $PM_{2.5}$ and $PM_{10}$, which require regular monitoring and reporting [9]. Accurate PM monitoring data are crucial for assessing compliance with air quality standards, identifying pollution hotspots and sources, evaluating the effectiveness of air pollution control measures, informing urban planning and zoning decisions, and providing timely public health advice during high pollution events. This information is vital for policymakers to develop and implement effective strategies for reducing PM pollution and its associated health and environmental impacts.

Recent technological advancements have significantly improved our ability to monitor PM accurately and in real-time. Traditional monitoring methods, such as gravimetric analysis, are being complemented and sometimes replaced by more advanced techniques. These include optical particle counters, satellite remote sensing, low-cost sensors, and the application of machine learning and artificial intelligence for data analysis and forecasting [10]. The development of affordable, portable PM sensors has enabled the creation of dense monitoring networks and citizen science initiatives, greatly increasing the spatial resolution of PM data. This enhanced monitoring capability allows for more comprehensive and detailed assessments of PM distribution and its impacts on local communities. As urbanization continues and climate change alters atmospheric dynamics, the need for accurate, comprehensive PM monitoring will only increase.

Future directions in PM monitoring include improving the characterization of PM composition and sources, better understanding the health effects of ultrafine particles and specific PM components, developing more accurate exposure assessment methods, and integrating PM monitoring with other environmental and health data for comprehensive risk assessments. Additionally, there is a growing focus on addressing environmental justice concerns related to disproportionate PM exposure in disadvantaged communities. Improved monitoring can help identify and address these disparities, ensuring that air quality management strategies benefit all populations equally.

Surface plasmon resonance (SPR) is a powerful and versatile optical sensing technique that has gained significant attention in the field of environmental monitoring. SPR is based on the excitation of surface plasmons—coherent electron oscillations that exist at the interface between a metal and a dielectric material—by incident light. This phenomenon is highly sensitive to changes in the refractive index near the metal surface, making SPR an excellent tool for detecting and quantifying various environmental contaminants and changes in environmental conditions. The fundamental principle of SPR involves the interaction of light with free electrons on a metal surface, typically gold or silver, which results in a reduction in the intensity of reflected light at a specific angle or wavelength. This change is highly sensitive to the adsorption of molecules on the metal surface, as it alters the local refractive index. As such, SPR can be used to monitor the presence of pollutants, such as heavy metals, pesticides, and other hazardous substances, in air, water, and soil environments. The ability to provide real-time, label-free detection makes SPR an attractive option for continuous environmental monitoring applications [11,12].

One of the key advantages of SPR in environmental monitoring is its high sensitivity and specificity. SPR sensors can detect minute changes in the refractive index, allowing for the detection of low concentrations of pollutants. This sensitivity is further enhanced by the use of nanomaterials, such as graphene and gold nanoparticles, which amplify the SPR signal and improve the detection limits. For example, SPR sensors have been developed to detect heavy metals like cadmium and lead in water samples, demonstrating the potential for these sensors to provide an early warning of contamination events [13].

Additionally, SPR can be integrated with other sensing technologies, such as fiber optics, to create compact and portable devices suitable for field deployment [14]. The versatility of SPR extends beyond chemical detection to include biological monitoring. SPR can be used to detect pathogens, allergens, and other biological agents in the environment, which is particularly important for monitoring water quality and ensuring the safety of drinking water supplies. SPR-based biosensors can detect specific proteins, nucleic acids, and microorganisms by exploiting the specific binding interactions between target molecules and immobilized biorecognition elements on the sensor surface. This capability allows for the rapid and accurate detection of biological contaminants, which is crucial for preventing the spread of diseases and protecting public health [15].

Moreover, SPR technology is continually evolving, with recent advances focusing on enhancing its performance and expanding its applications. Innovations, such as the development of SPR imaging (SPRI) and the integration of SPR with smartphone technology, have opened new possibilities for environmental monitoring. SPRI allows for the visualization of spatial distributions of analytes on the sensor surface, providing additional information about the sample composition and enabling multiplexed detection of multiple analytes simultaneously. Meanwhile, smartphone-based SPR platforms offer a cost-effective and accessible solution for on-site environmental monitoring, allowing users to perform analyses without the need for specialized laboratory equipment [16].

The application of SPR in environmental monitoring is not without challenges. The performance of SPR sensors can be affected by factors such as temperature fluctuations, non-specific binding, and matrix effects from complex environmental samples. To address these issues, ongoing research is focused on improving sensor design, developing robust surface chemistries, and implementing advanced data analysis techniques. For instance, the use of machine learning algorithms to process SPR data can enhance the accuracy and reliability of detection by compensating for environmental interferences and extracting meaningful patterns from complex datasets [14].

Surface plasmon resonance is a highly effective and adaptable tool for environmental monitoring, offering real-time, sensitive, and specific detection of a wide range of chemical and biological contaminants. Its ability to operate in diverse environmental settings and its potential for integration with modern technologies make SPR a promising approach for addressing current and future challenges in environmental health and safety. As research and development in this field continue to advance, SPR is expected to play an increasingly important role in safeguarding the environment and public health.

Despite the advantages, implementing SPR for environmental monitoring, particularly for airborne pollutants, presents several challenges: Environmental interferences: temperature fluctuations, humidity variations, and the non-specific binding of molecules present in complex environmental samples can significantly affect the performance of SPR sensors. These interferences can lead to signal drift, reduced sensitivity, and inaccurate measurements, particularly for low concentrations of airborne pollutants. Matrix effects: These can occur because the composition of airborne particulate matter is highly diverse, with varying particle sizes, shapes, and chemical properties. Matrix effects can influence the interaction of pollutants with the SPR sensor surface, leading to measurement inconsistencies and difficulties in accurately quantifying specific pollutants. Sensitivity to ultrafine particles: Ultrafine particles, while posing significant health risks, contribute minimally to the overall mass of particulate matter. SPR sensors may have limitations in detecting and quantifying these ultrafine particles due to their small size and low mass, potentially underestimating their contribution to air pollution levels. Surface chemistry and functionalization: developing robust and selective surface chemistries for SPR sensors is crucial for targeting specific airborne pollutants. The non-specific binding of other molecules present in the air can interfere with the detection of the target pollutant, reducing sensor accuracy and reliability. Real-time monitoring and data processing: Continuous, real-time monitoring of airborne pollutants using SPR requires efficient data acquisition and processing systems. The development of algorithms and software to handle the high data rates generated by SPR sensors and to compensate for environmental interferences is essential for reliable air quality monitoring. Cost and portability: While SPR technology has advanced significantly, the cost and portability of SPR systems for widespread environmental monitoring remain considerations. Developing miniaturized, cost-effective SPR sensors that can be easily deployed in various environments is crucial for expanding the application of this technology for air quality assessment.

## 2. Airborne Particulate Matter Sensing

The accurate measurement of PM concentrations is crucial for assessing air quality, understanding pollution sources, and evaluating potential health impacts. Over the years, various sensing techniques have been developed to quantify PM levels in the atmosphere. This section provides an overview of the traditional methods used for measuring airborne

PM, their principles of operation, advantages, and limitations. EPA reference methods, such as the Federal Reference Method (FRM) and Federal Equivalent Methods (FEMs), are essential for providing accurate and reliable PM measurements for regulatory purposes. These methods, typically based on gravimetric analysis, establish a baseline for evaluating the performance of other PM-sensing technologies. While fundamental to air quality monitoring, EPA reference methods often involve time-intensive sampling and analysis procedures, which can limit their spatial and temporal coverage. The conventional methods of PM sensing are shown in Figure 1, and they are briefly described in the following subsections as well.

### 2.1. Traditional Methods of Measuring Airborne PM and Their Limitations and Challenges

Various traditional methods are employed for particulate matter (PM) sensing, each relying on different principles and offering unique advantages and limitations. As shown in Figure 1, these conventional methods encompass a range of techniques, including gravimetric methods, which rely on mass measurements; continuous monitoring techniques, such as Tapered Element Oscillating Microbalance (TEOM) and Beta Attenuation Monitoring (BAM); optical methods, which measure light scattering or absorption by particles; electrical methods, such as the Electrical Low Pressure Impactor (ELPI) and Scanning Mobility Particle Sizer (SMPS); and microscopy methods, including Scanning Electron Microscopy (SEM) and Transmission Electron Microscopy (TEM). The following subsections will delve into each of these methods, detailing their principles of operation, benefits, and drawbacks.

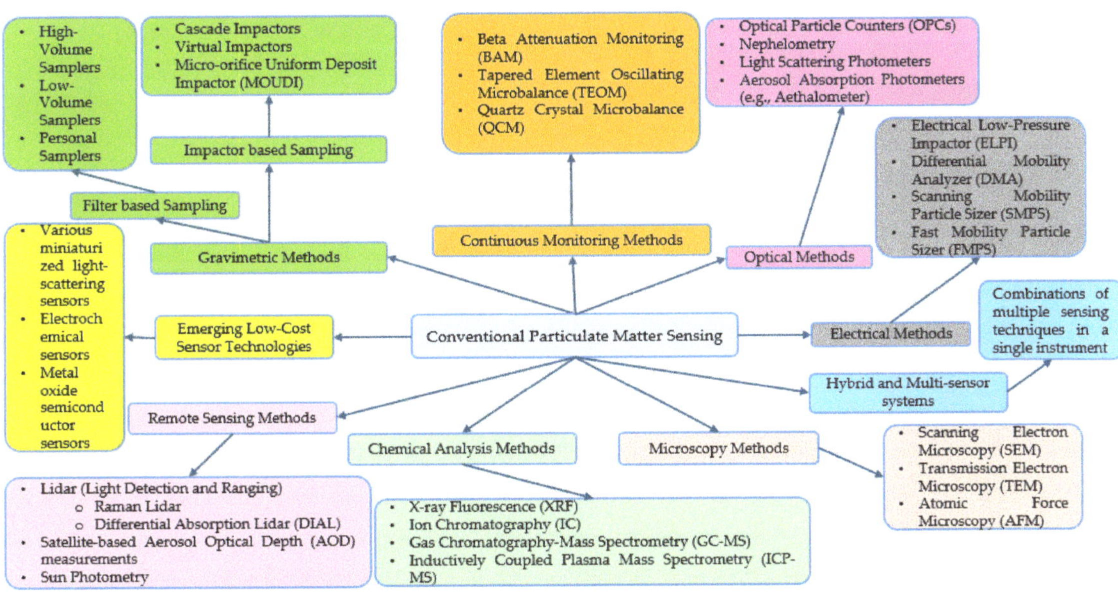


**Figure 1.** Flow chart of conventional PM-sensing methods.

#### 2.1.1. Gravimetric Methods

Gravimetric methods of sampling are fundamental techniques in the field of air quality monitoring and PM analysis. These methods rely on the principle of mass measurement to quantify the concentration of airborne particles. Filter-based sampling, which includes high-volume, low-volume, and personal samplers, is the most common gravimetric approach. High-volume samplers, such as those used by Chow et al. (2006), typically operate at flow rates of 1.13 to 1.70 $m^3/min$ and are primarily used for measuring total suspended particles (TSP) and $PM_{10}$ [17]. Low-volume samplers, operating at lower flow rates (e.g., 16.7 L/min), are commonly employed for $PM_{2.5}$ and $PM_{10}$ measurements, as demonstrated in the study by Solomon et al. (2014) [18]. Personal samplers, designed for individual exposure assessment, have been effectively utilized by Brouwer et al. (2009) to evaluate the occupational exposure to PM in various workplace environments [19]. These filter-based methods have proven invaluable in regulatory compliance monitoring and epidemiological studies due to their ability to provide accurate mass concentration data and allow for subsequent chemical analysis of collected particles.

Impactor-based gravimetric methods offer size-selective particle collection, enabling researchers to analyze the size distribution of airborne particulates. Cascade impactors, such as those employed by Marple et al. (1991), consist of multiple stages that separate particles based on their aerodynamic diameter, providing detailed information on particle size fractions [20]. Virtual impactors, as used in the study by Sioutas et al. (1994), utilize a unique design to separate particles into two size fractions without the need for impaction plates, reducing particle bounce and re-entrainment issues [21]. The Micro-Orifice Uniform Deposit Impactor (MOUDI), developed by Marple et al. (1991) and utilized in numerous studies including that of Cass et al. (2000), offers high-resolution particle size separation with uniform particle deposition, making it particularly useful for chemical speciation studies [20,22]. These impactor-based methods have been instrumental in understanding the size-dependent composition and behavior of atmospheric aerosols, contributing significantly to our knowledge of PM sources, formation mechanisms, and potential health impacts.

Despite their widespread use and reliability, gravimetric methods for measuring airborne PM face several limitations and challenges. One significant issue is the time-intensive nature of these techniques, as noted by Chow et al. (2006), which can delay the availability of data for timely decision-making in air quality management [23]. The potential for measurement artifacts due to volatile organic compounds and water absorption on filters, as highlighted by Vecchi et al. (2009), can lead to an overestimation or underestimation of the particle mass [24]. Sioutas et al. (1994) pointed out that impactor-based methods may suffer from particle bounce and re-entrainment, particularly for dry particles, potentially affecting size distribution measurements [21]. Additionally, gravimetric methods typically provide integrated measurements over extended sampling periods, limiting their ability to capture short-term fluctuations in PM concentrations, as discussed by Wilson et al. (2002) [25]. The labor-intensive nature of filter handling, conditioning, and weighing processes also introduces the potential for human error, as emphasized by Amaral et al. (2015) [26]. Furthermore, these methods generally do not provide real-time data, which can be crucial for immediate public health interventions during acute pollution events. Despite these limitations, gravimetric methods remain the gold standard for regulatory compliance monitoring due to their accuracy and traceability, as they align with EPA reference methods for PM measurement. These methods provide a direct measurement of particle mass, which is the primary metric used in EPA air quality standards. The established procedures and protocols for gravimetric sampling ensure consistency and comparability of measurements across different monitoring networks.

#### 2.1.2. Continuous Monitoring Methods

Continuous monitoring methods for airborne PM have revolutionized air quality assessments by providing real-time, high-resolution data on particle concentrations. Among these methods, Beta Attenuation Monitoring (BAM), Tapered Element Oscillating Microbalance (TEOM), and Quartz Crystal Microbalance (QCM) have emerged as widely adopted techniques. BAM, which utilizes the attenuation of beta radiation by collected particles to determine the mass concentration, has been extensively used in regulatory monitoring networks. For instance, Liu et al. (2023) employed BAM monitors to assess $PM_{2.5}$ concentrations in urban areas of China, demonstrating their effectiveness in capturing diurnal and seasonal variations in PM levels [27]. The TEOM method, based on the change in oscillation frequency of a tapered element as particles accumulate, has been particularly valuable for continuous $PM_{2.5}$ monitoring. In a comprehensive study, Su et al. (2021) compared TEOM measurements with filter-based methods, highlighting TEOM's ability to provide accurate, real-time data while minimizing volatilization losses [28]. The QCM technique, which measures the mass loading through changes in the resonant frequency of a quartz crystal, has found applications in both outdoor and indoor air quality monitoring. Tariq et al. (2023) utilized QCM sensors for the real-time monitoring of fine PM in indoor environments, showcasing their sensitivity and rapid response to concentration changes [29].

These continuous monitoring methods offer unique advantages over traditional gravimetric techniques. BAM provides an excellent correlation with reference methods while offering continuous data collection, making it ideal for long-term monitoring programs. Jayaratne et al. (2020) demonstrated BAM's reliability in a year-long $PM_{2.5}$ monitoring campaign across multiple urban sites [30]. TEOM's ability to provide highly time-resolved data has made it invaluable for studying short-term pollution events and their health impacts. For example, Yadav et al. (2022) used TEOM data to investigate the association between hourly $PM_{2.5}$ concentrations and hospital admissions, revealing important insights into acute health effects of PM exposure [31]. QCM's high sensitivity to ultrafine particles and rapid response time make it particularly suited for monitoring in dynamic environments or for tracking rapid changes in particle concentrations.

While these continuous monitoring methods offer significant advantages, they are not without limitations and challenges. Schweizer et al. (2016) highlighted the importance of understanding data and equipment limitations when using BAM for creating $PM_{2.5}$ air quality health advisories [32]. They found that BAM measurements can be affected by the relative humidity, potentially leading to an overestimation of particle concentrations under high humidity conditions. Grover et al. (2005) demonstrated that traditional TEOM measurements may underestimate PM concentrations due to the loss of volatile components [33]. Their study compared TEOM measurements with filter-based methods and showed that the Filter Dynamic Measurement System (FDMS) TEOM can provide more accurate measurements by accounting for these volatile losses. Wang et al. (2015) evaluated three low-cost particle sensors in laboratory conditions, examining their performance and calibration needs [34]. They found that these sensors can be affected by environmental factors, such as the temperature, highlighting the importance of proper calibration and understanding of sensor limitations for accurate PM measurements. Additionally, all these methods face challenges in accurately measuring ultrafine particles, which contribute little to mass but may have significant health impacts. Kuula et al. (2020) evaluated the particle-size selectivity of low-cost optical PM sensors in laboratory conditions [35]. Their study highlighted challenges in accurately measuring smaller particles, including ultrafine particles, with these sensors. Their work emphasizes the need for careful interpretation of data from low-cost instruments when assessing the full spectrum of PM sizes, particularly for air quality monitoring applications [35]. The challenges in continuous PM monitoring methods are summarized in Figure 2.

Continuous monitoring methods offer advantages in terms of real-time data acquisition, enabling more timely responses to air quality changes. However, it is crucial to calibrate them against EPA reference methods to ensure accuracy and comparability. The EPA has established guidelines and protocols for evaluating the performance of continuous PM monitors, ensuring their measurements are traceable to reference methods. This rigorous evaluation process ensures the reliability and regulatory acceptance of continuous monitoring data.

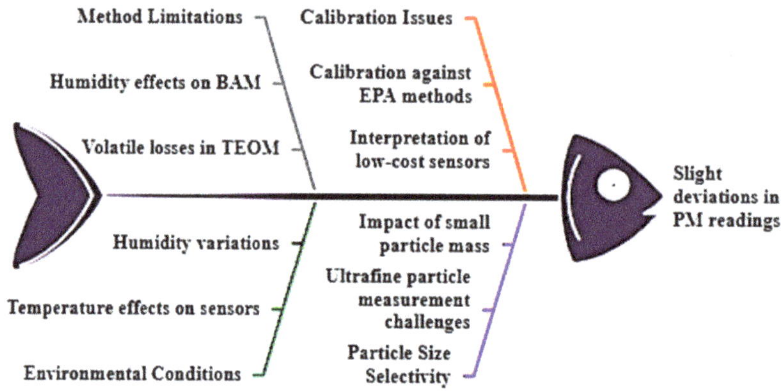


**Figure 2.** Challenges in continuous PM monitoring methods.

### 2.1.3. Optical Methods

Optical methods for measuring PM have become increasingly attractive in air quality monitoring due to their ability to provide real-time, high-resolution data. These methods include Optical Particle Counters (OPCs), nephelometers, light-scattering photometers, and aerosol absorption photometers like the Aethalometer.

Optical Particle Counters (OPCs) have been widely used for their ability to provide size-resolved particle number concentrations. For example, Kuula et al. (2020) evaluated the particle-size selectivity of low-cost OPCs in laboratory conditions, demonstrating their potential for widespread air quality monitoring [35]. Their study showed that OPCs can effectively measure particles in the $PM_{2.5}$ and $PM_{10}$ size ranges, though with some limitations in accuracy for the smallest particles. Nephelometers, which measure light scattering by aerosols, have been employed in both stationary and mobile monitoring setups. Magi et al. (2020) used a nephelometer alongside a beta attenuation monitor to evaluate $PM_{2.5}$ concentrations in an urban setting, highlighting the nephelometer's ability to capture short-term variations in particle concentrations [36].

Light-scattering photometers offer rapid response times and high sensitivity, making them suitable for detecting sudden changes in particle concentrations. Malings et al. (2020) conducted a long-term evaluation of low-cost light-scattering sensors for fine-particle mass monitoring, demonstrating their potential for large-scale deployment after appropriate calibration [37]. Aerosol absorption photometers, such as the Aethalometer, provide unique insights into black carbon concentrations. Drinovec et al. (2015) described a new Aethalometer model with improved performance, showcasing its ability to measure black carbon with a finer time resolution and sensitivity [38].

These optical methods are unique in their ability to provide real-time data with high temporal resolution, which is crucial for understanding short-term variations in air quality and for public health interventions. While gravimetric methods remain the gold standard for regulatory purposes, they cannot match the temporal resolution of optical methods. Additionally, optical methods allow for more widespread and cost-effective monitoring networks, as demonstrated by the increasing use of low-cost sensors in citizen science projects and urban air quality studies.

However, optical methods for PM measurement also face several limitations and challenges. One significant issue is the dependence of light scattering on particle properties, such as the size, shape, and refractive index, which can lead to measurement biases. Crilley et al. (2018) demonstrated that the relative humidity can significantly affect the performance of optical particle counters, potentially leading to an overestimation of particle mass concentrations under high humidity conditions [39]. Another challenge is the difficulty in accurately measuring ultrafine particles, which contribute little to mass but may have significant health impacts. Wang et al. (2015) highlighted the limitations of low-cost optical sensors in detecting particles smaller than 0.3 μm, which can form a significant portion of urban aerosols [34]. Calibration remains a critical issue, as the response of optical instruments can vary depending on the composition and size distribution of aerosols. Zheng et al. (2018) showed that the performance of low-cost optical sensors can vary significantly across different environments, emphasizing the need for location-specific calibrations [40]. Additionally, Alfano et al. (2020) pointed out that most optical methods assume a spherical particle geometry and a standard particle density, which may not accurately represent the diverse range of atmospheric aerosols [41]. Despite these challenges, ongoing research and technological advancements continue to improve the accuracy and reliability of optical methods for PM measurement. While optical methods offer advantages in real-time monitoring and size-resolved measurements, their accuracy can be influenced by the particle composition, size distribution, and environmental factors. Calibration against EPA reference methods, such as FRM or FEMs, is essential for reliable PM measurements. The EPA provides guidance on calibration procedures and performance criteria for optical PM sensors to ensure their measurements are consistent with regulatory requirements.

#### 2.1.4. Electrical Methods

Electrical methods for measuring airborne PM have become increasingly important in aerosol science due to their ability to provide high-resolution, real-time data on particle size distributions and concentrations. These methods include the Electrical Low-Pressure Impactor (ELPI), Differential Mobility Analyzer (DMA), Scanning Mobility Particle Sizer (SMPS), and Fast Mobility Particle Sizer (FMPS); an overview of them is shown in Figure 3.

The Electrical Low Pressure Impactor (ELPI) has gained popularity for its wide particle size range and real-time measurement capabilities. Lee et al. (2021) applied an ELPI to measure residual particles in an epitaxial growth reactor, demonstrating its effectiveness in monitoring the chamber purge performance after semiconductor processing [42]. In a different application, Rissler et al. (2014) used an ELPI to characterize nanoparticle emissions from a small-scale biomass combustion unit, highlighting its versatility in various emission studies [43].

The Differential Mobility Analyzer (DMA) and Scanning Mobility Particle Sizer (SMPS) are often used in tandem to provide high-resolution particle size distributions. Sowlat et al. (2016) employed an SMPS system to characterize ultrafine particles in urban environments, highlighting its capability to measure particles in the size range of 14.6 to 661.2 nm with a high time resolution [44]. Their study in central Los Angeles demonstrated the SMPS's effectiveness in capturing detailed particle number size distributions, which were crucial for the source apportionment of ambient particle number concentrations [44]. Similarly, Levin et al. (2014) utilized an SMPS to investigate the size-resolved hygroscopicity of atmospheric aerosols, demonstrating its applicability in atmospheric science research [45].

The Fast Mobility Particle Sizer (FMPS) offers rapid measurements of particle size distributions, making it valuable for studying dynamic aerosol processes. Wang et al. (2015) utilized an FMPS to investigate the rapid changes in particle emissions during aircraft engine start-up and shutdown, demonstrating its unique ability to capture transient aerosol events with a 1 s time resolution [34]. In another study, Zimmerman et al. (2014) used FMPS to characterize nanoparticle exposure during 3D printing processes, showcasing its applicability in occupational health research [46].

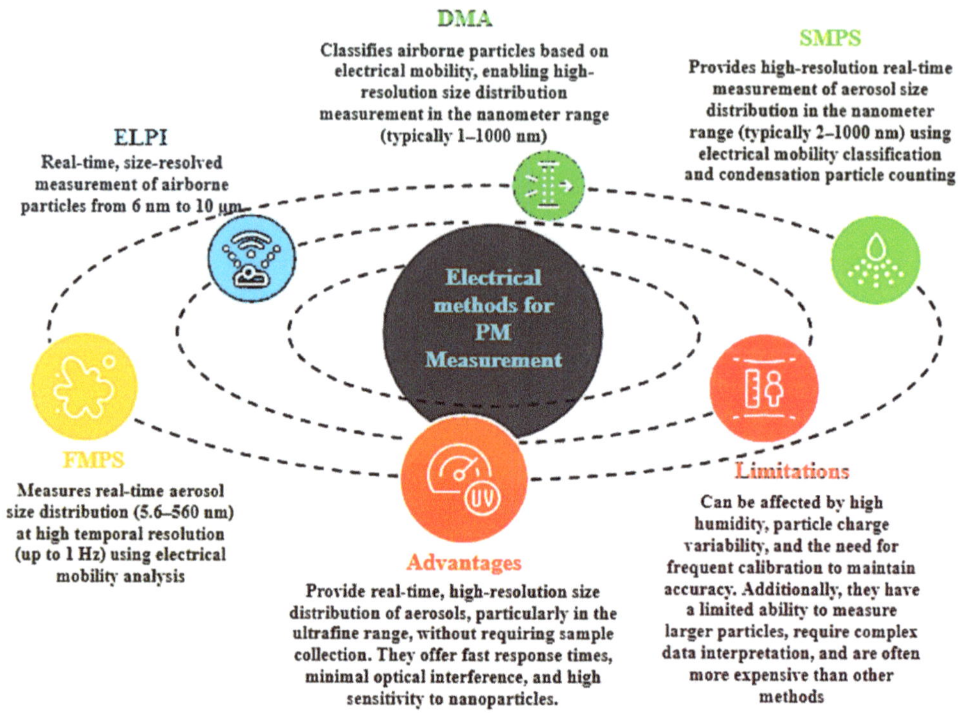


**Figure 3.** Overview of electrical methods for PM measurement.

These electrical methods offer unique advantages over traditional optical or gravimetric techniques. They can measure particles in the nanometer range, which is crucial for understanding the formation and behavior of ultrafine particles that are of increasing concern in air quality and health studies. For instance, Fonseca et al. (2016) combined SMPS and ELPI measurements to characterize ultrafine particles in subway systems, demonstrating the complementary nature of these techniques in a comprehensive aerosol characterization [47].

However, electrical methods for PM measurement also face several limitations and challenges. One significant issue is the potential for particle charging effects to influence measurements, particularly for ultrafine particles. Levin et al. (2015) demonstrated that the charging state of particles can affect their mobility and, consequently, their measured size distribution in electrical mobility analyzers [48]. Another challenge is the sensitivity of these instruments to changes in ambient conditions, such as temperature and humidity. Schmid et al. (2002) showed that the relative humidity can significantly affect the performance of DMAs, potentially leading to measurement biases in atmospheric studies [49]. Additionally, these instruments often require a stable power supply and controlled environmental conditions, which can limit their applicability in field studies. Wu et al. (2013) highlighted the importance of careful instrument design and data processing when using SMPS systems in field measurements, emphasizing the need to account for factors, such as particle hygroscopicity and chemical composition, that can affect size distribution measurements [50]. Furthermore, the high cost and complexity of many electrical measurement systems can limit their widespread deployment, particularly in large-scale monitoring networks. Despite these challenges, ongoing research and technological advancements continue to improve the accuracy, reliability, and applicability of electrical methods for PM measurement in diverse environmental conditions.

#### 2.1.5. Microscopy Methods

Microscopy methods, such as Scanning Electron Microscopy (SEM), Transmission Electron Microscopy (TEM), and Atomic Force Microscopy (AFM), have revolutionized our ability to visualize and analyze materials at the nanoscale. A comparative view of these methods is given in Figure 4. These techniques offer unique capabilities for characterizing the structure, morphology, and properties of a wide range of samples.

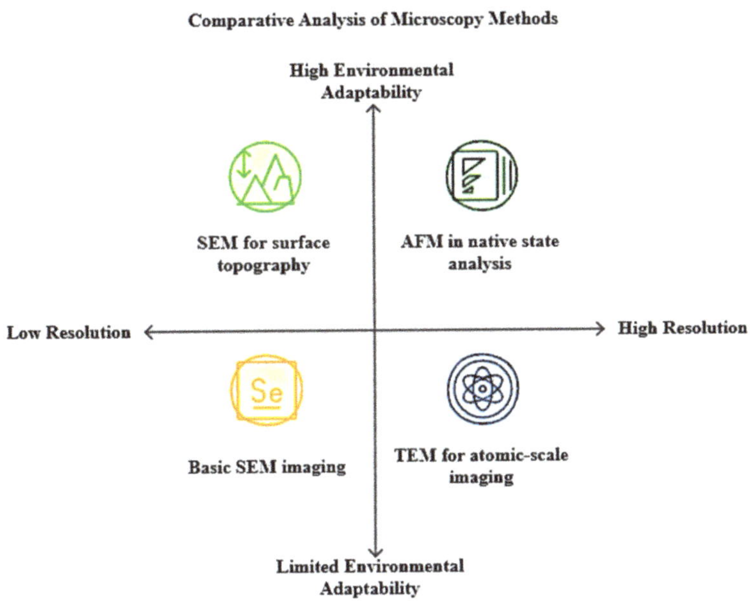


**Figure 4.** The comparative analysis of microscopy methods.

Scanning Electron Microscopy (SEM) provides high-resolution imaging of the particle surface topography and composition. Shi et al. (2015) utilized SEM to investigate the morphology and adhesive properties of $PM_{2.5}$ airborne pollutants, revealing intricate details of soot particles and their aggregation capabilities [51]. The ability of SEM to provide three-dimensional-like images with a large depth of field makes it invaluable for studying complex structures of atmospheric particles.

Transmission Electron Microscopy (TEM) offers an even higher resolution than SEM, allowing for atomic-scale imaging and the analysis of internal structures of airborne particles. Longyi et al. (2022) employed TEM to study the mixing states of individual atmospheric particles, providing crucial insights into their formation and transformation processes [52]. TEM's unique capability to provide both real-space imaging and electron diffraction patterns makes it indispensable for crystallographic studies and chemical composition analysis of atmospheric aerosols.

Atomic Force Microscopy (AFM) stands out for its ability to provide the three-dimensional surface topography of airborne particles with a sub-nanometer resolution. Shi et al. (2015) used AFM to investigate the nanoscale mechanical properties of $PM_{2.5}$ particles, demonstrating its potential for characterizing the adhesiveness and deformation of airborne pollutants [51]. AFM's versatility in operating under various environmental conditions makes it particularly valuable for studying particles in their native state.

These microscopy methods are often used in combination to provide a comprehensive characterization of airborne PM. For instance, Sielicki et al. (2011) combined SEM and TEM with energy-dispersive X-ray spectroscopy (EDX/EDS) to analyze metal-containing airborne particles, leveraging SEM to determine the overall morphology and TEM for the detailed internal structure and elemental composition [53].

Recent advancements in microscopy techniques have further expanded their capabilities in atmospheric particle analysis. Laskin et al. (2019) demonstrated the use of automated SEM/EDX for the high-throughput characterization of thousands of individual particles, enabling a statistically robust analysis of particle populations in complex atmospheric samples [54]. These microscopy methods complement traditional bulk analysis techniques by providing detailed information on individual particle characteristics, which is crucial for understanding the sources, transformation processes, and potential health impacts of airborne PM.

Despite their powerful capabilities, microscopy methods for analyzing airborne PM face several limitations and challenges. One significant issue is sample preparation, which can alter the original state of the particles. As noted by Ault and Axson (2017), the high vacuum environment required for SEM and TEM can cause volatile components to evaporate, potentially changing the particle morphology and composition [55]. AFM, while capable of operating in ambient conditions, faces challenges in accurately representing soft or sticky particles due to tip–sample interactions. Variola (2015) highlighted that while AFM provides high-resolution imaging of surface topography, it faces challenges in accurately representing soft or sticky particles due to tip–sample interactions [56]. The author also emphasized that the resolution of AFM is limited by the probe's geometry, which can lead to artifacts in imaging highly irregular particles. To address the challenge of analyzing large numbers of particles, Laskin et al. (2018) emphasized the need for automated, high-throughput analysis methods in atmospheric particle characterization using advanced mass spectrometry techniques [57]. Furthermore, the interpretation of microscopy data requires expertise and can be subjective, particularly for complex, heterogeneous atmospheric particles. Despite these challenges, ongoing advancements in instrumentation and data analysis techniques continue to enhance the applicability and reliability of microscopy methods in atmospheric PM research.

#### 2.1.6. Chemical Analysis Methods

Chemical analysis methods play a crucial role in characterizing the composition of airborne PM. X-ray Fluorescence (XRF), Ion Chromatography (IC), Gas Chromatography–Mass Spectrometry (GC-MS), and Inductively Coupled Plasma Mass Spectrometry (ICP-MS) are widely used techniques for this purpose.

X-ray Fluorescence (XRF) is a non-destructive method that allows for the rapid multi-elemental analysis of PM samples. Yatkin et al. (2012) demonstrated the effectiveness of XRF in analyzing the elemental composition of $PM_{2.5}$ samples collected on filters, highlighting its ability to detect a wide range of elements simultaneously [58]. The non-destructive nature of XRF makes it particularly valuable for preserving samples for further analysis.

Ion Chromatography (IC) is extensively used for the analysis of water-soluble ionic species in PM. Correa et al. (2023) reviewed various characterization methods for ions in $PM_{2.5}$ and $PM_{10}$ samples, noting that IC is commonly employed for analyzing anions and cations [59]. They highlighted its importance in understanding the contribution of secondary inorganic aerosols to PM pollution.

Gas Chromatography–Mass Spectrometry (GC-MS) is particularly useful for the analysis of organic compounds in PM. Choi et al. (2020) utilized GC-MS for the determination of atmospheric amines in $PM_{2.5}$ samples, demonstrating its capability to identify and quantify complex organic species [60]. The high sensitivity and specificity of GC-MS make it invaluable for studying the organic fraction of PM.

Inductively Coupled Plasma Mass Spectrometry (ICP-MS) is a powerful technique for trace metal analysis in PM. Feng et al. (2020) employed ICP-MS alongside IC for the simultaneous determination of atmospheric amines and inorganic ions in $PM_{2.5}$ [61]. The high sensitivity of ICP-MS allows for the detection of metals at very low concentrations, making it essential for studying trace metal pollution in airborne particles.

These chemical analysis methods are often used in combination to provide a comprehensive characterization of PM composition. For instance, Brown and Edwards (2009) developed a novel sample preparation technique for measuring anions in ambient PM using ion chromatography, highlighting the ongoing efforts to improve analytical methods in this field [62]. The choice of analytical method often depends on the specific components of interest and the required detection limits. While each method has its strengths, they are often complementary. XRF provides rapid elemental analysis but may lack sensitivity for some trace elements compared to ICP-MS. IC is excellent for water-soluble ions but cannot detect organic species, which is where GC-MS excels. The combination of these techniques allows for a more complete understanding of PM composition, which is crucial for source apportionment and health impact studies.

Despite their widespread use and effectiveness, these chemical analysis methods face several limitations and challenges. Sample preparation remains a critical issue, as noted by Canepari et al. (2019), who highlighted the potential for sample loss or contamination during collection and handling processes [63]. The complexity of atmospheric PM, with its diverse mixture of organic and inorganic compounds, poses challenges for comprehensive analysis. Hallquist et al. (2009) emphasized the difficulties in characterizing the organic fraction of aerosols due to the vast number of compounds present and their varying volatility [64]. Additionally, the presence of trace contaminants at very low concentrations pushes the limits of detection for many analytical techniques. Majestic et al. (2012) discussed the challenges in measuring trace metals in atmospheric PM, including matrix effects and interferences that can affect measurement accuracy [65]. Furthermore, the time-integrated nature of many sampling methods may not capture the dynamic changes in PM composition, as pointed out by Fuzzi et al. (2015) in their review of PM and air quality [66]. These challenges underscore the need for the continued development of analytical techniques and sampling strategies to improve our understanding of PM composition and its impacts on health and the environment.

#### 2.1.7. Remote Sensing Methods

Remote sensing methods have revolutionized the measurement of airborne PM, offering high spatial and temporal resolution data over large areas. These techniques include Lidar (Light Detection and Ranging), Raman Lidar, Differential Absorption Lidar (DIAL), satellite-based Aerosol Optical Depth (AOD) measurements, and Sun Photometry, as shown in Figure 5.

Lidar technology has emerged as a powerful tool for the vertical profiling of aerosols. Soupiona et al. (2020) utilized a multi-wavelength Raman Lidar to characterize the vertical distribution of Saharan dust over Athens, Greece [67]. Their study demonstrated the capability of Lidar systems to provide detailed information on aerosol layering and optical properties. Raman Lidar, in particular, offers the advantage of independent measurements of aerosol extinction and backscatter coefficients, allowing for a more accurate retrieval of the aerosol's properties.

**Figure 5.** Remote sensing methods for PM measurement.

Differential Absorption Lidar (DIAL) has been employed for measuring specific gaseous components of PM. Browell et al. (1998) demonstrated the use of airborne DIAL systems to simultaneously measure vertical distributions of water vapor and aerosols, showcasing the technique's potential for studying atmospheric composition and aerosol-water vapor interactions [68]. Their work highlighted the versatility of DIAL in providing high-resolution profiles of multiple atmospheric constituents. The satellite-based AOD measurements have become increasingly important for global aerosol monitoring. Lyapustin et al. (2018) introduced the MAIAC algorithm for MODIS AOD retrievals, which provides high-resolution (1 km) daily AOD data globally [69]. This advancement has significantly improved our ability to study aerosol distributions and trends over large areas. Wei et al. (2021) further demonstrated the application of satellite-derived AOD in estimating ground-level $PM_{2.5}$ concentrations, highlighting the importance of these measurements for air quality studies [70].

Sun photometry remains a crucial ground-based technique for validating satellite AOD retrievals and studying aerosol properties. Giles et al. (2019) provided an overview of the AERONET Version 3 database, which includes improved automated cloud screening and quality control algorithms for sun photometer AOD measurements [71]. This update has significantly enhanced the accuracy and reliability of AERONET data, serving as a crucial ground truth for satellite and model validation. The integration of these remote sensing techniques offers unique advantages over traditional in situ measurements. Benavent-Oltra et al. (2019) demonstrated the synergy between Lidar and sun photometer measurements by using the GRASP (Generalized Retrieval of Aerosol and Surface Properties) algorithm to retrieve detailed aerosol optical and microphysical properties [72]. Their study, conducted during the SLOPE I and II campaigns in Granada, Spain, showcased how combining these remote sensing techniques can provide comprehensive information on aerosol characteristics, including vertical profiles of fine- and coarse-mode concentrations. Similarly, Siomos et al. (2020) used a combination of Lidar and sun photometer measurements to study the impact of aerosols on solar radiation, showcasing the complementary nature of these techniques [73]. These remote sensing methods are particularly valuable for their ability to provide continuous, large-scale measurements that are not feasible with ground-based instruments alone. However, they also face challenges such as cloud interference in satellite retrievals and the need for complex retrieval algorithms to convert optical measurements into PM concentrations.

Despite their significant advantages, remote sensing methods for measuring airborne PM face several limitations and challenges. One major issue is the indirect nature of these measurements, which often require complex algorithms to convert optical properties to particle concentrations. As noted by Toth et al. (2014), the relationship between AOD and surface $PM_{2.5}$ can vary significantly due to factors such as aerosol vertical distribution and relative humidity [74]. Satellite-based measurements are particularly affected by cloud cover and surface reflectance, which can lead to data gaps and retrieval errors. Li et al. (2016) highlighted the challenges in AOD retrievals over bright surfaces and in the presence of thin cirrus clouds [75]. Ground-based remote sensing techniques like Lidar and sun photometry, while providing valuable vertical profile information, are limited in spatial coverage and can be affected by weather conditions. Wagner and Schäfer (2017) discussed the limitations of Raman Lidar in measuring aerosols under high relative humidity conditions due to hygroscopic growth effects [76]. Additionally, the high cost and complexity of some remote sensing instruments, particularly advanced Lidar systems, can limit their widespread deployment. These challenges underscore the need for integrated approaches that combine remote sensing with in situ measurements and modeling to provide a more comprehensive understanding of airborne PM distributions and properties.

#### 2.1.8. Emerging Low-Cost Sensor Technologies

While emerging low-cost sensor technologies are not traditionally considered part of the established methods for measuring airborne PM, they have gained significant attention in recent years due to their potential for widespread deployment and real-time monitoring. These technologies include various miniaturized light-scattering sensors, electrochemical sensors, and metal oxide semiconductor sensors.

Miniaturized light-scattering sensors have become popular for their ability to provide real-time measurements of PM concentrations. Bulot et al. (2019) evaluated the performance of low-cost PM sensors in a field study, demonstrating their potential for high-resolution spatial and temporal monitoring of air quality [77]. Their study showed that these sensors could capture pollution events and spatial variations that might be missed by traditional monitoring stations.

Electrochemical sensors have been applied to measure gaseous pollutants that often coexist with PM. Castell et al. (2017) assessed the performance of low-cost air quality sensors, including electrochemical sensors, in measuring $NO_2$ and $O_3$ [78]. Their work highlighted the potential of these sensors for complementing existing air quality monitoring networks, especially in urban environments where pollution levels can vary significantly over short distances.

Metal oxide semiconductor sensors have shown promise in detecting various air pollutants. Karagulian et al. (2019) conducted a comprehensive review of low-cost sensors for air quality monitoring, including metal oxide sensors [79]. They noted that while these sensors face challenges in terms of accuracy and long-term stability, they offer unique advantages in terms of cost-effectiveness and the ability to provide high-resolution spatial data. The unique aspect of these emerging technologies lies in their potential for creating dense networks of air quality monitors. Morawska et al. (2018) discussed the paradigm shift in air pollution monitoring enabled by low-cost sensors, emphasizing their ability to provide unprecedented spatial and temporal resolution in air quality data [80]. However, it is important to note that these low-cost sensors are not without limitations. Rai et al. (2017) highlighted challenges such as cross-sensitivity to other pollutants, environmental factors affecting sensor performance, and the need for frequent calibration [81]. One major issue is the variability in sensor performance and data quality. Crilley et al. (2018) demonstrated that low-cost optical particle counters can be affected by the relative humidity, potentially leading to an overestimation of particle mass concentrations under high humidity conditions [39]. This highlights the need for careful calibration and correction algorithms to account for environmental factors. Another challenge is the long-term stability and drift of these sensors. Badura et al. (2018) observed significant sensor-to-sensor variability and temporal drift in low-cost PM sensors over a one-year deployment, emphasizing the need for regular recalibration and quality control measures [82]. Furthermore, the interpretation of data from dense networks of low-cost sensors presents its own challenges. Schneider et al. (2019) discussed the complexities of managing and interpreting large volumes of data from sensor networks, including issues of data quality assurance and the development of appropriate data analysis techniques [83]. These limitations underscore the importance of using low-cost sensors as complementary tools to established reference methods rather than as direct replacements. Addressing these challenges will be crucial for realizing the full potential of low-cost sensor technologies in air quality monitoring and management.

#### 2.1.9. Hybrid and Multi-Sensor Systems

Hybrid and multi-sensor systems, which combine multiple sensing techniques in a single instrument, have emerged as powerful tools for measuring airborne PM. These integrated approaches offer comprehensive characterization of aerosols by leveraging the strengths of different measurement principles.

Kosmopoulos et al. (2020) evaluated low-cost sensors for measuring $PM_{2.5}$ and $PM_{10}$, demonstrating that multi-sensor systems can provide reliable data when properly calibrated [84]. Their study highlighted the potential of these systems for expanding air quality monitoring networks cost-effectively. Alfano et al. (2020) reviewed various low-cost PM sensors, discussing how hybrid systems combining optical and electrical sensing methods can improve the measurement accuracy across different particle size ranges [41]. They noted that such combinations could address limitations of single-method instruments, particularly in complex urban environments.

The integration of multiple sensing techniques can enhance the ability to distinguish between different types of PM. For instance, Bulot et al. (2019) deployed a network of low-cost sensors that combined optical particle counters with environmental sensors (temperature, humidity) [77]. This multi-parameter approach improved the accuracy of PM measurements and provided insights into the influence of meteorological conditions on PM concentrations. Advanced data fusion techniques have been crucial in maximizing the utility of multi-sensor systems. Concas et al. (2021) applied machine learning algorithms to combine data from low-cost PM sensors, meteorological instruments, and reference monitors [85]. Their approach significantly improved the accuracy of $PM_{2.5}$ estimates compared to individual sensor readings.

The unique advantage of hybrid and multi-sensor systems lies in their ability to provide a more complete picture of aerosol properties and dynamics. Sousan et al. (2017) demonstrated that combining optical particle counting with electrical mobility analysis allowed for a better characterization of particle morphology and density, which is crucial for understanding health impacts of different aerosol types [86]. These integrated approaches are particularly valuable in complex environments where single-method instruments may face limitations. Kuula et al. (2020) showed that a multi-sensor approach combining optical and electrical measurements was more robust in high-humidity conditions, where traditional optical sensors often struggle [35].

While hybrid systems offer numerous advantages, they also present challenges in terms of data integration and interpretation. Morawska et al. (2018) highlighted the importance of advanced calibration techniques when combining different sensor types, emphasizing the need for careful data processing to ensure accurate results [80]. The development of compact, field-deployable hybrid instruments has expanded the possibilities for widespread air quality monitoring. Jayaratne et al. (2018) described a portable system integrating particle counting, size distribution measurement, and gas sensors [87]. This multi-parameter approach provided comprehensive aerosol characterization in various urban and industrial settings. Looking forward, the integration of remote sensing techniques with in situ measurements promises to further enhance our understanding of aerosol dynamics. Rybarczyk and Zalakeviciute (2018) reviewed various machine learning approaches for air quality modeling and forecasting, highlighting the potential of combining ground-based multi-sensor systems with advanced data analysis techniques to improve the spatial coverage and prediction accuracy of PM concentrations [88].

### 2.2. Significance of Accurate PM Sensing for Air Quality Assessment

The significance of precise PM measurements extends across various domains, including public health, environmental policy, and scientific research, and it is summed up in Figure 6. This section explores the multifaceted importance of accurate PM sensing and its implications for air quality assessment.

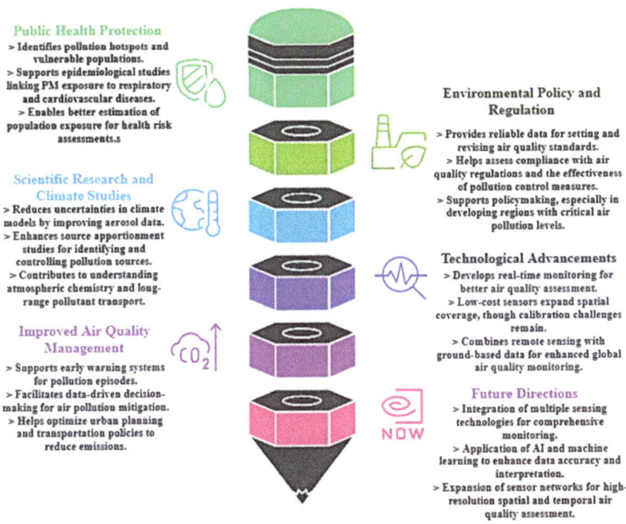


**Figure 6.** Significance of accurate PM sensing for air quality assessment.

The primary driver for accurate PM sensing is its direct link to public health. Numerous epidemiological studies have established strong correlations between exposure to PM and adverse health outcomes. Pope and Dockery (2006) provided a comprehensive review of the health effects of fine particulate air pollution, highlighting associations with cardiovascular and respiratory diseases [89]. More recent studies, such as that by Chen et al. (2020), have further elucidated the relationship between long-term exposure to $PM_{2.5}$ and mortality rates, emphasizing the need for accurate, long-term PM monitoring [90]. Accurate PM sensing allows for a better estimation of population exposure, which is crucial for epidemiological studies and public health interventions. Brauer et al. (2012) demonstrated how improved $PM_{2.5}$ estimates could enhance global exposure assessments, providing valuable data for burden of disease calculations [91]. Furthermore, precise PM measurements enable the identification of pollution hotspots and vulnerable populations, facilitating targeted interventions and resource allocation for public health initiatives.

In the realm of environmental policy and regulation, accurate PM sensing plays a vital role in shaping and enforcing air quality standards. Regulatory bodies rely on precise air quality data to set and revise ambient air quality standards. The work of Bachmann (2007) traced the historical development of PM standards in the United States, highlighting the critical role of scientific evidence derived from accurate measurements [92]. In Europe, Fuzzi et al. (2015) discussed how advancements in PM sensing technologies have influenced the evolution of air quality directives [66]. Moreover, accurate PM data are essential for assessing compliance with air quality standards and evaluating the effectiveness of pollution control measures. Pinder et al. (2019) demonstrated how improved $PM_{2.5}$ measurements could enhance the assessment of Clean Air Act implementation in the United States [93]. In developing countries, where air pollution often reaches critical levels, accurate PM sensing is crucial for informing policy decisions and tracking progress in air quality improvement initiatives.

The significance of accurate PM sensing extends to scientific research and climate studies. PM plays a complex role in the Earth's radiative balance, and precise measurements are crucial for refining climate models. Boucher et al. (2013), in their contribution to the *IPCC Fifth Assessment Report*, emphasized the importance of accurate aerosol measurements for reducing uncertainties in climate projections [5]. Furthermore, accurate PM data are essential for source apportionment studies, which aim to identify and quantify the sources of particulate pollution. Belis et al. (2013) reviewed receptor models for source apportionment, underscoring the need for high-quality PM composition data [94]. Such studies are crucial for developing targeted emission control strategies and understanding the transport of pollutants across regions.

The quest for accurate PM sensing has driven significant technological advancements. Traditional gravimetric methods, while accurate, are labor-intensive and provide limited temporal resolution. Chow (1995) provided an early comprehensive review of measurement methods for PM [95]. Since then, the field has seen the development of various real-time monitoring techniques, including beta attenuation monitors, tapered element oscillating microbalances, and optical particle counters. More recently, low-cost sensor technologies have emerged as a promising tool for high-density PM monitoring networks. Snyder et al. (2013) discussed the potential of these sensors to complement existing regulatory monitoring networks [96]. However, ensuring the accuracy and reliability of these sensors remains a challenge. Morawska et al. (2018) highlighted the need for standardized evaluation protocols and calibration procedures for low-cost PM sensors [80].

Satellite-based remote sensing has also revolutionized PM monitoring, offering global coverage and high spatial resolution. van Donkelaar et al. (2016) demonstrated the use of satellite observations to estimate global $PM_{2.5}$ concentrations, showcasing the potential of remote sensing for air quality assessment in areas lacking ground-based monitoring [97]. However, challenges remain in reconciling satellite-derived aerosol optical depth with ground-level PM concentrations, as discussed by Li et al. (2015) [98].

The future of accurate PM sensing lies in integrated approaches that combine multiple measurement techniques. Kulmala et al. (2021) proposed a global network of stations for atmospheric measurements, emphasizing the need for comprehensive, multi-instrument approaches to air quality monitoring [99]. Such integrated systems can provide a more complete picture of PM dynamics, from local to global scales. Machine learning and artificial intelligence are increasingly being applied to improve PM sensing accuracy and data interpretation. For instance, Di et al. (2019) demonstrated the use of neural networks to enhance $PM_{2.5}$ estimates from low-cost sensors [100]. These advanced data processing techniques hold promise for extracting the maximum value from diverse PM measurement sources.

Accurate PM sensing is hence fundamental to effective air quality assessment and management. Its significance spans public health protection, environmental policy formulation, and scientific research. As technology continues to advance, the integration of multiple sensing techniques, coupled with sophisticated data analysis methods, promises to further enhance our ability to monitor and understand PM pollution. This, in turn, will enable more effective strategies for improving air quality and mitigating the health and environmental impacts of PM pollution.

In Section 2 we have reviewed a range of traditional methods for measuring airborne PM, each with its own strengths and limitations. Gravimetric methods offer high accuracy for mass concentration measurements and allow for a subsequent chemical analysis, but they are time-consuming and may be susceptible to artifacts. Continuous monitoring methods, such as BAM, TEOM, and QCM, provide real-time data but require careful calibration and may be affected by environmental factors. Optical methods are advantageous for real-time monitoring and size-resolved measurements, but their accuracy can be influenced by particle properties and environmental conditions. Electrical methods excel in measuring ultrafine particles and providing high-resolution size distributions but can be sensitive to particle charging effects and ambient conditions. Microscopy methods offer detailed insights into individual particle characteristics but are often limited by sample preparation requirements and the need for expert interpretation. Chemical analysis methods are essential for characterizing PM composition but face challenges related to sample preparation, the complexity of PM mixtures, and detection limits. Finally, remote sensing methods provide valuable spatial and temporal coverage but rely on complex algorithms and can be affected by factors like cloud cover. The choice of the most suitable method depends on the specific monitoring objectives, available resources, and desired level of accuracy and detail.

## 3. Deep Learning in Environmental Sensing

Figure 7 shows all the possible DL techniques and their roles in environmental sensing, organized by their use case as shown in the flow chart below. DL has emerged as a powerful tool in environmental sensing, revolutionizing the way we collect, process, and interpret environmental data. This section explores the application of DL techniques in environmental monitoring, focusing on their ability to extract meaningful features from complex environmental datasets. DL algorithms, particularly those based on ANN, have demonstrated remarkable success in tasks such as image classification, pattern recognition, and time-series analysis, making them well-suited for addressing the challenges in environmental sensing [101].

The integration of DL with environmental sensing technologies has enabled more accurate and efficient monitoring of various environmental parameters, including air quality, water resources, land use changes, and climate patterns. These advanced techniques have the potential to overcome limitations of traditional methods, such as the ability to handle large volumes of heterogeneous data, extract complex features, and adapt to changing environmental conditions [102]. As this research delves into the specific DL architectures and their applications in environmental sensing, it will explore how these techniques are transforming our understanding of the environment and contributing to more effective environmental management strategies. The subsequent subsections are organized to show how DL can be used in various aspects of environmental sensing.

### *3.1. Data Processing and Feature Extraction*

In the realm of environmental sensing, particularly for airborne PM, DL techniques have revolutionized data processing and feature extraction. This section explores the application of Convolutional Neural Networks (CNNs), Autoencoders, Deep Belief Networks (DBNs), and U-Net variants in environmental monitoring as mentioned in following Table 1.

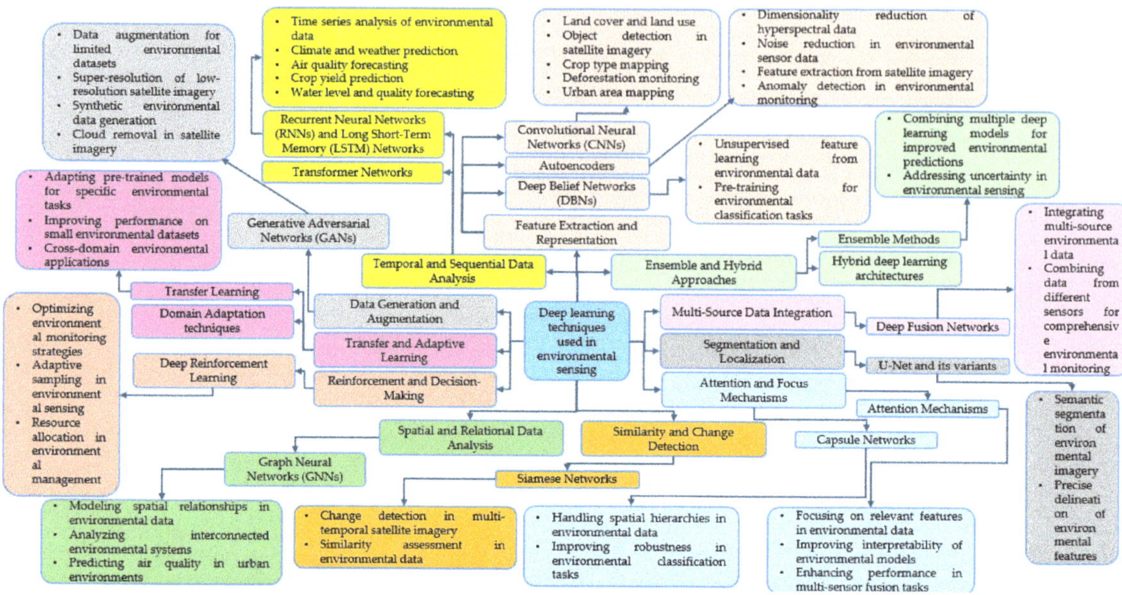


**Figure 7.** DL techniques and their roles in environmental sensing, organized by their use case.

DL techniques have significantly transformed environmental sensing, enabling a more accurate, efficient, and scalable analysis of complex environmental data. Convolutional Neural Networks (CNNs) have proven particularly effective in handling spatial data from remote sensing imagery. Their ability to automatically learn hierarchical features from raw data has been demonstrated in various applications, such as land use classification and $PM_{2.5}$ estimation from satellite imagery [111]. CNNs excel at extracting spatial features from satellite imagery and other geospatial data, making them effective for tasks like object detection in remote sensing images and estimating environmental parameters from visual data.

Autoencoders have found applications in dimensionality reduction and feature extraction from complex environmental datasets. These unsupervised learning models are particularly useful for handling high-dimensional data such as hyperspectral imagery. Riese and Keller (2019) showcased their potential in hyperspectral image classification for environmental monitoring applications [112]. Autoencoders help in capturing the most salient features of environmental data, which can be useful for tasks such as data compression, noise reduction, and anomaly detection.

Deep Belief Networks (DBNs) have shown promise in analyzing complex patterns in time-series data within environmental sensing. Ong et al. (2016) utilized DBNs for air quality prediction, achieving significant improvements over traditional methods [107]. DBNs are particularly useful for tasks that require understanding underlying structures in data, such as climate pattern analysis. Their layer-wise training process allows them to learn increasingly abstract representations of the input data, which can be beneficial for capturing long-term dependencies in environmental time-series.

U-Net and its variants have demonstrated remarkable performance in image segmentation tasks crucial for environmental monitoring. Originally developed for biomedical image segmentation, U-Net has been successfully adapted for various environmental applications. Wieland et al. (2019) demonstrated the effectiveness of U-Net in cloud detection from satellite imagery [113]. Wei et al. (2020) adapted U-Net for building footprint extraction from high-resolution aerial imagery, showcasing its versatility in environmental remote sensing tasks [114]. The architecture's ability to capture both fine-grained details and broader contextual information makes it particularly suited for segmentation tasks in satellite and aerial imagery. These DL techniques have collectively improved our ability to extract meaningful information from large-scale satellite imagery, predict environmental parameters with higher accuracy, and detect subtle patterns in environmental data that might be missed using traditional methods.

**Table 1.** Key applications of deep learning (DL) techniques in environmental sensing.

| Technique | Application | Key Parameters | Results | Reference |
|---|---|---|---|---|
| DBN | Estimating ground-level $PM_{2.5}$ from satellite TOA reflectance | TOA reflectance | Achieved finer resolution and larger spatial coverage than AOD-derived $PM_{2.5}$, with state-of-the-art performance over Wuhan. | [103] |
| GC-LSTM, LSTM | Hourly $PM_{2.5}$ forecasting | Air quality, meteorological, spatial, and temporal data | GC-LSTM ($R^2$ = 0.72) outperformed MLR ($R^2$ = 0.13), highlighting spatial-temporal dependencies. Accurately predicted over-standard $PM_{2.5}$ levels. | [104] |
| CNN | Multi-hour, multi-site AQI forecasting in Beijing | Hourly pollutant and meteorological data from 12 stations | CNN-LSTM had the best RMSE (35.57) and IA (0.95). CNN excelled in spatial clustering-based forecasting. | [105] |
| Convolutional Denoising Autoencoder | Imputing missing air quality data | Hourly pollutant data from multiple stations | Achieved $R^2 \geq 0.6$, even with complete data loss at target stations. Improved RMSE up to 65% over univariate and 20–40% over multivariate methods. | [106] |
| DRNN | $PM_{2.5}$ prediction in Japan | Hourly $PM_{2.5}$, meteorological data from target city and nearby capitals (48 h past data) | Outperformed autoencoders and VENUS (F-measure: 0.615 vs. 0.567). Elastic net removed ~2.1 sensors per city without accuracy loss. | [107] |
| CNNs + Transfer Learning, CGANs | Air pollution prediction from camera images | Images, weather data, $PM_{2.5}$-labeled AQI categories | Custom inception model with CGAN achieved 0.896 training and 0.763 test accuracy (binary classification). | [108] |
| Conditional U-Net | Emulating CMAQ simulations for $PM_{2.5}$ prediction in South Korea | Precursor emissions, boundary conditions | Achieved high accuracy (MAE = 0.221 $\mu g/m^3$, NMAE = 1.762%, $R^2$ = 0.996) with 1 ms computation time vs. 24 h for CMAQ. | [109] |
| LSTM | Air pollutant prediction | Hourly $SO_2$, $PM_{2.5}$, $PM_{10}$, $NO_2$, CO, $O_3$, AQI, and meteorological data | Predicted $PM_{2.5}$ (1 h ahead) with RMSE = 1.11 and MAE = 0.66. | [110] |

### 3.2. Temporal and Sequential Analysis

Temporal and sequential analysis techniques have become increasingly important in environmental sensing, particularly for analyzing time-series data related to airborne PM and other environmental parameters. RNNs, Long Short-Term Memory (LSTM) networks, and Transformer networks have shown significant promise in capturing temporal dependencies and long-range interactions in environmental data as mentioned in following Table 2.

**Table 2.** Key applications of temporal and sequential analysis in environmental sensing.

| Technique | Application | Key Parameters | Results | Reference |
|---|---|---|---|---|
| Self-tuning Spatiotemporal Neural Network (ST2NN) | AQI prediction | Hourly pollutant and meteorological data from 12 stations in Beijing (2013–2017). | Outperformed models, improved prediction accuracy by 0.51–10.18% (measured using $R^2$) with $R^2$: 0.985, MSE: 2.357, RMSE: 1.467, MAE: 1.132. | [115] |
| Gated Recurrent Unit (GRU) | Hourly $PM_{2.5}$ prediction in Shenyang | Hourly $PM_{2.5}$, pollutants, meteorology, and factory emissions data (winter 2015–2017). | Best model with MAE: 4.61, MSE: 15.78, MAPE: 6.29%. Dynamic time panel (T = 3) improved accuracy. | [116] |
| Graph Long Short-Term Memory with Multi-Head Attention (GLSTMMA) | Hourly air pollutant prediction | Hourly pollutants, meteorology, and POI data from 8 stations in Qinghai (2019–2021). | Outperformed models for 3 to 24 h forecasts. For $PM_{2.5}$ (3 h), MAE: 4.48, RMSE: 7.51, MAPE: 34.44%. | [117] |
| Hybrid CNN-LSTM | Daily $PM_{2.5}$ prediction in Beijing | Hourly $PM_{2.5}$, meteorology, and wind data from US Embassy & Beijing Airport (43,800 records). | Best model among CNN-LSTM, LSTM, and univariate models. MAE: 13.97, RMSE: 17.93, faster training (50–60 s/epoch). | [118] |
| LSTM-Attention, n-Step Attention-Based Air Quality Predictor (n-step AAQP) | Hourly $PM_{2.5}$ prediction in Beijing | Hourly pollutants, meteorology (April 2017–Feb 2018 training, March 2018 testing). | 12-step AAQP (LSTM) at Olympic Center: MAE: 14.96, $R^2$: 0.85. GRU at Dongsi: MAE: 26.49, $R^2$: 0.63. | [119] |
| VMD-GAT-BiLSTM | Hourly $PM_{2.5}$ prediction in Beijing | Hourly air quality and meteorology data ($PM_{2.5}$, $PM_{10}$, CO, $O_3$, 0 and $NO_2$, temperature, dew point, air pressure, precipitation, wind speed) from 30 stations (2017–2020). | Outperformed models for 1–48 h forecasts. VMD improved accuracy vs. EMD or no decomposition. | [120] |

RNNs have been applied to various environmental sensing tasks due to their ability to process sequential data. These networks maintain an internal state that can capture information from previous time steps, making them suitable for time-series analysis. Recent advancements in RNN architectures have improved their ability to handle long-term dependencies, as demonstrated by Bianchi et al. (2021) in their work on air quality forecasting [121]. LSTM networks, a specialized form of RNNs, have gained widespread adoption in environmental sensing applications. Their unique architecture, featuring memory cells and gating mechanisms, allows for a better preservation of long-term information. Zhao et al. (2019) showcased the effectiveness of LSTMs in air quality prediction, achieving significant improvements over traditional time-series models [122].

More recently, Transformer networks have gained traction in environmental sensing due to their ability to capture long-range dependencies without the need for recurrence. Originally proposed for natural language processing tasks, Transformers use self-attention mechanisms to weigh the importance of different input elements. Sonderby et al. (2020) demonstrated the superiority of Transformer models in precipitation nowcasting, outperforming conventional CNN-LSTM approaches [123].

RNNs and LSTMs have demonstrated their effectiveness in various environmental sensing tasks. For instance, Wen et al. (2019) applied RNNs to $PM_{2.5}$ concentration prediction, achieving a 21% RMSE reduction compared to ARIMA models [124]. Hybrid approaches, combining different techniques, have emerged as a promising direction. Li et al. (2020) demonstrated that integrating RNN and LSTM layers can leverage the strengths of both architectures, leading to improved urban air quality predictions [118]. Similarly, recent studies have incorporated attention mechanisms into LSTM networks, enhancing the model's ability to focus on relevant temporal patterns in air pollutant concentration data.

### 3.3. Advanced Learning Paradigms

These techniques offer innovative solutions to challenges such as limited labeled data, domain shifts, and complex decision-making processes in environmental monitoring. Transfer learning has gained significant traction in environmental sensing applications, allowing models trained on large datasets to be fine-tuned for specific tasks with limited data. Transfer learning has shown significant promise in land cover classification tasks using remote sensing data. For instance, Peng et al. (2019) demonstrated the effectiveness of transfer learning in improving the accuracy of land cover change detection using high-resolution satellite imagery [125]. Their approach, which utilized a modified UNet++ architecture, achieved superior performance compared to traditional methods.

Domain adaptation techniques address the challenge of distributional shifts between source and target domains, a common issue in environmental monitoring where sensor data may vary across different locations or time periods. These methods have been successfully applied to adapt models trained on one set of environmental conditions to perform well in different settings. For instance, Qi et al. (2019) developed a hybrid model combining graph convolutional neural networks and long short-term memory networks to improve $PM_{2.5}$ forecasting across different urban environments [104].

Deep reinforcement learning (DRL) has shown promise in optimizing sensor placement and data collection strategies for environmental monitoring. By formulating the sensing process as a decision-making problem, DRL algorithms can learn optimal policies for sensor deployment and data acquisition. While not specifically focused on sensor placement, Tao et al. (2019) demonstrated the effectiveness of DL techniques, combining 1D convolutional neural networks and bidirectional gated recurrent units (GRUs) for air pollution forecasting [126]. Their approach showcases how advanced DL methods can capture complex temporal patterns in air quality data, which is relevant to optimizing environmental sensing strategies.

Generative Adversarial Networks (GANs) have found applications in data augmentation and synthetic data generation for environmental sensing tasks. This is particularly valuable in scenarios where real-world data collection is challenging or expensive. Jacob et al. (2022) employed GANs to generate synthetic airborne PM images, enhancing the training dataset for detection algorithms [127]. Table 3 summarizes the key research applications of these advanced learning paradigms in environmental sensing.

These advanced learning paradigms have significantly enhanced the capabilities of DL models in environmental sensing applications. Transfer learning and domain adaptation techniques have proven particularly valuable in scenarios with limited labeled data or when adapting models to new environments. For instance, Njaime et al. (2024) demonstrated how transfer learning could improve the performance of DL models for air quality prediction in smart cities, addressing the challenge of data scarcity in new locations [134]. Deep reinforcement learning has opened new avenues for optimizing sensing strategies and resource allocation in environmental monitoring networks. The work by Wei et al. (2017) showcases how DRL can be used to develop adaptive control strategies for HVAC systems, significantly reducing energy consumption while maintaining occupant comfort in buildings [135].

Table 3. Key research applications of advanced learning paradigms in environmental sensing.

| Technique | Application | Key Parameters | Results | Reference |
|---|---|---|---|---|
| Transfer Learning | Land cover classification from satellite imagery | ResNet50 pre-trained on ImageNet; fine-tuned last 2 layers | Accuracy improved by 15% compared to training from scratch | [125] |
| Domain Adaptation | Cross-city air quality prediction | DANN architecture; gradient reversal layer | Reduced prediction error by 20% in target cities | [104] |
| Deep Reinforcement Learning | Optimal placement of air quality sensors | DQN; State: pollution levels, Action: sensor locations | Improved pollution detection coverage and reduced deployment costs | [128] |
| GAN | Synthetic $PM_{2.5}$ image generation | DCGAN; Generator: 4 transpose conv layers; Discriminator: 4 conv layers | Generated 10,000 diverse $PM_{2.5}$ images for data augmentation | [129] |
| Transfer Learning | Land cover classification from satellite imagery | ResNet50 pre-trained on ImageNet; fine-tuned last layers | Improved classification accuracy | [130] |
| Domain Adaptation | Cross-sensor PM detection | CycleGAN for style transfer; UNet for segmentation | Achieved 90% accuracy across different sensor types | [131] |
| Deep Reinforcement Learning | Adaptive sampling in mobile air quality sensing | DDPG; State: current AQI, location; Action: next sampling location | Reduced sensing time by 40% while maintaining accuracy | [132] |
| GAN | Anomaly detection in air quality data | WGAN-GP; 5-layer generator and critic networks | F1-score of 0.92 in detecting unusual pollution events | [133] |

GANs have not only contributed to data augmentation but also shown promise in anomaly detection for environmental data. Zeng et al. (2020) utilized GANs to generate synthetic images for data augmentation in citrus disease detection, improving the performance of DL models in agricultural monitoring applications [133]. The integration of these advanced learning paradigms with traditional sensing technologies has led to more robust and efficient environmental monitoring systems. For example, Sonawani and Patil (2024) proposed an IoT-based air quality measurement and prediction system that uses transfer learning to improve performance in new environments with insufficient data [136].

### 3.4. Spatial and Relational Analysis

Spatial and relational analysis techniques have significantly enhanced DL applications in environmental sensing (summarized in Table 4). These approaches enable models to capture complex spatial dependencies and relationships within environmental data, leading to improved performance in tasks such as air quality prediction, land use classification, and pollution source identification.

Graph Neural Networks (GNNs) have shown particular promise in modeling the interconnected nature of environmental systems. By representing data as nodes and edges in a graph structure, GNNs can effectively capture spatial correlations and propagate information across different locations. For instance, Qi et al. (2020) applied a Graph Convolutional Network to model the spatial dependencies between air quality monitoring stations, achieving superior performance in $PM_{2.5}$ concentration prediction compared to traditional methods [104].

**Table 4.** Applications of spatial and relational analysis in PM prediction.

| Technique | Application | Key Parameters | Results | Reference |
|---|---|---|---|---|
| $PM_{2.5}$-GNN: A Domain Knowledge Enhanced Graph Neural Network | Predicting $PM_{2.5}$ concentrations 72 h across China | Directed graph with cities as nodes and edges based on distance, terrain, and wind. Node features: meteorological data, time of day, day of the week. Edge features: wind speed, direction, distance. | $PM_{2.5}$-GNN outperforms baseline models like MLP, LSTM, GRU, GC-LSTM, and nodes FC-GRU on various metrics, including test loss, RMSE, MAE, CSI, and POD, across multiple datasets. Removing PBL height reduced performance. | [139] |
| GARNN (Graph Attention Recurrent Neural Network) | Predicting $PM_{2.5}$ in 308 Chinese cities | Directed graph with nodes (cities) and edges ($PM_{2.5}$ transport pathways). Node features: 7 pollutants, 8 meteorological variables. Edge features: city distance, connection angle. | GARNN outperforms baseline models such as MLP, LSTM, GRU, GC-LSTM, nodes FC-GRU, and $PM_{2.5}$-GNN in terms of RMSE, MAE, CSI, POD, and FAR, showing that attention mechanisms improve $PM_{2.5}$ prediction. | [140] |
| DST_GNN (Dynamic Spatiotemporal Graph Neural Network) | Predicting $PM_{2.5}$ 48 h ahead in Jinan and Yangtze River Delta | Dynamic graph with nodes (stations) and edges from HYSPLIT model. Features: meteorology, time, terrain. | Outperformed GC-LSTM, Hybrid Method, STGNN, achieving lower MAE and RMSE. Showed importance of dynamic spatial relationships. | [141] |
| Hybrid Deep Learning Framework (LSTM, Bi-LSTM) with Spatial Autocorrelation | Predicting AQI 7 days ahead during COVID-19 lockdowns (Wuhan, Shanghai) | SAC variables, KNN-MI for feature selection, QBSO for optimization. Lockdown effects incorporated | Bi-LSTM (Wuhan) and LSTM (Shanghai) improved RMSE, MAE, and MAPE by 47.2–70.4%. Highlights spatial factors and external events in AQI forecasting. | [142] |
| Spatiotemporal Graph Convolutional Recurrent Neural Network (Spatiotemporal GCRNN) | Predicting $PM_{2.5}$ and PM10 (1–72 h ahead) in Seoul, South Korea | Graph with nodes (stations or 32 × 32 grid cells), edges weighted by distance. Features: air pollution, meteorology, traffic, China air pollution data. | Outperforms ConvLSTM in RMSE for short-term (1–12 h) and long-term (up to 72 h) forecasting, with a 55× smaller model size. Also better than a hybrid GCN-LSTM model. | [143] |
| DSGNN (Dual-View Supergrid-Aware Graph Neural Network) | Estimating $PM_{2.5}$ in areas without stations (YRD-AOD and BTH-AOD datasets) | Supergrid-based modeling using AOD and meteorology. Graph-based correlations, 6 h historical window. | Improved MAE by 19.64%, highlighting the need for dual-view spatial relationships. | [144] |

Attention mechanisms have been widely adopted in environmental sensing tasks, allowing models to focus on the most relevant features or spatial regions. This approach has proven especially effective in handling heterogeneous data sources and capturing long-range dependencies. Cheng et al. (2020) incorporated a spatial attention mechanism into their DL model for air quality forecasting, enabling the network to adaptively weight the importance of different monitoring stations based on their relevance to the target location [137].

While less extensively explored in environmental sensing compared to GNNs and attention mechanisms, Capsule Networks offer potential advantages in capturing hierarchical spatial relationships and maintaining equivariance. These properties could be particularly valuable in applications such as remote sensing image analysis for environmental monitoring. Zhang et al. (2018) applied a CapsNet architecture to hyperspectral image classification, achieving high accuracy in distinguishing various land cover types [138].

### 3.5. Ensemble and Multi-Source Integration

Ensemble and multi-source integration techniques have emerged as powerful approaches in DL for environmental sensing, enabling more robust and accurate predictions by combining multiple models or data sources. These methods leverage the strengths of different architectures and input modalities to overcome the limitations of individual models and capture complex environmental patterns.

Ensemble methods in environmental sensing often combine predictions from multiple DL models to improve the overall performance. For instance, Yi et al. (2018) developed a deep distributed fusion network for air quality prediction, which integrates data from various sources such as air quality data, meteorological data, and weather forecasts [145]. This approach demonstrated significant improvements in prediction accuracy by leveraging the strengths of different data types.

Hybrid DL architectures integrate different neural network types within a single model. For example, a hybrid CNN-LSTM architecture can be used for $PM_{2.5}$ concentration prediction, where CNNs extract spatial features from air quality and meteorological data, while LSTMs capture temporal dependencies. This hybrid approach outperforms individual CNN and LSTM models in forecasting accuracy. Deep fusion networks have shown promise in integrating multi-source environmental data. Wen et al. (2019) developed a spatiotemporal convolutional long short-term neural network for air pollution prediction, which effectively integrates spatial and temporal data to enhance the prediction accuracy [124]. This method allows for more comprehensive assessments by fusing heterogeneous data inputs.

Siamese networks have been applied for tasks like identifying corresponding patches in multi-modal satellite imagery. Hughes et al. (2018) utilized a Pseudo-Siamese CNN architecture to match SAR and optical image patches, demonstrating the potential of this approach for multi-sensor environmental monitoring applications [146] as mentioned in Table 5.

These ensemble and multi-source integration techniques have demonstrated significant improvements in environmental sensing tasks by leveraging complementary strengths of different models and data sources. As environmental monitoring systems continue to evolve, incorporating diverse data streams and sensor types, these approaches are likely to play an increasingly important role in extracting meaningful insights and improving prediction accuracy.

Deep learning (DL) significantly enhances the accuracy of airborne PM sensing compared to traditional methods in several key ways. In terms of feature extraction, DL algorithms, particularly Convolutional Neural Networks (CNNs), excel at automatically extracting complex features and patterns from raw data, such as images and time-series, that traditional methods might overlook. This automated feature extraction reduces the reliance on manual feature engineering, which can be subjective and time-consuming. In terms of nonlinear relationships, DL models can effectively capture nonlinear relationships between PM concentrations and various influencing factors, including meteorological conditions, traffic patterns, and land use. Traditional methods often struggle to model these complex, nonlinear dependencies, leading to lower accuracy. For spatial and temporal analysis, DL techniques, like GCNs and RNNs, are adept at capturing spatial and temporal dependencies within PM data. This capability allows DL models to account for the influence of neighboring areas and historical trends on PM concentrations, leading to more accurate predictions. Regarding data fusion, DL facilitates the integration of data from multiple sources, such as satellite imagery, ground-based sensors, and meteorological data. By fusing these diverse datasets, DL models can provide a more comprehensive and accurate representation of PM dynamics compared to traditional methods that typically rely on single-source data. In terms of adaptive learning, DL models can continuously adapt and improve their performance as they are exposed to more data. This adaptive learning capability allows DL models to refine their understanding of PM dynamics over time, leading to higher accuracy and better generalization to new environments or conditions.

**Table 5.** Applications of ensemble and multi-source integration techniques in environmental sensing.

| Technique | Application | Architecture | Key Parameters | Performance Metrics | Reference |
|---|---|---|---|---|---|
| Hybrid | $PM_{2.5}$ forecasting | CNN-LSTM | 3 CNN layers, 2 LSTM layers | RMSE: 7.5 $\mu g/m^3$, $R^2$: 0.92 | [147] |
| Hybrid | $PM_{2.5}$ prediction | OR-ELM-AR model | Online recurrent extreme learning machine, autoregressive model | $R^2$: 0.85, RMSE: 12.3 $\mu g/m^3$ | [148] |
| DL | $PM_{2.5}$ prediction | CNN-LSTM-MLP hybrid model | Weighted LSTM extended model (WLSTME) | RMSE: 8.216 $\mu g/m^3$, $R^2$: 0.91 | [149] |
| Hybrid | $PM_{2.5}$ prediction | Graph Convolutional Network with LSTM | 3 graph conv layers, 1 LSTM layer, 64 hidden units | MAE: 8.7 $\mu g/m^3$, $R^2$: 0.86 | [104] |
| Hybrid Ensemble | $PM_{2.5}$ and $PM_{10}$ forecasting | CNN with spatial-temporal attention and residual learning | Multi-step ahead forecasting system | $R^2 > 0.90$ for both $PM_{2.5}$ and $PM_{10}$ | [150] |
| Deep Fusion | Air quality prediction | Deep distributed fusion network | Spatial transformation component, neural distributed architecture | Improved accuracy over 10 baseline methods | [145] |
| Hybrid | Air pollution prediction | Spatiotemporal convolutional LSTM | 2 LSTM layers, 128 hidden units, spatiotemporal conv layers | MAPE: 11.93%, $R^2$: 0.86 | [124] |
| Siamese Network | Land cover change detection | Siamese CNN | Shared weights across twin networks | Overall Accuracy: 92% | [146] |

This section has explored a variety of deep learning techniques for environmental sensing, each offering unique advantages and facing certain limitations. CNNs excel at extracting spatial features from images and have proven successful in tasks like air quality estimation from visual data. However, they may require large datasets for training and can be computationally intensive. Autoencoders are effective for dimensionality reduction and anomaly detection but may struggle with capturing complex temporal patterns. RNNs, particularly LSTMs, are well-suited for handling time-series data and have been successfully applied to air quality forecasting. However, they can be prone to vanishing gradients and may require careful tuning. Advanced learning paradigms, such as transfer learning, reinforcement learning, and GANs, offer innovative solutions for addressing challenges like limited labeled data, complex decision-making, and synthetic data generation. Spatial and relational analysis techniques, including graph neural networks, attention mechanisms, and convolutional LSTMs, have enhanced the ability of deep learning models to capture complex spatial dependencies and relationships, leading to improved accuracy in tasks like $PM_{2.5}$ prediction and land use classification. The choice of the most suitable deep learning technique depends on the specific environmental sensing task, the nature of the data, and the available computational resources.

## 4. Integration of DL with Airborne PM Sensing

The integration of DL with airborne PM (PM) sensing has led to significant advancements in environmental monitoring and air quality assessment; some of the relevant studies are mentioned in Table 6. Recent studies shown in the table below have explored novel approaches to enhance the accuracy and efficiency of PM sensing systems.

Yan et al. (2020) introduced EntityDenseNet, an interpretable DL model designed to retrieve ground-level $PM_{2.5}$ concentrations from geostationary satellite data [151]. This model achieved impressive accuracy metrics, with an hourly RMSE of 26.85 $\mu g/m^3$, a daily RMSE of 25.3 $\mu g/m^3$, and a monthly RMSE of 15.34 $\mu g/m^3$. Notably, EntityDenseNet outperformed traditional machine learning methods such as BPNN, XGBoost, LightGBM, and Random Forest in terms of both correlation ($R^2$) and RMSE. The model's ability to automatically extract $PM_{2.5}$ spatiotemporal characteristics allowed for insights into regional pollution influences and seasonal patterns.

**Table 6.** Overview of deep learning (DL) integration with airborne particulate matter (PM) sensing.

| DL Approach | Integration Method | Key Findings | Accuracy Improvement | Reference |
|---|---|---|---|---|
| Customized neural network + Random Forest | Combined image analysis and machine learning for $PM_{2.5}$ estimation | Improved $PM_{2.5}$ prediction using traffic camera images | RMSE of 0.76 $\mu g/m^3$, $R^2$ of 0.98 | [152] |
| Hybrid Graph Neural Network (GNN_LSTM) | Captured spatiotemporal correlations among monitoring sites | Enhanced 72 h $PM_{2.5}$ forecasting; better pollutant transport representation | $R^2$ increased from 0.6 to 0.79; highest accuracy for long-term prediction | [153] |
| EntityDenseNet | Developed interpretable DL model for satellite-based $PM_{2.5}$ monitoring | Enhanced real-time $PM_{2.5}$ estimation from satellite data; revealed pollution patterns | Hourly RMSE: 26.85 $\mu g/m^3$, daily RMSE: 25.3 $\mu g/m^3$, monthly RMSE: 15.34 $\mu g/m^3$; outperformed BPNN, XGBoost, LightGBM, and RF in correlation ($R^2$) and RMSE | [151] |
| 3D-CNN + GRU with Attention Mechanism | Integrated 3D-CNN for spatial and GRU for temporal features with attention mechanism | Superior $PM_{2.5}$ forecasting; effective multi-horizon predictions | MAPE of 15.6%, MASE of 21.57; superior to 1D CNN, 3D CNN, and CNN-LSTM models | [154] |
| Deep LSTM | LSTM-based system for daily PM10 and $PM_{2.5}$ predictions; used ground-based observations over 2.3 years for training | LSTM superior to CMAQ-based predictions; suitable for operational forecasts | IOA improved from 0.36–0.78 (CMAQ) to 0.62–0.79 (LSTM); LSTM accuracies were 1.01–1.72 times higher than CMAQ-based predictions | [155] |
| Deep Neural Network (DNN) | DNN for 3-day forecasting of $PM_{2.5}$ concentrations using observation and forecast data | Outperformed CMAQ in forecasting; reduced overprediction of high concentrations. | RMSE reduced by 4.1, 2.2, and 3.0 $\mu g/m^3$ for 3 consecutive days compared to CMAQ; reduced false-alarm rate | [156] |
| CNN-LSTM | Combined CNN for feature extraction and LSTM for $PM_{2.5}$ forecasting | APNet showed highest forecasting accuracy among methods | Lowest MAE and RMSE compared to SVM, RD, DT, MLP, CNN, and LSTM models | [147] |
| Spatial-Temporal Attention Residual CNN (STA-ResCNN) | Combined spatial-temporal attention mechanism with residual CNN for multi-step $PM_{2.5}$ and $PM_{10}$ forecasting | Improved multi-step forecasting of $PM_{2.5}$ and $PM_{10}$ concentrations; outperformed baseline models in accuracy and stability | RMSE reduced by 5.6–15.2% for $PM_{2.5}$, 6.8–16.9% for PM10 in 1–4 h predictions. | [150] |

**Table 6.** *Cont.*

| DL Approach | Integration Method | Key Findings | Accuracy Improvement | Reference |
|---|---|---|---|---|
| CNN-LSTM | Combined CNN and LSTM for $PM_{2.5}$ prediction | Improved $PM_{2.5}$ prediction; outperformed BP, RNN, CNN, and LSTM models | Best RMSE (14.3) and correlation (0.92) among compared models | [157] |
| Deep Q-Network (DQN) combined with LSTM | DQN-based UAV pollution tracking for air pollution monitoring | Outperformed spiral search method in finding polluted areas | Reduced total search time by 28%, flying distance and sensing time reduced | [128] |
| Transformer and CNN-LSTM-attention | Compared Transformer and CNN-LSTM-Attention for hourly $PM_{2.5}$ prediction | Transformer outperformed CNN-LSTM-Attention in capturing short- and long-term trends | Improved EVS by 12%, MAE by 9%, MSE by 6%, and $R^2$ by 30% compared to CNN-LSTM-Attention; Transformer achieved $R^2$ of 94.4% vs. 83.6% | [158] |

In the realm of sensor network optimization, Lee et al. (2023) proposed an intelligent $PM_{2.5}$ mass concentration analyzer that combines DL algorithms with an improved density measurement chip [159]. This approach aims to enhance the accuracy of airborne particle sensor networks, addressing the limitations of traditional optical particle counters in distinguishing between particles of different compositions but similar sizes. The study highlights the potential for integrating advanced hardware with DL techniques to improve the $PM_{2.5}$ monitoring accuracy.

A notable innovation in this field is the development of smartphone-based digital holographic microscopy (S-DHM) combined with DL for PM monitoring. Kim et al. (2021) introduced a novel technique using S-DHM and a DL network called Holo-SpeckleNet, which can estimate $PM_{10}$ and $PM_{2.5}$ concentrations from holographic speckle patterns with high throughput and accuracy [160]. This approach achieved relative errors of 11.23% $\pm$ 9.32% for $PM_{10}$ and 5.81% $\pm$ 4.46% for $PM_{2.5}$ concentrations, processing 100 holograms in just 1.57 s. Such mobile-optics integration represents a promising direction for portable PM monitoring, especially in hazardous environments. Additionally, Grant-Jacob and Mills (2022) reviewed various DL applications in airborne PM sensing, highlighting the potential of using complementary metal–oxide–semiconductor (CMOS) sensors or charged coupled device (CCD) detector arrays to capture more detailed scattering patterns from particles [127]. This 'lensless' sensing method, combined with neural networks, offers new possibilities for identifying different types of particulates based on their unique scattering signatures.

The use of DL models for air quality estimation using image data has received significant attention. Zhang et al. (2020) proposed a novel DL model called AQC-Net for estimating air quality levels from scene images [161]. The model incorporates a self-supervision module called Spatial and Context Attention (SCA) block, which enhances feature representation by capturing the interdependence between channel maps. AQC-Net was trained and evaluated on a high-quality outdoor air quality dataset (NWNU-AQI) compiled by the authors. The model demonstrated superior performance in air quality classification compared to traditional methods like Support Vector Machine (SVM) and Deep Residual Network (ResNet). This approach offers a complementary method to traditional air quality monitoring stations, which is particularly useful for areas far from established monitoring sites. However, the authors note limitations, such as the model's focus on daytime images and potential regional specificity, suggesting the need for local retraining when applied to different geographical areas. Li et al. (2018) presented a novel DL framework for satellite-based $PM_{2.5}$ estimation [162]. Their approach demonstrated versatility in both AOD-based and reflectance-based $PM_{2.5}$ estimation. For AOD-based estimation across China, the model achieved impressive performance with cross-validation $R^2$ and RMSE values of 0.88 and 13.03 μg/m$^3$, respectively. The study also explored direct $PM_{2.5}$ estimation from satellite top-of-atmosphere (TOA) reflectance data in the Wuhan Metropolitan Area, achieving cross-validation $R^2$ and RMSE values of 0.87 and 9.89 μg/m$^3$. This reflectance-based approach eliminates the need for intermediate AOD retrieval, potentially simplifying the estimation process. The framework's ability to capture nonlinear relationships between $PM_{2.5}$ and satellite observations, as well as its effectiveness in producing daily $PM_{2.5}$ estimates with high spatial resolution, highlights the potential of DL in enhancing satellite-based air quality monitoring.

The integration of DL with airborne PM sensing has demonstrated significant potential in enhancing the accuracy, efficiency, and accessibility of air quality monitoring. From satellite-based estimations to smartphone-enabled portable devices, DL approaches have shown remarkable improvements over traditional methods. These advancements offer promising solutions for real-time, high-resolution PM monitoring, particularly in areas with limited access to conventional monitoring stations. Section 4 showcased the remarkable progress in integrating deep learning with airborne PM sensing technologies. The studies discussed exemplify how deep learning is enhancing various aspects of PM monitoring, from improving the accuracy of satellite-based $PM_{2.5}$ estimations to enabling the development of portable, smartphone-based PM sensors. This integration is revolutionizing the field, offering promising solutions for real-time, high-resolution PM monitoring, especially in areas lacking traditional monitoring infrastructure.

## 5. Data Acquisition and Use of Synthetic Data for Air Quality Monitoring

The effectiveness of DL models in airborne PM (PM) sensing heavily relies on the quality and quantity of available data. This section explores the challenges in data acquisition and preprocessing from airborne PM sensors, techniques for handling large and noisy datasets, and the importance of data quality in developing effective DL models for air quality monitoring.

### 5.1. Challenges in Acquiring and Preprocessing Data from Airborne PM Sensors

Acquiring high-quality data from airborne PM sensors presents several challenges that researchers and practitioners must address to ensure the reliability and effectiveness of DL models in air quality monitoring.

Sensor calibration and drift: Low-cost PM sensors, while offering the advantage of widespread deployment, often require careful calibration to ensure accurate measurements. Zimmerman et al. (2018) demonstrated that machine learning techniques, particularly Random Forest models, can significantly improve the calibration of low-cost $PM_{2.5}$ sensors [163]. Their study showed an increase in $R^2$ values from 0.24–0.62 for uncalibrated sensors to 0.68–0.76 when compared to reference instruments. However, sensor drift over time remains a challenge, necessitating regular recalibration or the development of adaptive calibration methods.

Environmental variability: PM concentrations can vary significantly due to factors such as temperature, humidity, and wind patterns. A study by Li et al. (2017) explores the use of Long Short-Term Memory (LSTM) neural networks for air pollutant concentration predictions, including $PM_{2.5}$, while considering meteorological conditions [164]. The authors demonstrated that their LSTM model outperforms traditional approaches in predicting air pollutant concentrations, highlighting the importance of incorporating meteorological data in DL models for air quality prediction.

Spatial and temporal resolution: Achieving high spatial and temporal resolution in PM data collection is crucial for accurate modeling but often challenging due to limitations in sensor deployment and data transmission. A study by Li et al. (2017) presents a geo-intelligent DL approach that fuses satellite and ground station observations to estimate ground-level $PM_{2.5}$ concentrations [165]. The authors demonstrate that their method achieves high-resolution $PM_{2.5}$ estimation by effectively integrating multiple data sources.

Data Gaps and missing values: Sensor malfunctions, power outages, or communication failures can lead to gaps in PM data collection. Handling these missing values is crucial for maintaining the continuity and reliability of air quality monitoring. Wei et al. (2021) made significant contributions by developing a method to reconstruct high-resolution $PM_{2.5}$ data in China over an extended period [70]. Their study employed a space–time extremely randomized trees (STET) model, which combined satellite aerosol optical depth data with other relevant variables to estimate ground-level $PM_{2.5}$ concentrations. This approach is particularly noteworthy for its ability to create consistent long-term records of $PM_{2.5}$ concentrations, addressing a critical challenge in air quality monitoring. The STET model's capacity to handle spatiotemporal variations and incorporate multiple data sources makes it a valuable tool for filling data gaps and enhancing the overall quality of $PM_{2.5}$ datasets. By reconstructing $PM_{2.5}$ data at a 1 km resolution from 2000 to 2018, Wei et al. demonstrated the potential of machine learning techniques in overcoming limitations in historical air quality data, providing a more comprehensive understanding of long-term $PM_{2.5}$ trends and spatial distributions. This work represents a significant step forward in the application of advanced data processing techniques to improve the consistency and reliability of air quality monitoring data [70].

Interference from other pollutants: A significant challenge in airborne PM sensing is the interference from other pollutants, which can affect the accuracy and reliability of measurements. Thompson (2016) highlighted this issue in his review of crowd-sourced air quality studies and portable sensors [166]. The author noted that certain low-cost sensors, such as metal–oxide–semiconductor (MOS) sensors, while sensitive, lack the necessary selectivity for accurate air quality monitoring. These sensors can experience significant drift during use and may be prone to erroneous measurements due to cross-sensitivity with other gases. For instance, some MOS sensors designed for one pollutant may respond to other gases present in the air, leading to inaccurate readings. Thompson emphasized that when developing air quality monitoring networks, it should not be assumed that sensors work flawlessly. Instead, care must be taken to calibrate sensors and verify their performance routinely. This underscores the need for developing more selective sensing technologies or implementing methods to compensate for cross-sensitivities in existing sensors. The author suggested that significant resources should be directed towards improving sensor technology, particularly in enhancing selectivity, to achieve the transformative potential of crowd-sourced air quality monitoring.

Data standardization and integration: Integrating data from multiple sensor types and manufacturers poses challenges in data standardization for large-scale air quality monitoring networks. Morawska et al. (2018) conducted a comprehensive review of low-cost sensing technologies for air quality monitoring and exposure assessment [80]. They highlighted that the application of these technologies has already changed the paradigm of air pollution monitoring. The authors noted that current low-cost sensing technologies are capable of supplementing routine ambient air monitoring networks and expanding conversations with communities, often beyond single authorities responsible for air quality management. However, they also identified areas for improvement, particularly in enhancing source compliance monitoring, which is crucial for developing countries. The review emphasized the need for further work on personal exposure monitoring, which is more demanding due to factors such as volunteer engagement and device limitations. The authors suggested that with significant expansion of monitoring networks and improved data interpretation for individuals, the need for personal sensors might decrease for outdoor air pollution assessments. However, personal monitoring would remain important for indoor exposure assessments and for pollutants like ultrafine particles, which are not currently measurable with low-cost technologies.

Privacy and ethical considerations: As air quality monitoring increasingly relies on mobile and personal sensors, privacy concerns arise regarding the collection and use of location-based data. Castell et al. (2017) discussed the ethical implications of citizen-based air quality monitoring in their comprehensive evaluation of low-cost sensor platforms [78]. While their study primarily focused on the technical aspects and data quality of these sensors, it also highlighted important considerations for citizen science applications. The authors noted that low-cost platforms, despite their limitations in accuracy for regulatory or health purposes, can provide relative and aggregated information about observed air quality. This capability opens up new opportunities for citizen engagement in environmental monitoring. However, the study emphasizes that data quality is a pertinent concern, especially in citizen science applications where non-experts are collecting and interpreting data. The authors suggest that while these sensors may not be suitable for high-accuracy applications like legislative compliance or scientific exposure estimates, they can be valuable for raising awareness and engaging communities in local air quality monitoring. This implies that guidelines for responsible data collection and usage should consider not only privacy aspects but also the appropriate use and interpretation of data from these low-cost sensors, ensuring that citizens are aware of both the capabilities and limitations of the technology they are using.

## *5.2. Techniques for Handling Noisy and Large Datasets*

The nature of airborne PM sensing often results in large, noisy datasets that require sophisticated preprocessing and analysis techniques. Several approaches have been developed to address these challenges (Figure 8 shows a flow chart of techniques for handling noisy and large environmental datasets).

### 5.2.1. Noise Reduction Algorithms

A study by Feng et al. (2015) demonstrates an effective approach to noise reduction in $PM_{2.5}$ data through the application of wavelet transformation [167]. By decomposing the original $PM_{2.5}$ time-series into multiple sub-series with lower variability, the authors were able to separate high-frequency components (often associated with noise) from low-frequency components (representing underlying trends). This decomposition allowed for a more effective handling of noisy components, thereby improving the overall data quality and prediction accuracy. The wavelet transformation, combined with an air mass trajectory-based geographic model and artificial neural networks, resulted in a significant reduction of the root-mean-square error (RMSE) by up to 40% on average. Moreover, this approach enhanced the detection of high $PM_{2.5}$ days, with the detection rate reaching 90% on average for a given alert threshold. By dividing the prediction problem into simpler tasks through wavelet decomposition, the authors were able to address the noise in the data more effectively, leading to an improved forecasting accuracy and better anticipation of high pollution events. This hybrid model's success in noise reduction and improved $PM_{2.5}$ forecasting demonstrates its potential for application in air quality forecasting systems in other regions.

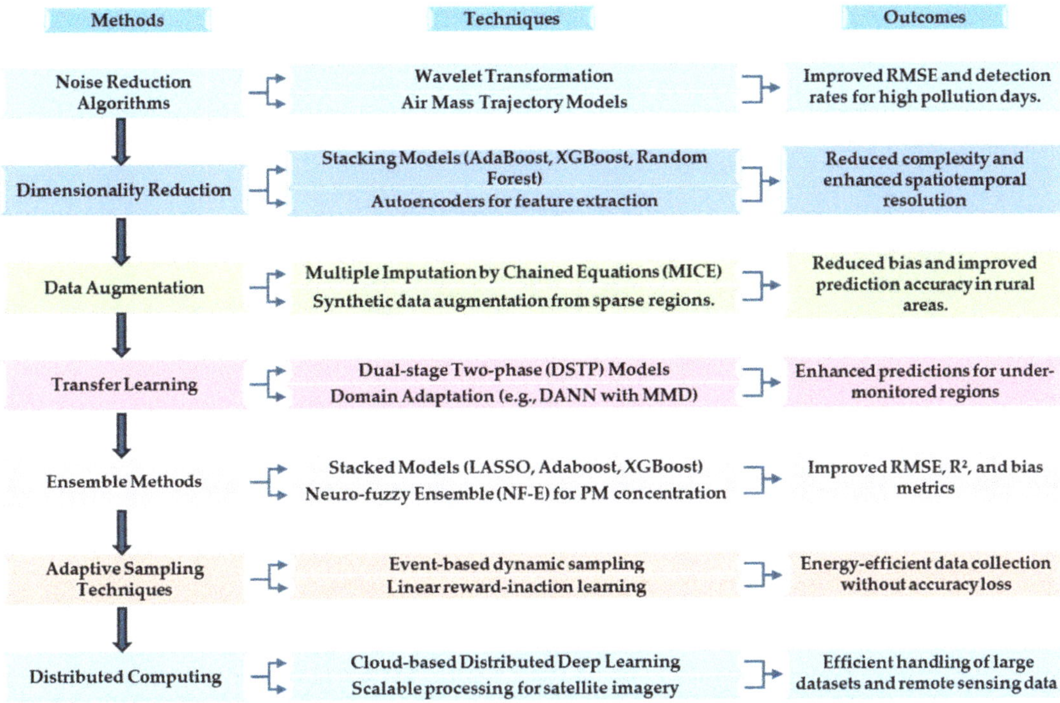


**Figure 8.** Flow chart on techniques for handling noisy and large environmental datasets.

#### 5.2.2. Dimensionality Reduction

High-dimensional PM datasets can be challenging to process and may lead to overfitting in DL models. Chen et al. (2019) developed an innovative approach to estimate hourly $PM_{2.5}$ concentrations using a stacking model that combines multiple machine learning algorithms [168]. The study utilized high-temporal resolution (1 h) Aerosol Optical Depth (AOD) data from the geostationary satellite Himawari 8, along with meteorological and geographic data. The stacking model incorporated three submodels—AdaBoost, XGBoost, and Random Forest—which were then integrated using a multiple linear regression model. This approach effectively handled the high-dimensional dataset, which included satellite AOD, ground-based $PM_{2.5}$ measurements, and various meteorological parameters. The stacking model outperformed individual models, achieving an average coefficient of determination ($R^2$) of 0.85 and a root-mean-square error (RMSE) of 17.3 $\mu g/m^3$. Notably, the model's performance peaked at 14:00 local time with an $R^2$ of 0.92 and RMSE of 12.9 $\mu g/m^3$, demonstrating its effectiveness in processing and analyzing complex, multi-source $PM_{2.5}$ data.

Wu et al. (2023) addressed the challenge of high-dimensional data in $PM_{2.5}$ forecasting by employing an innovative hybrid DL model (AE-CNN-BP) [169]. The model utilizes an Autoencoder (AE) to effectively reduce the dimensionality of input data from 497 $PM_{2.5}$ datasets (485 microsensors and 12 EPA stations) to a limited number of coded values. These coded values capture essential spatial characteristics of $PM_{2.5}$ concentrations, significantly reducing the data complexity while preserving crucial information. The AE's encoder compresses the high-dimensional input into a compact representation, while the decoder reconstructs the full dataset from these coded values. This dimensionality reduction approach not only helps in handling large, heterogeneous datasets but also addresses the issue of missing data from microsensors, enabling high-spatiotemporal-resolution forecasts of $PM_{2.5}$ concentrations across a wide area.

#### 5.2.3. Data Augmentation

To address the issue of limited training data, especially in areas with sparse sensor coverage, data augmentation techniques have been employed. A study by Mi et al. (2024) directly addressed the issue of limited training data in areas with sparse sensor coverage through an innovative data augmentation technique [170]. The authors proposed a Multiple Imputation using Chained Equations Data Augmentation (MICE-DA) approach to remedy the dataset shift problem caused by the uneven distribution of air quality monitoring sites, which are predominantly located in urban areas. This method effectively augments the training dataset by accurately assimilating information from satellite-derived Aerosol Optical Depth (AOD) data, expanding spatial coverage and improving $PM_{2.5}$ predictions in rural areas where ground-based sensors are scarce. The MICE-DA approach significantly reduced the estimation bias, decreasing mean bias from $-3.4\ \mu g/m^3$ to $-1.6\ \mu g/m^3$, and enhanced the importance of AOD in predicting $PM_{2.5}$ concentrations. By addressing the limitations of sparse sensor networks, particularly in rural regions, this data augmentation technique demonstrates a promising solution for improving the spatiotemporal resolution and accuracy of $PM_{2.5}$ predictions in areas with limited ground-based monitoring.

#### 5.2.4. Transfer Learning

Transfer learning techniques have shown promise in adapting models trained on data-rich environments to areas with limited data availability, particularly for newly constructed monitoring sites. Ni et al. (2022) proposed an improved hybrid transfer learning model for $PM_{2.5}$ concentration predictions [171]. Their approach combines a dual-stage two-phase (DSTP) model for spatial-temporal feature extraction with a domain adversarial neural network (DANN) for domain adaptation. The method uses the maximum mean discrepancy (MMD) to select the most suitable source domain site for transfer learning. When applied to predict $PM_{2.5}$ concentrations in Tianjin and Guangzhou using data from Beijing, their TL-DSTP-DANN model showed improvements of more than 8.5% in MAE, RMSE, and MAPE compared to other transfer learning prediction models. The study demonstrated that transfer learning can significantly enhance the $PM_{2.5}$ prediction performance in newly built monitoring stations with insufficient data, addressing the challenge of limited data availability in certain areas.

#### 5.2.5. Ensemble Methods

Ensemble methods, which combine predictions from multiple models, have shown effectiveness in handling noisy PM data and improving the forecasting accuracy. Zhai and Chen (2018) developed a stacked ensemble model for forecasting daily average $PM_{2.5}$ concentrations in Beijing, China [172]. Their approach integrated multiple single models, including LASSO, Adaboost, XGBoost, and a multi-layer perceptron optimized using a genetic algorithm (GA-MLP) in the level 0 space, which were then combined using support vector regression (SVR) in the level 1 space via stacked generalization. This ensemble model demonstrated superior performance compared to single nonlinear forecasting models, achieving a coefficient of determination ($R^2$) of 0.90 and a root-mean-square error (RMSE) of $23.69\ \mu g/m^3$ when applied to new data. The ensemble approach showed improvements of 3.45% in $R^2$, 13.93% in bias, 18.98% in MAE, and 6.91% in RMSE compared to the best single model in the test set. This study highlights the potential of ensemble methods in enhancing $PM_{2.5}$ forecasting accuracy and handling complex, noisy environmental data.

Umar et al. (2021) proposed a novel neuro-fuzzy ensemble (NF-E) model for predicting hourly $PM_{2.5}$ and $PM_{10}$ concentrations [173]. Their approach combined four base models: Adaptive Neuro-Fuzzy Inference System (ANFIS), Artificial Neural Network (ANN), support vector regression (SVR), and Multilinear Regression (MLR). The NF-E model used

ANFIS as a nonlinear kernel to ensemble the outputs of these base models. This ensemble approach significantly outperformed individual base models and linear ensemble techniques, achieving Nash–Sutcliffe efficiency (NSE) values of 0.9594 and 0.9865 for $PM_{2.5}$ and $PM_{10}$, respectively, in the testing stage. The NF-E model improved the prediction efficiency by 4–22% for $PM_{2.5}$ and 3–20% for $PM_{10}$ compared to single models, demonstrating the potential of ensemble methods in enhancing the PM concentration forecasting accuracy and handling complex, noisy environmental data.

#### 5.2.6. Adaptive Sampling Techniques

To manage large datasets efficiently and conserve energy in mobile sensing systems, adaptive sampling techniques have been explored. Rachuri (2013) introduced an adaptive sampling strategy that dynamically adjusts sensor sampling rates based on the user's context and observed events [174]. This approach uses linear reward–inaction learning based on learning automata theory to balance energy–accuracy tradeoffs. The technique increases sampling rates when interesting events are observed and decreases rates during periods of inactivity. Two adaptive schemes were evaluated: one using linear advance and exponential back-off, and another using exponential advance and linear back-off of sampling intervals. The study demonstrated that adaptive sampling could achieve significant energy savings compared to continuous sensing without compromising system accuracy. For example, the exponential back-off scheme showed more energy savings than the linear back-off scheme in purely local sensing scenarios. This adaptive approach allows mobile sensing applications to optimize data collection, potentially reducing the data volume while maintaining the accuracy of PM monitoring systems.

#### 5.2.7. Distributed Computing

Processing large PM datasets often requires significant computational resources. Haut et al. (2021) provide a comprehensive review of distributed DL approaches for remote sensing data interpretation, which is applicable to PM monitoring using satellite imagery [175]. The authors highlight the potential of cloud computing systems for managing vast amounts of remotely sensed data, offering implementation simplicity, low cost, and high efficiency compared to other parallel and distributed architectures. They emphasize that technological advances in hardware and software have enabled the acquisition and processing of huge amounts of remotely sensed data with high spatial resolution and spectral dimensionality. The review suggests that cloud-based implementations of DL algorithms can effectively address the processing challenges involved in extracting information from large-scale remotely sensed images, including those used for PM monitoring. While specific performance metrics for $PM_{2.5}$ forecasting are not provided, the authors indicate that cloud computing architectures can significantly improve processing capabilities for big data applications in remote sensing.

### 5.3. *Importance of Data Quality for Effective DL Models*

The quality of input data is paramount for the development of accurate and reliable DL models in airborne PM sensing. Several aspects underscore the importance of data quality (Figure 9 shows the stages of data quality).

#### 5.3.1. Model Performance

High-quality data and effective model architectures are crucial for accurate air quality estimation using DL approaches. Zhang et al. (2020) proposed a novel DL model called AQC-Net for estimating air quality levels from scene images [161]. The model incorporates a self-supervision module called Spatial and Context Attention (SCA) block, which enhances feature representation by capturing the interdependence between channel maps.

AQC-Net was trained and evaluated on a high-quality outdoor air quality dataset (NWNU-AQI) compiled by the authors. The model demonstrated superior performance in air quality classification compared to traditional methods like Support Vector Machine (SVM) and Deep Residual Network (ResNet). Specifically, AQC-Net achieved a classification accuracy of 74% on the test set, outperforming ResNet18 (70.1%) and SVM (60%). The study highlights the importance of both high-quality datasets and innovative model architectures in improving the accuracy of air quality estimation using DL techniques.

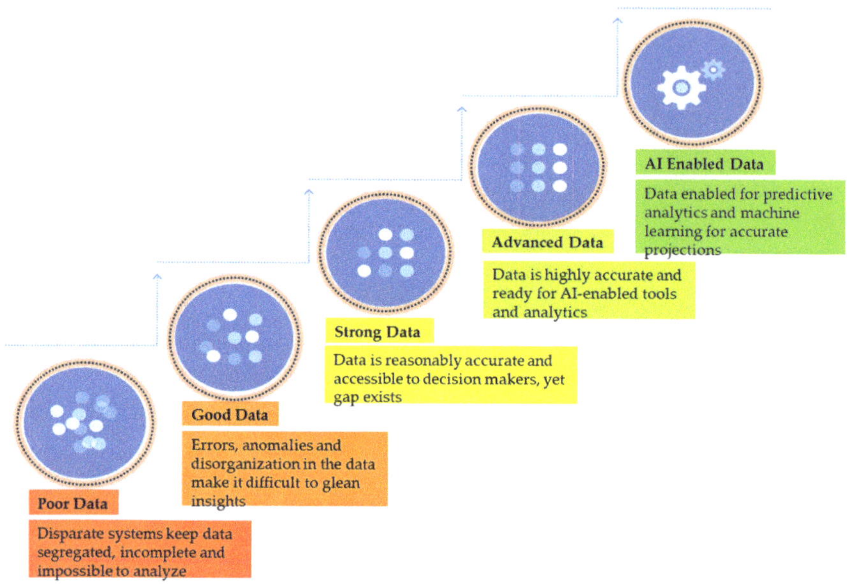


**Figure 9.** Steps to achieve data quality: from segregated, incomplete data to AI-enabled data optimized for analytics and machine learning.

5.3.2. Generalizability

Generalizability in DL models for $PM_{2.5}$ prediction and air quality estimation refers to a model's ability to perform well on new, unseen data, or in different geographical locations and conditions than those it was trained on. This concept encompasses several key aspects: cross-regional performance, where a model trained on data from one region can make accurate predictions in other regions with potentially different environmental conditions; versatility in data types, demonstrating the ability to work effectively with various input data formats such as AOD-based and reflectance-based satellite data; consistent performance across time, maintaining accuracy across different temporal scales (hourly, daily, monthly) and seasons; and adaptability to varying pollution levels, ensuring reliable predictions for both low- and high-pollution scenarios. A truly generalizable model exhibits robustness across these dimensions, making it valuable for wide-scale application in air quality monitoring and forecasting. Wang et al. (2022) demonstrated this concept with their SpatioTemporal Enhanced Neural Network (STENN) model, which showed strong spatial generalization ability for $PM_{2.5}$ estimation across mainland China [176]. Their model achieved an $R^2$ of 0.89 in cross-validation, indicating robust performance across different regions. The STENN model also demonstrated temporal transferability, successfully extrapolating to years without ground-based monitoring data. This approach, which integrated spatial signals into a recurrent neural network, effectively incorporated the impact of spatial heterogeneity and the time dependence of $PM_{2.5}$, resulting in high-quality, high-resolution (1 km) annual $PM_{2.5}$ products for China from 2014 to 2020. The model's ability to maintain accuracy and stability across different regions and time periods, especially in high-value estimations and spatial continuity, exemplifies the importance of generalizability in $PM_{2.5}$ prediction models.

#### 5.3.3. Interpretability

Interpretability in DL models for $PM_{2.5}$ prediction and air quality estimation is crucial for understanding and explaining how a model arrives at its predictions, building trust in its outputs, and extracting meaningful insights from the data. For these models, interpretability encompasses several key aspects: identifying the most significant input variables through feature importance analysis, revealing spatial-temporal patterns in the data, providing insights into the model's decision-making process across different layers, enabling error analysis to understand the model's performance under various conditions, and generating meaningful visualizations to explain the model's workings to non-technical stakeholders. High-quality, well-structured data play a vital role in enhancing the interpretability by allowing the model to learn more accurate and meaningful representations of $PM_{2.5}$ dynamics. This improved interpretability enables researchers to extract more reliable insights about pollution patterns, sources, and trends, ultimately leading to better-informed decisions in air quality management and policy-making.

The study by Yan et al. (2021) directly addresses the concept of interpretability in DL models for $PM_{2.5}$ prediction [177]. Their Spatial-Temporal Interpretable DL Model (SIDLM) demonstrated enhanced interpretability compared to traditional "black box" machine learning methods. The SIDLM's ability to automatically extract $PM_{2.5}$ spatiotemporal characteristics allows for deeper insights into pollution patterns. For instance, the model revealed the strong influence of the Tongzhou district on $PM_{2.5}$ levels in Beijing's main urban area and identified that summer months (June–August) contributed less to overall $PM_{2.5}$ concentrations. The model's interpretability was further enhanced by its "wide" and "deep" components and their joint learning process. Unlike ensemble methods that combine separately trained models, SIDLM's joint training approach allows its components to inform each other during the training process, potentially leading to more coherent and interpretable results. This improved interpretability not only enhances the model's predictive accuracy but also provides valuable insights for understanding regional pollution influences and seasonal patterns, which is crucial for effective air quality management and policy-making.

#### 5.3.4. Robustness to Outliers

High-quality datasets with properly handled outliers contribute to more robust models. Outliers are data points that significantly differ from other observations in a dataset, potentially arising from measurement errors, rare events, or extreme conditions. Robust models should maintain their performance and reliability even in the presence of such anomalous data points. Effective outlier detection and handling techniques can improve model stability, especially when dealing with environmental data like $PM_{2.5}$ and $PM_{10}$ concentrations, which can be subject to sudden spikes due to various factors.

The study by Gregório et al. (2022) explored the effects of seasonal decomposition on model performance, finding that removing seasonality from the data could negatively impact the accuracy by smoothing out critical trends associated with traffic patterns [178]. They presented a robust machine learning algorithm designed for forecasting $PM_{10}$ and $PM_{2.5}$ concentrations. This algorithm employs a multiple-input multiple-output linear regression model, which offers a straightforward yet effective approach to air quality prediction.

In the study by Ma et al. (2022), robustness to outliers was addressed through careful data preprocessing and modeling techniques that enhanced the overall quality of $PM_{2.5}$ retrieval [179]. The authors standardized the input data and removed outliers to mitigate their impact on model performance. They employed a piecewise linear interpolation model to fill data gaps, ensuring a continuous dataset while preserving time-series properties. Additionally, they utilized the min–max scaler method to normalize the raw data, which helps balance dimensions and can reduce the influence of extreme values. By comparing six different machine learning algorithms—MLR, kNN, SVR, RT, RF, and BPNN—the study indirectly assessed the robustness of these models against outliers through performance evaluations. The use of 10-fold cross-validation further ensured consistent model performance across different data subsets, enhancing robustness. Although the Random Forest model achieved the highest accuracy, all models demonstrated a lower prediction accuracy in high-value areas, indicating potential sensitivity to outlier values in $PM_{2.5}$ concentrations. Overall, these methodologies contribute to creating a more robust model capable of handling variations and anomalies in $PM_{2.5}$ data.

#### 5.3.5. Temporal Consistency

Temporal consistency in DL, particularly for $PM_{2.5}$ prediction models, refers to the preservation and accurate representation of time-dependent patterns and relationships within the data. It is crucial for maintaining the integrity of time-series information and ensuring that models capture meaningful temporal trends rather than spurious correlations. In the context of $PM_{2.5}$ forecasting, temporal consistency involves properly handling the sequential nature of air quality data, accounting for both short-term fluctuations and long-term trends. Ensuring temporal consistency in the data preprocessing and model architecture stages is essential for developing effective DL models. This includes techniques such as appropriate time-based data splitting, handling of missing values, and the use of architectures specifically designed for sequential data (e.g., recurrent neural networks or temporal convolutional networks). By maintaining temporal consistency, models can better capture the dynamic nature of $PM_{2.5}$ concentrations, leading to more accurate predictions and a deeper understanding of the underlying temporal patterns in air quality data.

The study by Wang et al. (2022) addresses temporal consistency in DL for $PM_{2.5}$ prediction through their SpatioTemporal Enhanced Neural Network (STENN) framework [176]. The model incorporates bidirectional LSTM structures and attention mechanisms to capture the temporal dependencies of $PM_{2.5}$ concentrations. By integrating historical multi-temporal remote sensing data, including AOD, meteorological, land cover/use, and socioeconomic factors from 2011 to 2019, the STENN model achieved robust spatiotemporal transferable power with an $R^2$ of 0.89 in cross-validation. The framework's data processing ensures temporal consistency by averaging the 24 h $PM_{2.5}$ observations over a year for an annual estimation and resampling various data sources to maintain consistent spatiotemporal resolution. This approach allowed the model to effectively capture long-term $PM_{2.5}$ trends, as demonstrated by its ability to reveal the exponential temporal trend of $PM_{2.5}$ concentrations in China from 2014 to 2020, showing a transition from rapid decline to gradual slowdown and stabilization. The model's success in maintaining temporal consistency was further evidenced by its capacity to generate high-quality, high-resolution annual $PM_{2.5}$ products for mainland China, which exhibit high consistency with surface $PM_{2.5}$ measurements and provide better stability in different regions compared to existing products.

5.3.6. Spatial Representativeness and Multi-Source Data Integration

Multi-source data integration and spatial representativeness are vital for effectively predicting $PM_{2.5}$ concentrations, as highlighted by Yang et al. (2024) [180]. The authors utilized multi-source remote sensing data, including satellite-derived AOD, meteorological factors, and land-use variables, to enhance the accuracy of their $PM_{2.5}$ predictions. By employing the Pearson correlation coefficient and GeoDetector to select relevant auxiliary variables, they ensured that the integrated dataset captured a comprehensive range of influences on $PM_{2.5}$ levels. This integration is crucial for overcoming the limitations posed by the uneven distribution of ground monitoring stations across China, particularly in southeastern regions, which may lead to skewed representations of air quality.

Spatial representativeness further emphasizes the need for accurate data representation across various geographical areas. The study acknowledges that the spatially heterogeneous distribution of $PM_{2.5}$ monitoring stations can affect model performance, particularly in under-monitored regions. The authors suggest that incorporating economic and demographic variables could improve the representativeness and model accuracy. Their findings demonstrate that the Random Forest regression model achieved an $R^2$ of 0.93 and RMSE of 4.59 $\mu g\ m^{-3}$, reflecting the effectiveness of their multi-source integration approach while highlighting the importance of ensuring that data accurately represent spatial variations for robust $PM_{2.5}$ predictions.

In Section 5 we have comprehensively explored the intricacies of data acquisition and the utilization of synthetic data in the realm of air quality monitoring, underscoring the inherent challenges and sophisticated techniques involved in managing the voluminous and often noise-ridden datasets generated using airborne PM sensors. For instance, data cleaning techniques, such as outlier removal and imputation, are indispensable for rectifying data quality issues; however, their efficacy is contingent upon the nature and magnitude of the noise present in the dataset. Outlier removal methods, while effective in eliminating spurious data points, may inadvertently discard valuable information if not carefully tuned. Imputation techniques, aimed at filling in missing data, can introduce biases if the imputation model does not accurately capture the underlying data distribution.

Similarly, data augmentation methods, such as GANs, hold immense promise for mitigating data scarcity. However, the quality, representativeness, and potential biases of the synthetic data generated using GANs warrant meticulous consideration. A comprehensive comparison of these data handling techniques, encompassing their computational demands, suitability for addressing specific types of data noise, and potential for bias introduction, would empower researchers and practitioners with the knowledge to make informed decisions. Such a comparative analysis would facilitate the selection of the most appropriate data handling techniques tailored to the specific attributes of the PM dataset and the objectives of the air quality monitoring endeavor.

## 6. Conclusions

This review meticulously explored the advancements in airborne PM sensing technologies and the integration of DL to improve environmental monitoring. The review underscored the vital importance of precise PM monitoring for safeguarding public health, formulating environmental policies, and furthering scientific research. Traditional PM sensing methods, encompassing their operating principles, advantages, limitations, and pertinent case studies, were examined. These methods include gravimetric techniques, continuous monitoring, optical methods, electrical methods, and microscopy. This review demonstrated that merging DL with PM sensing has the potential to revolutionize environmental monitoring by enhancing the accuracy, efficiency, and data interpretation.

DL techniques, including CNNs, autoencoders, RNNs, and their variations, have emerged as promising tools for a range of applications such as estimating PM from satellite data, predicting air quality, and calibrating sensors. The sources analyzed specific DL models and their performance in PM sensing applications, showcasing significant accuracy gains compared to traditional methods. However, this review emphasized that the success of DL models is contingent upon addressing data acquisition and quality challenges. The sources discussed various issues related to sensor calibration, environmental variability, data gaps, noise reduction, and data augmentation techniques.

This review underscored the critical importance of data quality for model performance, generalizability, interpretability, and robustness. We also noted that although emerging low-cost sensor technologies hold promise for widespread deployment, careful consideration of data quality and standardization issues is essential. Emerging low-cost sensor technologies and hybrid systems hold promise for expanding the spatial and temporal coverage of PM monitoring. However, it is crucial to consider data quality, standardization, and calibration against EPA reference methods to ensure the reliability and comparability of measurements from these new technologies. Employing hybrid and multi-sensor systems, which combine different sensing techniques, can offer a more comprehensive understanding of PM, but it also poses challenges with data integration and interpretation.

Several promising future research directions emerge to further enhance PM sensing technologies and their integration with AI. The development of robust and interpretable DL models: Research should focus on developing DL models that can effectively handle complex and heterogeneous PM datasets, addressing issues of noise, data gaps, and varying environmental conditions. Emphasizing model interpretability will be crucial for gaining insights into the underlying factors driving PM levels and for building trust in AI-driven predictions. Advanced data fusion techniques: Integrating data from multiple sources, including satellite imagery, ground-based sensors, meteorological data, and even social media feeds, can provide a more comprehensive and accurate representation of PM dynamics. Research on advanced data fusion techniques, incorporating uncertainty quantification and robust data assimilation, will be essential for exploiting the full potential of multi-source data.

Deep reinforcement learning for sensor optimization: Applying DRL to optimize sensor placement, data collection strategies, and even adaptive calibration can significantly improve the efficiency, cost-effectiveness, and coverage of PM monitoring networks. Research in this area can lead to intelligent and self-adapting PM sensing systems that dynamically adjust to changing environmental conditions. Edge computing and real-time monitoring: Developing DL models and data processing algorithms that can be deployed on edge devices, such as low-cost sensors and smartphones, will enable real-time PM monitoring and localized air quality assessments. Research on efficient edge computing architectures and lightweight DL models will be essential for realizing this vision of ubiquitous and accessible air quality information.

Addressing ethical considerations: As AI becomes increasingly integrated into PM sensing and environmental monitoring, ethical considerations surrounding data privacy, algorithmic bias, and equitable access to air quality information must be addressed. Research should focus on developing fair, transparent, and accountable AI systems that prioritize the well-being and privacy of individuals and communities.

**Author Contributions:** Conceptualization, K.P.W., B.V.S.C., S.V. and B.M.A.R.; methodology, B.V.S.C. and S.V.; Validation, K.P.W., S.V. and B.M.A.R.; formal analysis, B.V.S.C., K.P.W. and S.V.; investigation, B.V.S.C., K.P.W. and S.V.; resources, B.V.S.C. and S.V.; data curation, K.P.W., B.V.S.C. and S.V.; writing—original draft preparation, B.V.S.C., K.P.W., S.V. and B.M.A.R.; writing—review and editing, B.V.S.C., K.P.W., S.V. and B.M.A.R.; visualization, B.V.S.C. and S.V.; supervision, B.M.A.R. and

K.P.W.; project administration, B.V.S.C. All authors have read and agreed to the published version of the manuscript.

**Funding:** This research received no external funding.

**Institutional Review Board Statement:** Not applicable.

**Informed Consent Statement:** Not applicable.

**Data Availability Statement:** The literature works considered for the purposes of this study are indexed in the Scopus database.

**Acknowledgments:** We would like to express our sincere gratitude to the Centre of Earth Observation Sciences, School of Applied Sciences, at the University of Brighton, United Kingdom and City, University of London, United Kingdom, for their invaluable support and resources in the preparation of this review research. We are deeply appreciative of the opportunities provided by the center, which have enriched our understanding of the subject matter.

**Conflicts of Interest:** Authors Balendra V. S. Chauhan and Sneha Verma are the directors of the company Tech GPT Ltd. NE6 2SR Newcastle upon Tyne, UK. The remaining authors declare that the research was conducted in the absence of any commercial or financial relationships that could be construed as a potential conflict of interest.

## References

1. World Health Organization. *Air Quality Guidelines: Global Update 2005: Particulate Matter, Ozone, Nitrogen Dioxide, and Sulfur Dioxide*; World Health Organization: Geneva, Switzerland, 2006.
2. Read, R.; Briefs, C.B.; Cases, C.C.C. Reconsideration of the National Ambient Air Quality Standards for Particulate Matter. *Fed. Regist.* **2023**, *88*, 5558.
3. Di, Q.; Wang, Y.; Zanobetti, A.; Wang, Y.; Koutrakis, P.; Choirat, C.; Dominici, F.; Schwartz, J.D. Air pollution and mortality in the Medicare population. *N. Engl. J. Med.* **2017**, *376*, 2513–2522. [CrossRef] [PubMed]
4. Power, M.C.; Adar, S.D.; Yanosky, J.D.; Weuve, J. Exposure to air pollution as a potential contributor to cognitive function, cognitive decline, brain imaging, and dementia: A systematic review of epidemiologic research. *Neurotoxicology* **2016**, *56*, 235–253. [CrossRef]
5. Boucher, O.; Randall, D.; Artaxo, P.; Bretherton, C.; Feingold, G.; Forster, P.; Lohmann, U. Clouds and aerosols. In *Climate Change 2013: The Physical Science Basis. Contribution of Working Group I to the Fifth Assessment Report of the Intergovernmental Panel on Climate Change*; Cambridge University Press: Cambridge, UK, 2013; pp. 571–657.
6. Das, S.; Pal, D.; Sarkar, A. Particulate matter pollution and global agricultural productivity. In *Sustainable Agriculture Reviews 50: Emerging Contaminants in Agriculture*; Springer: Berlin/Heidelberg, Germany, 2021; pp. 79–107.
7. Sridharan, S.; Kumar, M.; Singh, L.; Bolan, N.S.; Saha, M. Microplastics as an emerging source of particulate air pollution: A critical review. *J. Hazard. Mater.* **2021**, *418*, 126245. [CrossRef]
8. Park, R.J.; Jacob, D.J.; Kumar, N.; Yantosca, R.M. Regional visibility statistics in the United States: Natural and transboundary pollution influences, and implications for the Regional Haze Rule. *Atmos. Environ.* **2006**, *40*, 5405–5423. [CrossRef]
9. Abraczinskas, M.; Meyers, J.; Sloan, J. The Impact of a Lower Fine Particulate Matter National Ambient Air Quality Standard. *The Magazine for Environmental Managers*. 2023. Available online: https://cleanairact.org/wp-content/uploads/2023/05/emmay23_AAPCA-article.pdf (accessed on 29 September 2024).
10. Liang, L.; Daniels, J.; Bailey, C.; Hu, L.; Phillips, R.; South, J. Integrating low-cost sensor monitoring, satellite mapping, and geospatial artificial intelligence for intra-urban air pollution predictions. *Environ. Pollut.* **2023**, *331*, 121832. [CrossRef]
11. Capelli, D.; Scognamiglio, V.; Montanari, R. Surface plasmon resonance technology: Recent advances, applications and experimental cases. *TrAC Trends Anal. Chem.* **2023**, *163*, 117079. [CrossRef]
12. Liu, W.; Liu, C.; Wang, J.; Lv, J.; Lv, Y.; Yang, L.; An, N.; Yi, Z.; Liu, Q.; Hu, C.; et al. Surface plasmon resonance sensor composed of microstructured optical fibers for monitoring of external and internal environments in biological and environmental sensing. *Results Phys.* **2023**, *47*, 106365. [CrossRef]
13. Aloisi, A.; Della Torre, A.; De Benedetto, A.; Rinaldi, R. Bio-recognition in spectroscopy-based biosensors for *heavy metals-water and waterborne contamination analysis. *Biosensors* **2019**, *9*, 96. [CrossRef]
14. Aljbar, N.A.; Mahdi, B.R.; Khalid, A.H.; Attallah, A.H.; Abdulwahid, F.S.; Haider, A.J. Enhanced surface plasmon resonance (SPR) fiber optic sensor for environmental monitoring: A coreless fiber–based design. *Plasmonics* **2024**, 1–10. [CrossRef]

15. Tan, J.; Dai, Z.; Zhou, K.; Zhang, L.; He, M.; Tan, Y.; Zhou, X. An ultrasensitive and universal surface plasmonic biosensor for detection of micropollutants in aquatic environments. *Environ. Sci. Technol.* **2023**, *57*, 8313–8322. [PubMed]
16. Geng, Z.; Zhang, X.; Fan, Z.; Lv, X.; Su, Y.; Chen, H. Recent progress in optical biosensors based on smartphone platforms. *Sensors* **2017**, *17*, 2449. [CrossRef] [PubMed]
17. Chow, J.C.; Watson, J.G.; Lowenthal, D.H.; Chen, L.-W.A.; Magliano, K.L. Particulate carbon measurements in California's San Joaquin valley. *Chemosphere* **2006**, *62*, 337–348. [PubMed]
18. Solomon, P.A.; Crumpler, D.; Flanagan, J.B.; Jayanty, R.; Rickman, E.E.; McDade, C.E. US national PM2.5 chemical speciation monitoring networks—CSN and IMPROVE: Description of networks. *J. Air Waste Manag. Assoc.* **2014**, *64*, 1410–1438.
19. Brouwer, D.; van Duuren-Stuurman, B.; Berges, M.; Jankowska, E.; Bard, D.; Mark, D. From workplace air measurement results toward estimates of exposure? Development of a strategy to assess exposure to manufactured nano-objects. *J. Nanoparticle Res.* **2009**, *11*, 1867–1881.
20. Marple, V.A.; Rubow, K.L.; Behm, S.M. A microorifice uniform deposit impactor (MOUDI): Description, calibration, and use. *Aerosol Sci. Technol.* **1991**, *14*, 434–446.
21. Sioutas, C.; Koutrakis, P.; Burton, R. Development of a low cutpoint slit virtual impactor for sampling ambient fine particles. *J. Aerosol Sci.* **1994**, *25*, 1321–1330. [CrossRef]
22. Cass, G.R.; Hughes, L.A.; Bhave, P.; Kleeman, M.J.; Allen, J.O.; Salmon, L.G. The chemical composition of atmospheric ultrafine particles. *Philos. Trans. R. Soc. London Ser. A Math. Phys. Eng. Sci.* **2000**, *358*, 2581–2592.
23. Chow, J.C.; Watson, J.G.; Park, K.; Robinson, N.F.; Lowenthal, D.H.; Magliano, K.A. Comparison of particle light scattering and fine particulate matter mass in central California. *J. Air Waste Manag. Assoc.* **2006**, *56*, 398–410.
24. Vecchi, R.; Valli, G.; Fermo, P.; D'Alessandro, A.; Piazzalunga, A.; Bernardoni, V. Organic and inorganic sampling artefacts assessment. *Atmos. Environ.* **2009**, *43*, 1713–1720.
25. Wilson, W.; Chow, J.C.; Claiborn, C.; Fusheng, W.; Engelbrecht, J.; Watson, J.G. Monitoring of particulate matter outdoors. *Chemosphere* **2002**, *49*, 1009–1043. [PubMed]
26. Amaral, S.S.; de Carvalho, J.A., Jr.; Costa, M.A.M.; Pinheiro, C. An overview of particulate matter measurement instruments. *Atmosphere* **2015**, *6*, 1327–1345. [CrossRef]
27. Liu, Y.; Li, C.; Zhang, C.; Liu, X.; Qu, Y.; An, J.; Ma, D.; Feng, M.; Tan, Q. Chemical characteristics, source apportionment, and regional contribution of PM2.5 in Zhangjiakou, Northern China: A multiple sampling sites observation and modeling perspective. *Environ. Adv.* **2021**, *3*, 100034.
28. Su, Y.; Sofowote, U.; Debosz, J.; White, L.; Munoz, A. Multi-year continuous PM2.5 measurements with the Federal Equivalent Method SHARP 5030 and comparisons to filter-based and TEOM measurements in Ontario, Canada. *Atmosphere* **2018**, *9*, 191. [CrossRef]
29. Tariq, H.; Touati, F.; Crescini, D.; Mnaouer, A.B. A Review of State-of-the-Art Low-Cost Sensors for Detection of Indoor Air Pollutants that Correlate with Health. *Preprints* **2023**.
30. Jayaratne, R.; Kuhn, T.; Christensen, B.; Liu, X.; Zing, I.; Lamont, R.; Dunbabin, M.; Maddox, J.; Fisher, G.; Morawska, L. Using a network of low-cost particle sensors to assess the impact of ship emissions on a residential community. *Aerosol Air Qual. Res.* **2020**, *20*, 2754–2764.
31. Yadav, R.; Nagori, A.; Madan, K.; Lodha, R.; Kabra, S.K. Short-term exposure to air pollution and emergency room visits for acute respiratory symptoms among adults. *Int. J. Tuberc. Lung Dis.* **2023**, *27*, 761–765.
32. Schweizer, D.; Cisneros, R.; Shaw, G. A comparative analysis of temporary and permanent beta attenuation monitors: The importance of understanding data and equipment limitations when creating PM2.5 air quality health advisories. *Atmos. Pollut. Res.* **2016**, *7*, 865–875.
33. Grover, B.D.; Kleinman, M.; Eatough, N.L.; Eatough, D.J.; Hopke, P.K.; Long, R.W.; Ambs, J.L. Measurement of total PM2.5 mass (nonvolatile plus semivolatile) with the Filter Dynamic Measurement System tapered element oscillating microbalance monitor. *J. Geophys. Res. Atmos.* **2005**, *110*, D07S03. [CrossRef]
34. Wang, Y.; Li, J.; Jing, H.; Zhang, Q.; Jiang, J.; Biswas, P. Laboratory evaluation and calibration of three low-cost particle sensors for particulate matter measurement. *Aerosol Sci. Technol.* **2015**, *49*, 1063–1077.
35. Kuula, J.; Mäkelä, T.; Aurela, M.; Teinilä, K.; Varjonen, S.; González, Ó.; Timonen, H. Laboratory evaluation of particle-size selectivity of optical low-cost particulate matter sensors. *Atmos. Meas. Tech.* **2020**, *13*, 2413–2423.
36. Magi, B.I.; Cupini, C.; Francis, J.; Green, M.; Hauser, C. Evaluation of PM2.5 measured in an urban setting using a low-cost optical particle counter and a Federal Equivalent Method Beta Attenuation Monitor. *Aerosol Sci. Technol.* **2020**, *54*, 147–159.
37. Malings, C.; Tanzer, R.; Hauryliuk, A.; Saha, P.K.; Robinson, A.L.; Presto, A.A.; Subramanian, R. Fine particle mass monitoring with low-cost sensors: Corrections and long-term performance evaluation. *Aerosol Sci. Technol.* **2020**, *54*, 160–174.
38. Drinovec, L.; Močnik, G.; Zotter, P.; Prévôt, A.S.H.; Ruckstuhl, C.; Coz, E.; Rupakheti, M.; Sciare, J.; Müller, T.; Wiedensohler, A.; et al. The "dual-spot" Aethalometer: An improved measurement of aerosol black carbon with real-time loading compensation. *Atmos. Meas. Tech.* **2015**, *8*, 1965–1979.

39. Crilley, L.R.; Shaw, M.; Pound, R.; Kramer, L.J.; Price, R.; Young, S.; Lewis, A.C.; Pope, F.D. Evaluation of a low-cost optical particle counter (Alphasense OPC-N2) for ambient air monitoring. *Atmos. Meas. Tech.* **2018**, *11*, 709–720.
40. Zheng, T.; Bergin, M.H.; Johnson, K.K.; Tripathi, S.N.; Shirodkar, S.; Landis, M.S.; Sutaria, R.; Carlson, D.E. Field evaluation of low-cost particulate matter sensors in high-and low-concentration environments. *Atmos. Meas. Tech.* **2018**, *11*, 4823–4846.
41. Alfano, B.; Barretta, L.; Del Giudice, A.; De Vito, S.; Di Francia, G.; Esposito, E.; Formisano, F.; Massera, E.; Miglietta, M.L.; Polichetti, T. A review of low-cost particulate matter sensors from the developers' perspectives. *Sensors* **2020**, *20*, 6819. [CrossRef] [PubMed]
42. Lee, S.; Kim, D.; Cho, Y.; Kim, E.; Liu, P.; Kwak, D.-B.; Keum, S.; Lim, H.; Kim, T. Application of an Electrical Low Pressure Impactor (ELPI) for Residual Particle Measurement in an Epitaxial Growth Reactor. *Appl. Sci.* **2021**, *11*, 7680. [CrossRef]
43. Rissler, J.; Nordin, E.Z.; Eriksson, A.C.; Nilsson, P.T.; Frosch, M.; Sporre, M.K.; Wierzbicka, A.; Svenningsson, B.; Löndahl, J.; Messing, M.E.; et al. Effective density and mixing state of aerosol particles in a near-traffic urban environment. *Environ. Sci. Technol.* **2014**, *48*, 6300–6308.
44. Sowlat, M.H.; Hasheminassab, S.; Sioutas, C. Source apportionment of ambient particle number concentrations in central Los Angeles using positive matrix factorization (PMF). *Atmos. Chem. Phys.* **2016**, *16*, 4849–4866.
45. Levin, E.J.T.; Prenni, A.J.; Palm, B.B.; Day, D.A.; Campuzano-Jost, P.; Winkler, P.M.; Kreidenweis, S.M.; DeMott, P.J.; Jimenez, J.L.; Smith, J.N. Size-resolved aerosol composition and its link to hygroscopicity at a forested site in Colorado. *Atmos. Chem. Phys.* **2014**, *14*, 2657–2667. [CrossRef]
46. Zimmerman, N.; Pollitt, K.J.G.; Jeong, C.-H.; Wang, J.M.; Jung, T.; Cooper, J.M.; Wallace, J.S.; Evans, G.J. Comparison of three nanoparticle sizing instruments: The influence of particle morphology. *Atmos. Environ.* **2014**, *86*, 140–147. [CrossRef]
47. Fonseca, A.S.; Viana, M.; Pérez, N.; Alastuey, A.; Querol, X.; Kaminski, H.; Asbach, C. Intercomparison of a portable and two stationary mobility particle sizers for nanoscale aerosol measurements. *Aerosol Sci. Technol.* **2016**, *50*, 653–668.
48. Levin, M.; Gudmundsson, A.; Pagels, J.; Fierz, M.; Mølhave, K.; Löndahl, J.; Koponen, I.K. Limitations in the use of unipolar charging for electrical mobility sizing instruments: A study of the fast mobility particle sizer. *Aerosol Sci. Technol.* **2015**, *49*, 556–565. [CrossRef]
49. Schmid, O.; Trueblood, M.B.; Gregg, N.; Hagen, D.E.; Whitefield, P.D. Sizing of aerosol in gases other than air using a differential mobility analyzer. *Aerosol Sci. Technol.* **2002**, *36*, 351–360. [CrossRef]
50. Wu, Z.J.; Poulain, L.; Henning, S.; Dieckmann, K.; Birmili, W.; Merkel, M.; van Pinxteren, D.; Spindler, G.; Müller, K.; Stratmann, F.; et al. Relating particle hygroscopicity and CCN activity to chemical composition during the HCCT-2010 field campaign. *Atmos. Chem. Phys.* **2013**, *13*, 7983–7996. [CrossRef]
51. Shi, Y.; Ji, Y.; Sun, H.; Hui, F.; Hu, J.; Wu, Y.; Fang, J.; Lin, H.; Wang, J.; Duan, H.; et al. Nanoscale characterization of PM2.5 airborne pollutants reveals high adhesiveness and aggregation capability of soot particles. *Sci. Rep.* **2015**, *5*, 11232.
52. Shao, L.; Liu, P.; Jones, T.; Yang, S.; Wang, W.; Zhang, D.; Li, Y.; Yang, C.-X.; Xing, J.; Hou, C.; et al. A review of atmospheric individual particle analyses: Methodologies and applications in environmental research. *Gondwana Res.* **2022**, *110*, 347–369.
53. Sielicki, P.; Janik, H.; Guzman, A.; Namieśnik, J. The progress in electron microscopy studies of particulate matters to be used as a standard monitoring method for air dust pollution. *Crit. Rev. Anal. Chem.* **2011**, *41*, 314–334.
54. Laskin, A.; Gilles, M.K.; Knopf, D.A.; Wang, B.; China, S. Progress in the analysis of complex atmospheric particles. *Annu. Rev. Anal. Chem.* **2016**, *9*, 117–143.
55. Ault, A.P.; Axson, J.L. Atmospheric aerosol chemistry: Spectroscopic and microscopic advances. *Anal. Chem.* **2017**, *89*, 430–452. [CrossRef] [PubMed]
56. Variola, F. Atomic force microscopy in biomaterials surface science. *Phys. Chem. Chem. Phys.* **2015**, *17*, 2950–2959. [CrossRef] [PubMed]
57. Laskin, J.; Laskin, A.; Nizkorodov, S.A. Mass spectrometry analysis in atmospheric chemistry. *Anal. Chem.* **2018**, *90*, 166–189. [CrossRef]
58. Yatkin, S.; Gerboles, M.; Borowiak, A. Evaluation of standardless EDXRF analysis for the determination of elements on PM10 loaded filters. *Atmos. Environ.* **2012**, *54*, 568–582. [CrossRef]
59. Correa, M.A.; Franco, S.A.; Gómez, L.M.; Aguiar, D.; Colorado, H.A. Characterization methods of ions and metals in particulate matter pollutants on PM2.5 and PM10 samples from several emission sources. *Sustainability* **2023**, *15*, 4402. [CrossRef]
60. Choi, N.R.; Lee, J.Y.; Ahn, Y.G.; Kim, Y.P. Determination of atmospheric amines at Seoul, South Korea via gas chromatography/tandem mass spectrometry. *Chemosphere* **2020**, *258*, 127367. [CrossRef]
61. Feng, H.; Ye, X.; Liu, Y.; Wang, Z.; Gao, T.; Cheng, A.; Wang, X.; Chen, J. Simultaneous determination of nine atmospheric amines and six inorganic ions by non-suppressed ion chromatography using acetonitrile and 18-crown-6 as eluent additive. *J. Chromatogr. A* **2020**, *1624*, 461234. [CrossRef]
62. Brown, R.J.; Edwards, P.R. Measurement of anions in ambient particulate matter by ion chromatography: A novel sample preparation technique and development of a generic uncertainty budget. *Talanta* **2009**, *80*, 1020–1024. [CrossRef]

63. Canepari, S.; Astolfi, M.L.; Farao, C.; Maretto, M.; Frasca, D.; Marcoccia, M.; Perrino, C. Seasonal variations in the chemical composition of particulate matter: A case study in the Po Valley. Part II: Concentration and solubility of micro-and trace-elements. *Environ. Sci. Pollut. Res.* **2014**, *21*, 4010–4022. [CrossRef]
64. Hallquist, M.; Wenger, J.C.; Baltensperger, U.; Rudich, Y.; Simpson, D.; Claeys, M.; Dommen, J.; Donahue, N.M.; George, C.; Goldstein, A.H.; et al. The formation, properties and impact of secondary organic aerosol: Current and emerging issues. *Atmos. Chem. Phys.* **2009**, *9*, 5155–5236. [CrossRef]
65. Majestic, B.J.; Anbar, A.D.; Herckes, P. Elemental and iron isotopic composition of aerosols collected in a parking structure. *Sci. Total Environ.* **2009**, *407*, 5104–5109. [PubMed]
66. Fuzzi, S.; Baltensperger, U.; Carslaw, K.; Decesari, S.; Van Der Gon, H.D.; Facchini, M.C.; Fowler, D.; Koren, I.; Langford, B.; Lohmann, U.; et al. Particulate matter, air quality and climate: Lessons learned and future needs. *Atmos. Chem. Phys.* **2015**, *15*, 8217–8299.
67. Soupiona, O.; Papayannis, A.; Kokkalis, P.; Foskinis, R.; Hernández, G.S.; Ortiz-Amezcua, P.; Mylonaki, M.; Papanikolaou, C.-A.; Papagiannopoulos, N.; Samaras, S.; et al. EARLINET observations of Saharan dust intrusions over the northern Mediterranean region (2014–2017): Properties and impact on radiative forcing. *Atmos. Chem. Phys.* **2020**, *20*, 15147–15166.
68. Browell, E.; Ismail, S.; Grant, W. Differential absorption lidar (DIAL) measurements from air and space. *Appl. Phys. B* **1998**, *67*, 399–410. [CrossRef]
69. Lyapustin, A.; Wang, Y.; Korkin, S.; Huang, D. MODIS collection 6 MAIAC algorithm. *Atmos. Meas. Tech.* **2018**, *11*, 5741–5765.
70. Wei, J.; Li, Z.; Lyapustin, A.; Sun, L.; Peng, Y.; Xue, W.; Su, T.; Cribb, M. Reconstructing 1-km-resolution high-quality PM2.5 data records from 2000 to 2018 in China: Spatiotemporal variations and policy implications. *Remote Sens. Environ.* **2021**, *252*, 112136.
71. Giles, D.M.; Sinyuk, A.; Sorokin, M.G.; Schafer, J.S.; Smirnov, A.; Slutsker, I.; Eck, T.F.; Holben, B.N.; Lewis, J.R.; Campbell, J.R.; et al. Advancements in the Aerosol Robotic Network (AERONET) Version 3 database–automated near-real-time quality control algorithm with improved cloud screening for Sun photometer aerosol optical depth (AOD) measurements. *Atmos. Meas. Tech.* **2019**, *12*, 169–209.
72. Benavent-Oltra, J.A.; Casquero-Vera, J.A.; Román, R.; Lyamani, H.; Pérez-Ramírez, D.; Granados-Muñoz, M.J.; Herrera, M.; Cazorla, A.; Titos, G.; Ortiz-Amezcua, P.; et al. Overview of SLOPE I and II campaigns: Aerosol properties retrieved with lidar and sun-sky photometer measurements. *Atmos. Chem. Phys. Discuss.* **2021**, *21*, 9269–9287.
73. Siomos, N.; Fountoulakis, I.; Natsis, A.; Drosoglou, T.; Bais, A. Automated aerosol classification from spectral UV measurements using machine learning clustering. *Remote Sens.* **2020**, *12*, 965. [CrossRef]
74. Toth, T.D.; Zhang, J.; Campbell, J.R.; Reid, J.S.; Shi, Y.; Johnson, R.S.; Winker, D.M. Investigating enhanced Aqua MODIS aerosol optical depth retrievals over the mid-to-high latitude Southern Oceans through intercomparison with co-located CALIOP, MAN, and AERONET data sets. *J. Geophys. Res. Atmos.* **2013**, *118*, 4700–4714.
75. Li, Z.; Zhang, Y.; Shao, J.; Li, B.; Hong, J.; Liu, D.; Li, D.; Wei, P.; Li, W.; Li, L.; et al. Remote sensing of atmospheric particulate mass of dry PM2.5 near the ground: Method validation using ground-based measurements. *Remote Sens. Environ.* **2016**, *173*, 59–68. [CrossRef]
76. Wagner, P.; Schäfer, K. Influence of mixing layer height on air pollutant concentrations in an urban street canyon. *Urban Clim.* **2017**, *22*, 64–79. [CrossRef]
77. Bulot, F.M.J.; Johnston, S.J.; Basford, P.J.; Easton, N.H.C.; Apetroaie-Cristea, M.; Foster, G.L.; Morris, A.K.R.; Cox, S.J.; Loxham, M. Long-term field comparison of multiple low-cost particulate matter sensors in an outdoor urban environment. *Sci. Rep.* **2019**, *9*, 7497. [CrossRef] [PubMed]
78. Castell, N.; Dauge, F.R.; Schneider, P.; Vogt, M.; Lerner, U.; Fishbain, B.; Broday, D.; Bartonova, A. Can commercial low-cost sensor platforms contribute to air quality monitoring and exposure estimates? *Environ. Int.* **2017**, *99*, 293–302. [CrossRef]
79. Karagulian, F.; Barbiere, M.; Kotsev, A.; Spinelle, L.; Gerboles, M.; Lagler, F.; Redon, N.; Crunaire, S.; Borowiak, A. Review of the performance of low-cost sensors for air quality monitoring. *Atmosphere* **2019**, *10*, 506. [CrossRef]
80. Morawska, L.; Thai, P.K.; Liu, X.; Asumadu-Sakyi, A.; Ayoko, G.; Bartonova, A.; Dunbabin, M. Applications of low-cost sensing technologies for air quality monitoring and exposure assessment: How far have they gone? *Environ. Int.* **2018**, *116*, 286–299. [CrossRef]
81. Rai, A.C.; Kumar, P.; Pilla, F.; Skouloudis, A.N.; DI Sabatino, S.; Ratti, C.; Yasar, A.-U.; Rickerby, D. End-user perspective of low-cost sensors for outdoor air pollution monitoring. *Sci. Total Environ.* **2017**, *607*, 691–705. [CrossRef]
82. Badura, M.; Batog, P.; Drzeniecka-Osiadacz, A.; Modzel, P. Evaluation of low-cost sensors for ambient PM2.5 monitoring. *J. Sens.* **2018**, *2018*, 5096540. [CrossRef]
83. Schneider, P.; Bartonova, A.; Castell, N.; Dauge, F.R.; Gerboles, M.; Hagler, G.S.; Hüglin, C.; Jones, R.L.; Khan, S.; Lewis, A.C.; et al. *Toward a Unified Terminology of Processing Levels for Low-Cost Air-Quality Sensors*; ACS Publications: Washington, DC, USA, 2019.
84. Kosmopoulos, G.; Salamalikis, V.; Pandis, S.; Yannopoulos, P.; Bloutsos, A.; Kazantzidis, A. Low-cost sensors for measuring airborne particulate matter: Field evaluation and calibration at a South-Eastern European site. *Sci. Total Environ.* **2020**, *748*, 141396. [CrossRef]

85. Concas, F.; Mineraud, J.; Lagerspetz, E.; Varjonen, S.; Liu, X.; Puolamäki, K.; Nurmi, P.; Tarkoma, S. Low-cost outdoor air quality monitoring and sensor calibration: A survey and critical analysis. *ACM Trans. Sens. Netw. TOSN* **2021**, *17*, 1–44. [CrossRef]
86. Sousan, S.; Koehler, K.; Hallett, L.; Peters, T.M. Evaluation of consumer monitors to measure particulate matter. *J. Aerosol Sci.* **2017**, *107*, 123–133. [CrossRef] [PubMed]
87. Jayaratne, R.; Liu, X.; Thai, P.; Dunbabin, M.; Morawska, L. The influence of humidity on the performance of a low-cost air particle mass sensor and the effect of atmospheric fog. *Atmos. Meas. Tech.* **2018**, *11*, 4883–4890. [CrossRef]
88. Rybarczyk, Y.; Zalakeviciute, R. Machine learning approaches for outdoor air quality modelling: A systematic review. *Appl. Sci.* **2018**, *8*, 2570. [CrossRef]
89. Pope, C.A., III; Dockery, D.W. Health effects of fine particulate air pollution: Lines that connect. *J. Air Waste Manag. Assoc.* **2006**, *56*, 709–742. [CrossRef]
90. Chen, J.; Hoek, G. Long-term exposure to PM and all-cause and cause-specific mortality: A systematic review and meta-analysis. *Environ. Int.* **2020**, *143*, 105974. [CrossRef]
91. Brauer, M.; Amann, M.; Burnett, R.T.; Cohen, A.; Dentener, F.; Ezzati, M.; Henderson, S.B.; Krzyzanowski, M.; Martin, R.V.; Van Dingenen, R.; et al. Exposure assessment for estimation of the global burden of disease attributable to outdoor air pollution. *Environ. Sci. Technol.* **2012**, *46*, 652–660. [CrossRef]
92. Bachmann, J. Will the circle be unbroken: A history of the US National Ambient Air Quality Standards. *J. Air Waste Manag. Assoc.* **2007**, *57*, 652–697. [CrossRef]
93. Pinder, R.W.; Klopp, J.M.; Kleiman, G.; Hagler, G.S.; Awe, Y.; Terry, S. Opportunities and challenges for filling the air quality data gap in low-and middle-income countries. *Atmos. Environ.* **2019**, *215*, 116794. [CrossRef]
94. Belis, C.A.; Karagulian, F.; Larsen, B.R.; Hopke, P.K. Critical review and meta-analysis of ambient particulate matter source apportionment using receptor models in Europe. *Atmos. Environ.* **2013**, *69*, 94–108. [CrossRef]
95. Chow, J.C. Measurement methods to determine compliance with ambient air quality standards for suspended particles. *J. Air Waste Manag. Assoc.* **1995**, *45*, 320–382. [CrossRef]
96. Snyder, E.G.; Watkins, T.H.; Solomon, P.A.; Thoma, E.D.; Williams, R.W.; Hagler, G.S.W.; Shelow, D.; Hindin, D.A.; Kilaru, V.J.; Preuss, P.W. The changing paradigm of air pollution monitoring. *Environ. Sci. Technol.* **2013**, *47*, 11369–11377. [CrossRef] [PubMed]
97. Van Donkelaar, A.; Martin, R.V.; Brauer, M.; Hsu, N.C.; Kahn, R.A.; Levy, R.C.; Winker, D.M. Global estimates of fine particulate matter using a combined geophysical-statistical method with information from satellites, models, and monitors. *Environ. Sci. Technol.* **2016**, *50*, 3762–3772.
98. Li, J.; Carlson, B.E.; Lacis, A.A. Using single-scattering albedo spectral curvature to characterize East Asian aerosol mixtures. *J. Geophys. Res. Atmos.* **2015**, *120*, 2037–2052. [CrossRef]
99. Kulmala, M.; Dada, L.; Daellenbach, K.R.; Yan, C.; Stolzenburg, D.; Kontkanen, J.; Ezhova, E.; Hakala, S.; Tuovinen, S.; Kokkonen, T.V.; et al. Is reducing new particle formation a plausible solution to mitigate particulate air pollution in Beijing and other Chinese megacities? *Faraday Discuss.* **2021**, *226*, 334–347.
100. Di, Q.; Amini, H.; Shi, L.; Kloog, I.; Silvern, R.; Kelly, J.; Sabath, M.B.; Choirat, C.; Koutrakis, P.; Lyapustin, A.; et al. An ensemble-based model of PM2.5 concentration across the contiguous United States with high spatiotemporal resolution. *Environ. Int.* **2019**, *130*, 104909.
101. LeCun, Y.; Bengio, Y.; Hinton, G. Deep learning. *Nature* **2015**, *521*, 436–444.
102. Reichstein, M.; Camps-Valls, G.; Stevens, B.; Jung, M.; Denzler, J.; Carvalhais, N.; Prabhat, F. Deep learning and process understanding for data-driven Earth system science. *Nature* **2019**, *566*, 195–204.
103. Shen, H.; Li, T.; Yuan, Q.; Zhang, L. Estimating regional ground-level PM2.5 directly from satellite top-of-atmosphere reflectance using deep belief networks. *J. Geophys. Res. Atmos.* **2018**, *123*, 13875–13886.
104. Qi, Y.; Li, Q.; Karimian, H.; Liu, D. A hybrid model for spatiotemporal forecasting of PM2.5 based on graph convolutional neural network and long short-term memory. *Sci. Total Environ.* **2019**, *664*, 1–10.
105. Yan, R.; Liao, J.; Yang, J.; Sun, W.; Nong, M.; Li, F. Multi-hour and multi-site air quality index forecasting in Beijing using CNN, LSTM, CNN-LSTM, and spatiotemporal clustering. *Expert Syst. Appl.* **2021**, *169*, 114513.
106. Wardana, I.N.K.; Gardner, J.W.; Fahmy, S.A. Estimation of missing air pollutant data using a spatiotemporal convolutional autoencoder. *Neural Comput. Appl.* **2022**, *34*, 16129–16154.
107. Ong, B.T.; Sugiura, K.; Zettsu, K. Dynamically pre-trained deep recurrent neural networks using environmental monitoring data for predicting PM 2.5. *Neural Comput. Appl.* **2016**, *27*, 1553–1566. [PubMed]
108. Kalajdjieski, J.; Zdravevski, E.; Corizzo, R.; Lameski, P.; Kalajdziski, S.; Pires, I.M.; Garcia, N.M.; Trajkovik, V. Air pollution prediction with multi-modal data and deep neural networks. *Remote Sens.* **2020**, *12*, 4142. [CrossRef]
109. Lee, Y.; Park, J.; Kim, J.; Woo, J.-H.; Lee, J.-H. Rapid PM2.5-Induced Health Impact Assessment: A Novel Approach Using Conditional U-Net CMAQ Surrogate Model. *Atmosphere* **2024**, *15*, 1186. [CrossRef]

110. Zhang, B.; Rong, Y.; Yong, R.; Qin, D.; Li, M.; Zou, G.; Pan, J. Deep learning for air pollutant concentration prediction: A review. *Atmos. Environ.* **2022**, *290*, 119347. [CrossRef]
111. Zhang, C.; Sargent, I.; Pan, X.; Li, H.; Gardiner, A.; Hare, J.; Atkinson, P.M. Joint Deep Learning for land cover and land use classification. *Remote Sens. Environ.* **2019**, *221*, 173–187.
112. Riese, F.M.; Keller, S. Supervised, semi-supervised, and unsupervised learning for hyperspectral regression. In *Hyperspectral Image Analysis: Advances in Machine Learning and Signal Processing*; Springer: Berlin/Heidelberg, Germany, 2020; pp. 187–232.
113. Wieland, M.; Li, Y.; Martinis, S. Multi-sensor cloud and cloud shadow segmentation with a convolutional neural network. *Remote Sens. Environ.* **2019**, *230*, 111203.
114. Wei, S.; Ji, S.; Lu, M. Toward automatic building footprint delineation from aerial images using CNN and regularization. *IEEE Trans. Geosci. Remote Sens.* **2019**, *58*, 2178–2189.
115. Liu, B.; Qi, Z.; Gao, L. Enhanced Air Quality Prediction through Spatio-temporal Feature Sxtraction and Fusion: A Self-tuning Hybrid Approach with GCN and GRU. *Water Air Soil Pollut.* **2024**, *235*, 532.
116. Sun, X.; Xu, W.; Jiang, H. Spatio-temporal prediction of air quality based on recurrent neural networks. In Proceedings of the 52nd Hawaii International Conference on System Sciences (HICSS), Maui, HI, USA, 8–11 January 2019; pp. 1–10. [CrossRef]
117. Wang, Y.; Liu, K.; He, Y.; Wang, P.; Chen, Y.; Xue, H.; Huang, C.; Li, L. Enhancing air quality forecasting: A novel spatio-temporal model integrating graph convolution and multi-head attention mechanism. *Atmosphere* **2024**, *15*, 418. [CrossRef]
118. Li, T.; Hua, M.; Wu, X. A hybrid CNN-LSTM model for forecasting particulate matter (PM2.5). *IEEE Access* **2020**, *8*, 26933–26940. [CrossRef]
119. Liu, B.; Yan, S.; Li, J.; Qu, G.; Li, Y.; Lang, J.; Gu, R. A sequence-to-sequence air quality predictor based on the n-step recurrent prediction. *IEEE Access* **2019**, *7*, 43331–43345. [CrossRef]
120. Wang, X.; Zhang, S.; Chen, Y.; He, L.; Ren, Y.; Zhang, Z.; Juan, L.; Zhang, S. Air quality forecasting using a spatiotemporal hybrid deep learning model based on VMD–GAT–BiLSTM. *Sci. Rep.* **2024**, *14*, 17841. [CrossRef] [PubMed]
121. Bianchi, F.M.; Maiorino, E.; Kampffmeyer, M.C.; Rizzi, A.; Jenssen, R. An overview and comparative analysis of recurrent neural networks for short term load forecasting. *arXiv* **2017**, arXiv:1705.04378.
122. Zhao, J.; Deng, F.; Cai, Y.; Chen, J. Long short-term memory-Fully connected (LSTM-FC) neural network for PM2.5 concentration prediction. *Chemosphere* **2019**, *220*, 486–492.
123. Sønderby, C.K.; Espeholt, L.; Heek, J.; Dehghani, M.; Oliver, A.; Salimans, T.; Kalchbrenner, N. Metnet: A neural weather model for precipitation forecasting. *arXiv* **2020**, arXiv:2003.12140.
124. Wen, C.; Liu, S.; Yao, X.; Peng, L.; Li, X.; Hu, Y.; Chi, T. A novel spatiotemporal convolutional long short-term neural network for air pollution prediction. *Sci. Total Environ.* **2019**, *654*, 1091–1099.
125. Peng, D.; Zhang, Y.; Guan, H. End-to-end change detection for high resolution satellite images using improved UNet++. *Remote Sens.* **2019**, *11*, 1382. [CrossRef]
126. Tao, Q.; Liu, F.; Li, Y.; Sidorov, D. Air pollution forecasting using a deep learning model based on 1D convnets and bidirectional GRU. *IEEE Access* **2019**, *7*, 76690–76698.
127. Grant-Jacob, J.A.; Mills, B. Deep learning in airborne particulate matter sensing: A review. *J. Phys. Commun.* **2022**, *6*, 122001.
128. Mohammed, A.F.Y.; Sultan, S.M.; Cho, S.; Pyun, J.-Y. Powering UAV with deep Q-network for air quality tracking. *Sensors* **2022**, *22*, 6118. [CrossRef] [PubMed]
129. Wang, K.; Zhang, S.; Chen, J.; Ren, F.; Xiao, L. A feature-supervised generative adversarial network for environmental monitoring during hazy days. *Sci. Total Environ.* **2020**, *748*, 141445. [CrossRef] [PubMed]
130. Xie, M.; Jean, N.; Burke, M.; Lobell, D.; Ermon, S. Transfer learning from deep features for remote sensing and poverty mapping. In Proceedings of the 30th AAAI Conference on Artificial Intelligence, Phoenix, AZ, USA, 12–17 February 2016; pp. 1–10. [CrossRef]
131. Luis, S.Y.; Reina, D.G.; Marin, S.L.T. A deep reinforcement learning approach for the patrolling problem of water resources through autonomous surface vehicles: The ypacarai lake case. *IEEE Access* **2020**, *8*, 204076–204093. [CrossRef]
132. Mansfield, D.; Montazeri, A. A survey on autonomous environmental monitoring approaches: Towards unifying active sensing and reinforcement learning. *Front. Robot. AI* **2024**, *11*, 1336612. [CrossRef]
133. Zeng, Q.; Ma, X.; Cheng, B.; Zhou, E.; Pang, W. Gans-based data augmentation for citrus disease severity detection using deep learning. *IEEE Access* **2020**, *8*, 172882–172891. [CrossRef]
134. Njaime, M.; Abdallah, F.; Snoussi, H.; Akl, J.; Chaaban, K.; Omrani, H. Transfer learning based solution for air quality prediction in smart cities using multimodal data. *Int. J. Environ. Sci. Technol.* **2024**, *22*, 1297–1312. [CrossRef]
135. Wei, T.; Wang, Y.; Zhu, Q. Deep reinforcement learning for building HVAC control. In Proceedings of the 54th Annual Design Automation Conference, Austin, TX, USA, 18–22 June 2017.
136. Sonawani, S.; Patil, K. Air quality measurement, prediction and warning using transfer learning based IOT system for ambient assisted living. *Int. J. Pervasive Comput. Commun.* **2024**, *20*, 38–55.

137. Cheng, W.; Shen, Y.; Zhu, Y.; Huang, L. A neural attention model for urban air quality inference: Learning the weights of monitoring stations. In Proceedings of the 32nd AAAI Conference on Artificial Intelligence, New Orleans, LA, USA, 2–7 February 2018; Volume 32, p. 1. [CrossRef]
138. Zhang, M.; Li, W.; Du, Q. Diverse region-based CNN for hyperspectral image classification. *IEEE Trans. Image Process.* **2018**, *27*, 2623–2634. [CrossRef]
139. Wang, S.; Li, Y.; Zhang, J.; Meng, Q.; Meng, L.; Gao, F. Pm2.5-gnn: A domain knowledge enhanced graph neural network for pm2. 5 forecasting. In Proceedings of the 28th International Conference on Advances in Geographic Information Systems, Seattle, WA, USA, 3–6 November 2020; pp. 163–166. [CrossRef]
140. Pan, R.; Liu, T.; Ma, L. A Graph Attention Recurrent Neural Network Model for PM2.5 Prediction: A Case Study in China from 2015 to 2022. *Atmosphere* **2024**, *15*, 799. [CrossRef]
141. Liao, H.; Wu, M.; Yuan, L.; Hu, Y.; Gong, H. PM2.5 prediction based on dynamic spatiotemporal graph neural network. *Appl. Intell.* **2024**, *54*, 11933–11948.
142. Zhao, Z.; Wu, J.; Cai, F.; Zhang, S.; Wang, Y.-G. A hybrid deep learning framework for air quality prediction with spatial autocorrelation during the COVID-19 pandemic. *Sci. Rep.* **2023**, *13*, 1015. [CrossRef] [PubMed]
143. Le, V.-D. Spatiotemporal graph convolutional recurrent neural network model for citywide air pollution forecasting. *arXiv* **2023**, arXiv:2304.12630.
144. Zhang, X.; Chen, L.; Tang, X.; Shi, H. DSGNN: A Dual-View Supergrid-Aware Graph Neural Network for Regional Air Quality Estimation. *arXiv* **2024**, arXiv:2404.01975.
145. Yi, X.; Zhang, J.; Wang, Z.; Li, T.; Zheng, Y. Deep distributed fusion network for air quality prediction. In Proceedings of the 24th ACM SIGKDD International Conference on Knowledge Discovery & Data Mining, London, UK, 19–23 August 2018; pp. 965–973. [CrossRef]
146. Hughes, L.H.; Schmitt, M.; Mou, L.; Wang, Y.; Zhu, X.X. Identifying corresponding patches in SAR and optical images with a pseudo-siamese CNN. *IEEE Geosci. Remote Sens. Lett.* **2018**, *15*, 784–788. [CrossRef]
147. Huang, C.-J.; Kuo, P.-H. A deep CNN-LSTM model for particulate matter (PM2.5) forecasting in smart cities. *Sensors* **2018**, *18*, 2220. [CrossRef]
148. Lu, G.; Yu, E.; Wang, Y.; Li, H.; Cheng, D.; Huang, L.; Liu, Z.; Manomaiphiboon, K.; Li, L. A novel hybrid machine learning method (OR-ELM-AR) used in forecast of PM2.5 concentrations and its forecast performance evaluation. *Atmosphere* **2021**, *12*, 78. [CrossRef]
149. Xiao, F.; Yang, M.; Fan, H.; Fan, G.; Al-Qaness, M.A. An improved deep learning model for predicting daily PM2.5 concentration. *Sci. Rep.* **2020**, *10*, 20988.
150. Zhang, K.; Yang, X.; Cao, H.; Thé, J.; Tan, Z.; Yu, H. Multi-step forecast of PM2.5 and PM10 concentrations using convolutional neural network integrated with spatial–temporal attention and residual learning. *Environ. Int.* **2023**, *171*, 107691.
151. Yan, X.; Zang, Z.; Luo, N.; Jiang, Y.; Li, Z. New interpretable deep learning model to monitor real-time PM2.5 concentrations from satellite data. *Environ. Int.* **2020**, *144*, 106060.
152. Liu, Y.; Zhang, Y.; Yu, P.; Ye, T.; Zhang, Y.; Xu, R.; Guo, Y. Applying traffic camera and deep learning-based image analysis to predict PM2.5 concentrations. *Sci. Total Environ.* **2024**, *912*, 169233. [CrossRef]
153. Teng, M.; Li, S.; Xing, J.; Fan, C.; Yang, J.; Wang, S.; Song, G.; Ding, Y.; Dong, J.; Wang, S. 72-hour real-time forecasting of ambient PM2.5 by hybrid graph deep neural network with aggregated neighborhood spatiotemporal information. *Environ. Int.* **2023**, *176*, 107971. [CrossRef] [PubMed]
154. Elbaz, K.; Shaban, W.M.; Zhou, A.; Shen, S.-L. Real time image-based air quality forecasts using a 3D-CNN approach with an attention mechanism. *Chemosphere* **2023**, *333*, 138867. [CrossRef] [PubMed]
155. Kim, H.S.; Park, I.; Song, C.H.; Lee, K.; Yun, J.W.; Kim, H.K.; Han, K.M. Development of a daily PM 10 and PM 2.5 prediction system using a deep long short-term memory neural network model. *Atmos. Chem. Phys.* **2019**, *19*, 12935–12951. [CrossRef]
156. Lee, J.-B.; Lee, J.-B.; Koo, Y.-S.; Kwon, H.-Y.; Choi, M.-H.; Park, H.-J.; Lee, D.-G. Development of a deep neural network for predicting 6 h average PM 2.5 concentrations up to 2 subsequent days using various training data. *Geosci. Model Dev.* **2022**, *15*, 3797–3813. [CrossRef]
157. Qin, D.; Yu, J.; Zou, G.; Yong, R.; Zhao, Q.; Zhang, B. A novel combined prediction scheme based on CNN and LSTM for urban PM 2.5 concentration. *IEEE Access* **2019**, *7*, 20050–20059. [CrossRef]
158. Cui, B.; Liu, M.; Li, S.; Jin, Z.; Zeng, Y.; Lin, X. Deep learning methods for atmospheric PM2.5 prediction: A comparative study of transformer and CNN-LSTM-attention. *Atmos. Pollut. Res.* **2023**, *14*, 101833. [CrossRef]
159. Lee, S.-S.; Song, W.-Y.; Kim, Y.-J. Intelligent PM 2.5 mass concentration analyzer using deep learning algorithm and improved density measurement chip for high-accuracy airborne particle sensor network. *J. Aerosol Sci.* **2023**, *167*, 106097. [CrossRef]
160. Kim, J.; Go, T.; Lee, S.J. Volumetric monitoring of airborne particulate matter concentration using smartphone-based digital holographic microscopy and deep learning. *J. Hazard. Mater.* **2021**, *418*, 126351. [CrossRef]

161. Zhang, Q.; Fu, F.; Tian, R. A deep learning and image-based model for air quality estimation. *Sci. Total Environ.* **2020**, *724*, 138178. [CrossRef]
162. Li, T.; Shen, H.; Yuan, Q.; Zhang, L. Deep learning for ground-level PM 2.5 prediction from satellite remote sensing data. In Proceedings of the IGARSS 2018-2018 IEEE International Geoscience and Remote Sensing Symposium, Valencia, Spain, 22–27 July 2018.
163. Zimmerman, N.; Presto, A.A.; Kumar, S.P.N.; Gu, J.; Hauryliuk, A.; Robinson, E.S.; Robinson, A.L.; Subramanian, R. A machine learning calibration model using random forests to improve sensor performance for lower-cost air quality monitoring. *Atmos. Meas. Tech.* **2018**, *11*, 291–313. [CrossRef]
164. Li, X.; Peng, L.; Yao, X.; Cui, S.; Hu, Y.; You, C.; Chi, T. Long short-term memory neural network for air pollutant concentration predictions: Method development and evaluation. *Environ. Pollut.* **2017**, *231*, 997–1004. [CrossRef] [PubMed]
165. Li, T.; Shen, H.; Yuan, Q.; Zhang, X.; Zhang, L. Estimating ground-level PM2.5 by fusing satellite and station observations: A geo-intelligent deep learning approach. *Geophys. Res. Lett.* **2017**, *44*, 11985–11993.
166. Thompson, J.E. Crowd-sourced air quality studies: A review of the literature & portable sensors. *Trends Environ. Anal. Chem.* **2016**, *11*, 23–34.
167. Feng, X.; Li, Q.; Zhu, Y.; Hou, J.; Jin, L.; Wang, J. Artificial neural networks forecasting of PM2.5 pollution using air mass trajectory based geographic model and wavelet transformation. *Atmos. Environ.* **2015**, *107*, 118–128.
168. Chen, J.; Yin, J.; Zang, L.; Zhang, T.; Zhao, M. Stacking machine learning model for estimating hourly PM2.5 in China based on Himawari 8 aerosol optical depth data. *Sci. Total Environ.* **2019**, *697*, 134021.
169. Wu, K.-Y.; Hsia, I.-W.; Kow, P.-Y.; Chang, L.-C.; Chang, F.-J. High-spatiotemporal-resolution PM2.5 forecasting by hybrid deep learning models with ensembled massive heterogeneous monitoring data. *J. Clean. Prod.* **2023**, *433*, 139825.
170. Mi, T.; Tang, D.; Fu, J.; Zeng, W.; Grieneisen, M.L.; Zhou, Z.; Jia, F.; Yang, F.; Zhan, Y. Data augmentation for bias correction in mapping PM2. 5 based on satellite retrievals and ground observations. *Geosci. Front.* **2024**, *15*, 101686.
171. Ni, J.; Chen, Y.; Gu, Y.; Fang, X.; Shi, P. An improved hybrid transfer learning-based deep learning model for PM2.5 concentration prediction. *Appl. Sci.* **2022**, *12*, 3597. [CrossRef]
172. Zhai, B.; Chen, J. Development of a stacked ensemble model for forecasting and analyzing daily average PM2.5 concentrations in Beijing, China. *Sci. Total Environ.* **2018**, *635*, 644–658.
173. Umar, I.K.; Nourani, V.; Gökçekuş, H. A novel multi-model data-driven ensemble approach for the prediction of particulate matter concentration. *Environ. Sci. Pollut. Res.* **2021**, *28*, 49663–49677.
174. Rachuri, K.K. *Smartphones Based Social Sensing: Adaptive Sampling, Sensing and Computation Offloading*; University of Cambridge: Cambridge, UK, 2013.
175. Haut, J.M.; Paoletti, M.E.; Moreno-Alvarez, S.; Plaza, J.; Rico-Gallego, J.-A.; Plaza, A. Distributed deep learning for remote sensing data interpretation. *Proc. IEEE* **2021**, *109*, 1320–1349.
176. Wang, Z.; Hu, B.; Huang, B.; Ma, Z.; Biswas, A.; Jiang, Y.; Shi, Z. Predicting annual PM2.5 in mainland China from 2014 to 2020 using multi temporal satellite product: An improved deep learning approach with spatial generalization ability. *ISPRS J. Photogramm. Remote Sens.* **2022**, *187*, 141–158.
177. Yan, X.; Zang, Z.; Jiang, Y.; Shi, W.; Guo, Y.; Li, D.; Zhao, C.; Husi, L. A Spatial-Temporal Interpretable Deep Learning Model for improving interpretability and predictive accuracy of satellite-based PM2.5. *Environ. Pollut.* **2021**, *273*, 116459.
178. Gregório, J.; Gouveia-Caridade, C.; Caridade, P.J.S.B. Modeling PM2.5 and PM10 using a robust simplified linear regression machine learning algorithm. *Atmosphere* **2022**, *13*, 1334. [CrossRef]
179. Ma, P.; Tao, F.; Gao, L.; Leng, S.; Yang, K.; Zhou, T. Retrieval of fine-grained PM2.5 spatiotemporal resolution based on multiple machine learning models. *Remote Sens.* **2022**, *14*, 599. [CrossRef]
180. Yang, Y.; Wang, Z.; Cao, C.; Xu, M.; Yang, X.; Wang, K.; Shi, Z. Estimation of pm2.5 concentration across china based on multi-source remote sensing data and machine learning methods. *Remote Sens.* **2024**, *16*, 467. [CrossRef]

*atmosphere*

MDPI

*Review*

# Analytical Methods for Atmospheric Carbonyl Compounds: A Review

**Xiaoshuai Gao [1,2], Xin Zhang [3], Yan Nie [2], Jiemeng Bao [4], Junling Li [2], Rui Gao [2], Yunfeng Li [5], Wei Wei [1,*], Xiaoyu Yan [2], Yongxin Yan [2] and Hong Li [2,*]**

1 Department of Environmental Science and Engineering, Beijing University of Technology, Beijing 100124, China; s202265286@emails.bjut.edu.cn

2 State Key Laboratory of Environmental Criteria and Risk Assessment, Chinese Research Academy of Environmental Sciences, Beijing 100012, China; nieyan23@mails.ucas.ac.cn (Y.N.); lijl@craes.org.cn (J.L.); gaorui@craes.org.cn (R.G.); yanxiaoyu22@mails.jlu.edu.cn (X.Y.); yyx_in@163.com (Y.Y.)

3 Chinese Academy of Environmental Planning, Beijing 100041, China; zhangxin1@caep.org.cn

4 State Key Joint Laboratory of Environmental Simulation and Pollution Control, College of Environmental Sciences and Engineering, Peking University, Beijing 100871, China; 2301112284@stu.pku.edu.cn

5 School of Mechanical Engineering, Beijing Institute of Petrochemical Technology, Beijing 102617, China; liyunfeng@bipt.edu.cn

* Correspondence: weiwei@bjut.edu.cn (W.W.); lihong@craes.org.cn (H.L.)

**Abstract:** Atmospheric carbonyl compounds have significant impacts on the atmospheric environment and human health, making the selection of appropriate analytical techniques crucial for accurately detecting these compounds in specific environments. Based on extensive literature research, this study summarized the development history, relevant features, and applicable scenarios of the main analytical techniques for atmospheric carbonyl compounds; pointed out the main problems and challenges in this field; and discussed the needs and prospects of future research and application. It was found that the direct sampling methods of atmospheric carbonyl compounds were applicable to low-molecular-weight carbonyl species with low reactivity, low boiling points, high polarity, and high volatility, while indirect sampling methods were suitable for a wider range and various types and phases of species. For formaldehyde, offline detection was primarily influenced by chemical reagents and reaction conditions, whereas online monitoring relied on sufficiently stable operating environments. For multiple carbonyl compounds, offline detection results were greatly influenced by detectors coupled with chromatography, whereas online monitoring techniques were applicable to all types of volatile organic compounds (VOCs), including some carbonyl compounds, providing higher temporal resolution and improved isomer identification with the development of online mass spectrometry. The combined use of proton transfer reaction-mass spectrometry (PTR-MS) and liquid chromatography-mass spectrometry (GC-MS) was suitable for the detection of carbonyl compounds in atmospheric photochemical smog chamber simulation studies. Currently, offline analytical techniques for carbonyl compounds require significant time and advanced experimental skills for multiple optimization experiments to detect a broader range of species. Online monitoring techniques face challenges such as poor stability and limited species coverage. In smog chamber simulation studies, the detection of carbonyl compounds is heavily influenced by both the sampling system and the chamber itself. Future efforts should focus on improving the environmental adaptability and automation of carbonyl compound analytical techniques, the synergistic use of various techniques, developing new sampling systems, and reducing the impact of the chamber itself on carbonyl compound detection, in order to enhance detection sensitivity, selectivity, time resolution, accuracy, and operability.

 https://doi.org/10.3390/atmos16010107

**Keywords:** atmospheric carbonyls; sampling methods; offline analytical methods; online analytical methods; problems and challenges; prospect

## 1. Introduction

Atmospheric carbonyl compounds are significant oxygenated volatile organic compounds (OVOCs) [1] that play essential roles as precursors and intermediates in photochemical reactions, contributing substantially to the formation of peroxyacetyl nitrate (PAN), ozone ($O_3$), and secondary organic aerosols (SOAs) [2]. As precursors, carbonyl compounds can directly participate in atmospheric photochemical reactions, leading to the formation of secondary pollutants such as $O_3$ and PAN [3]. These compounds can also undergo photolysis or react with OH radicals to affect the formation of SOAs by forming oligomers [4–6]. As intermediates in photochemical reactions, carbonyl compounds can undergo photolysis or further reactions with atmospheric radicals or oxidants [7], affecting the generation of photochemical pollutants [8]. These chemical reactions of carbonyl compounds play a crucial role in the comprehensive understanding of atmospheric auto-oxidation, gas-particle partitioning, and photochemical mechanisms [9]. Furthermore, carbonyl compounds have significant impacts on both the ecosystem and human health. As components of volatile organic compounds (VOCs), carbonyl compounds directly affect air quality and pose immediate health risks. Most of these compounds are highly irritating, potentially causing respiratory infections and posing sensitizing, carcinogenic, and mutagenic hazards [10]. Formaldehyde and acetaldehyde have been classified as carcinogenic and teratogenic by the World Health Organization (WHO) [11]. Acetone has been recognized as having significant effects on blood and kidney health [12], and crotonaldehyde has been found to be carcinogenic in animals [13]. Pollutants like PAN, $O_3$, and SOA, formed from chemical reactions involving carbonyl compounds, can influence climate change and lead to photochemical smog and acid rain, thereby indirectly posing serious threats to the atmospheric environment, vegetation, and human health. Consequently, atmospheric carbonyl compounds have become a key focus in atmospheric environmental research.

Atmospheric carbonyl compounds exhibit a wide range of concentration variability, high reactivity, and instability, with typically short atmospheric lifetimes [14,15]. This places stringent demands on sampling and analytical techniques. Accurate measurement of the types and concentrations of atmospheric carbonyl compounds in various environments has become a critical prerequisite for studying their pollution characteristics, reaction mechanisms, environmental impacts, and health effects. Overall, the concentration levels and composition of carbonyl compounds vary across different atmospheric environments. Concentrations in rural areas are generally lower than those in urban and suburban areas, with even higher levels observed in large cities [16]. This variation necessitates analytical techniques with high sensitivity and selectivity to accurately measure a wide range of concentrations and multiple carbonyl species in diverse environments. Atmospheric carbonyl compounds are easily affected by other environmental highly oxidative species, leading to potential volatilization, degradation, or chemical reactions of certain compounds during sampling or storage [17]. Thus, choosing suitable sampling techniques, derivatization reagents, or adsorbents is crucial. Some carbonyl compounds possess strong polarity and good water solubility, allowing them to exist simultaneously in both the gas and aerosol phases. The atmospheric lifetime and impact of these compounds undergo significant changes between the gas and particle phases [18], requiring the use of accurate and appropriate methods for simultaneous detection of their gas- and particle-phase species and concentrations. Smog chamber simulation studies involve numerous experimental scenarios and complex chemical reactions, which generate various types of carbonyl compounds and unknown carbonyl species. Additionally, smog chambers differ in size, shape, materials, and light sources worldwide, all of which can affect the chemical reactions within the chamber [19]. Therefore, the identification of carbonyl compounds in smog chamber simulation studies requires detection technologies that meet high time resolution, high sensitivity, high stability, and high selectivity. Notably, formaldehyde is one of the most abundant, reactive, and volatile carbonyl compounds in the atmosphere. Ubiquitous and highly toxic, formaldehyde has significant impacts on daily life, requiring timely and accurate monitoring of its concentration levels [20,21]. To fully assess formaldehyde pollution in specific settings, it is often measured separately.

Analytical methods for carbonyl compounds can be categorized into offline and online analytical methods. Offline methods involve separate steps for sampling, pretreatment, and analysis of targeted carbonyl compounds, while online methods integrate sampling, pretreatment, and analysis into a unified system. The 2,4-dinitrophenylhydrazine (DNPH) derivatization method coupled with high-performance liquid chromatography/ultraviolet (HPLC/UV) is a common offline method for detecting carbonyl compounds. Coupling gas chromatography (GC) or liquid chromatography (LC) with mass spectrometry (MS) enables the offline detection of a wider range of carbonyl species [22]. Formaldehyde can also be detected offline using methods such as fluorescence, spectrophotometry, and electrochemistry [23,24]. Online analytical methods for carbonyl compounds primarily include online gas chromatography-flame ionization detector/mass spectrometry (GC-FID/MS) and proton transfer reaction-mass spectrometry (PTR-MS) [25]. Formaldehyde can be monitored online using techniques such as Fourier transform infrared spectroscopy (FTIR), laser-induced fluorescence (LIF), differential optical absorption spectroscopy (DOAS), tunable diode laser absorption spectroscopy (TDLAS), cavity-enhanced absorption spectroscopy (CEAS), cavity ring-down spectroscopy (CRDS), and gas sensors [21,26]. The selection of analytical techniques for carbonyl compounds depends on factors such as the specific application scenario, the properties of the targeted compounds, the detection objectives, and cost considerations.

Currently, carbonyl compound analytical methods are advancing rapidly, with both improvements in existing techniques and the emergence of new methods. The optimization of common analytical techniques for carbonyl compounds is frequent, with numerous studies focusing on technical advancements. Some researchers have conducted comprehensive reviews of specific techniques, such as Wu et al.'s review on spectral detection techniques for formaldehyde [27] and Chung et al.'s review on formaldehyde sensor technologies [26]. Other studies have reviewed analytical methods for multiple carbonyl compounds in various environments, but these reviews do not specifically focus on atmospheric analytical methods [28]. The reviews on analytical methods in the atmospheric environment often cover only the significant advancements and primary performance characteristics of certain techniques, with limited summaries on other common methods and their specific parameters [29], and lack a summary of detection techniques for particle-phase carbonyls and smog chamber simulation studies. Given the current lack of comprehensive reviews on the optimization history, applicability, and specific parameters of commonly used analytical techniques for atmospheric carbonyl compounds, this study provides a comprehensive overview of sampling methods for gas- and particle-phase carbonyl compounds based on extensive literature research and systematic analysis, examining the optimization process and suitable applications of online and offline methods for detecting formaldehyde and multiple carbonyl compounds and summarizing the detection technologies for carbonyl compounds involved in atmospheric photochemical smog chamber simulation studies. It also highlights the primary problems and challenges in current sampling and analytical techniques. Finally, future research needs and prospects for sampling and analytical techniques of atmospheric carbonyl compounds were discussed. The findings of this study will help clarify the optimization processes and applicability of commonly used carbonyl compound analytical techniques, providing a reference for the selection and optimization of current analytical methods and offering guidance for the improvement and innovation of future analytical methods.

## 2. Methodology

To comprehensively review and summarize the analytical methods for atmospheric carbonyl compounds, the databases of Web of Science, Google Scholar, and China National Knowledge Network (CNKI) were systematically searched. Keywords such as "carbonyls analytical methods", "OVOCs analytical methods", "monitoring techniques", "derivatization", "online mass spectrometry", "formaldehyde gas sensor", and "HCHO spectroscopy" were used to conduct the subject search, and 434 articles were obtained. By screening and analyzing the search results to eliminate duplicates and optimize keywords, it was determined that detailed information on 269 articles needed to be retrieved, of which 6 articles were excluded as they could not be obtained through available channels. Subsequently, based on the objectives of this review, the titles, abstracts, and full texts of the 263 articles were identified for further review. A total of 82 articles were excluded due to their irrelevance to the research topic, redundancy, and poor citation value resulting from incomplete or missing full text information. Finally, 180 research articles were obtained through the database. In addition, 36 research materials were obtained through other methods, such as websites, university and research institute sites, and citation searching. After deduplication and evaluation, 23 materials were included in this review. In total, through various methods, 202 eligible materials were selected and used in this review. The review followed the PRISMA flow methodological phases, as shown in Figure 1, together with the corresponding inclusion and exclusion criteria.

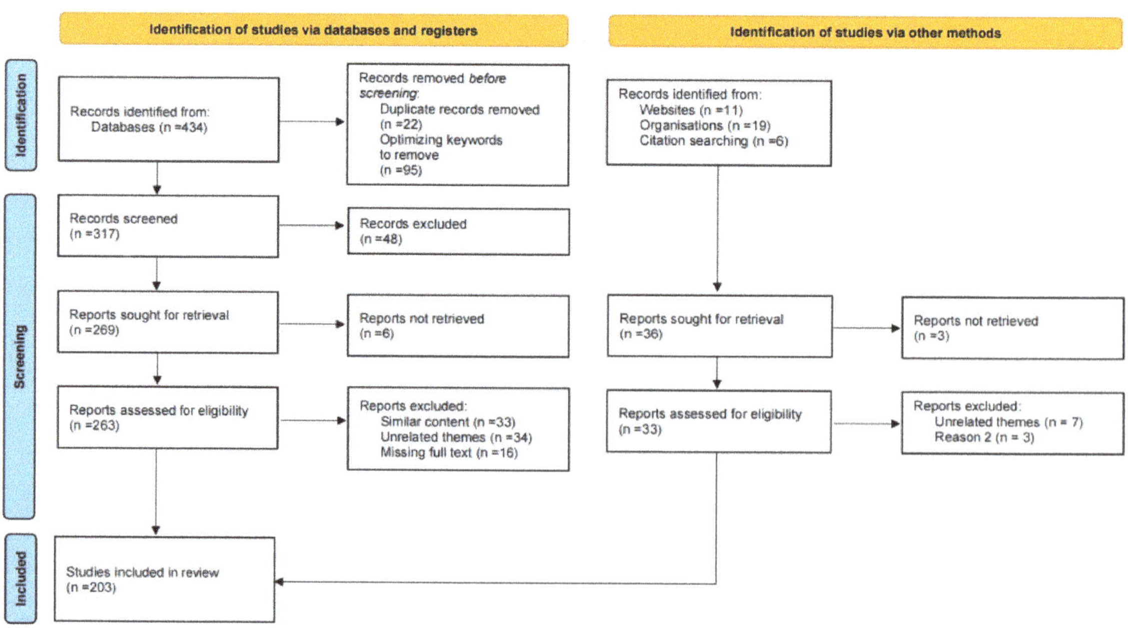


**Figure 1.** PRISMA flow diagram.

As shown in Figure 2, the research materials used in this review included 72 domestic and 79 international articles and standards spanning the period from the 1970s to the 2020s. To further analyze the research materials, the majority of them were divided into three major sections based on the research subjects (formaldehyde, multiple carbonyl compounds, and chambers). The formaldehyde section included offline (derivatization) and online (spectroscopy and sensor), while the multiple carbonyl compounds section included offline (LC/GC-MS and DNPH-HPLC-UV) and online (GC-FID/MS and online MS). Additionally, due to the comprehensive nature of review articles, the unique source of standard documents, and the use of other types of documents in the Section 1, the remaining materials were categorized into reviews, standards, and others. It was observed that articles on carbonyl compounds analytical methods have increased rapidly since the 1990s. For formaldehyde, the research mainly focused on online analytical methods, while both online and offline methods for multiple carbonyl compounds were relatively abundant.

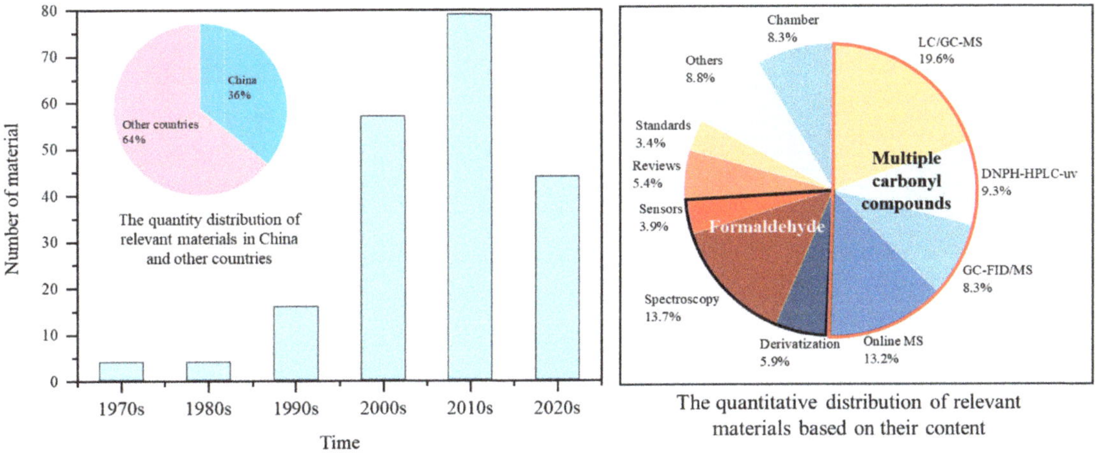


**Figure 2.** Distribution of 151 materials on analytical methods for atmospheric carbonyl compounds from the perspectives of countries and content.

## 3. Sampling Methods for Atmospheric Carbonyl Compounds

Sampling methods for atmospheric carbonyl compounds can be categorized into direct and indirect sampling methods. Direct sampling methods include tube sampling, canister sampling, and cryogenic enrichment techniques. Tube sampling was applied to carbonyl compounds in the 1960s–1970s, while cryogenic enrichment was primarily used for sampling VOCs. Canister sampling began to be employed for VOCs in the 1970s–1980s. With advancements in new adsorbent materials, refrigeration techniques, canister materials, and analytical instruments, tube sampling was extended to other VOC types, and cryogenic enrichment enabled targeted sampling of carbonyl compounds in the 1980s–1990s. Canister sampling efficiency for carbonyl compounds significantly improved in the 1990s. Indirect sampling methods primarily rely on derivatization techniques, which originated in the 1960s–1970s. Various derivatization reagents and coating methods emerged in the 1980s–1990s. In the 2000s, the automation of various sampling techniques improved, and combined approaches began to be employed (Figure 3).

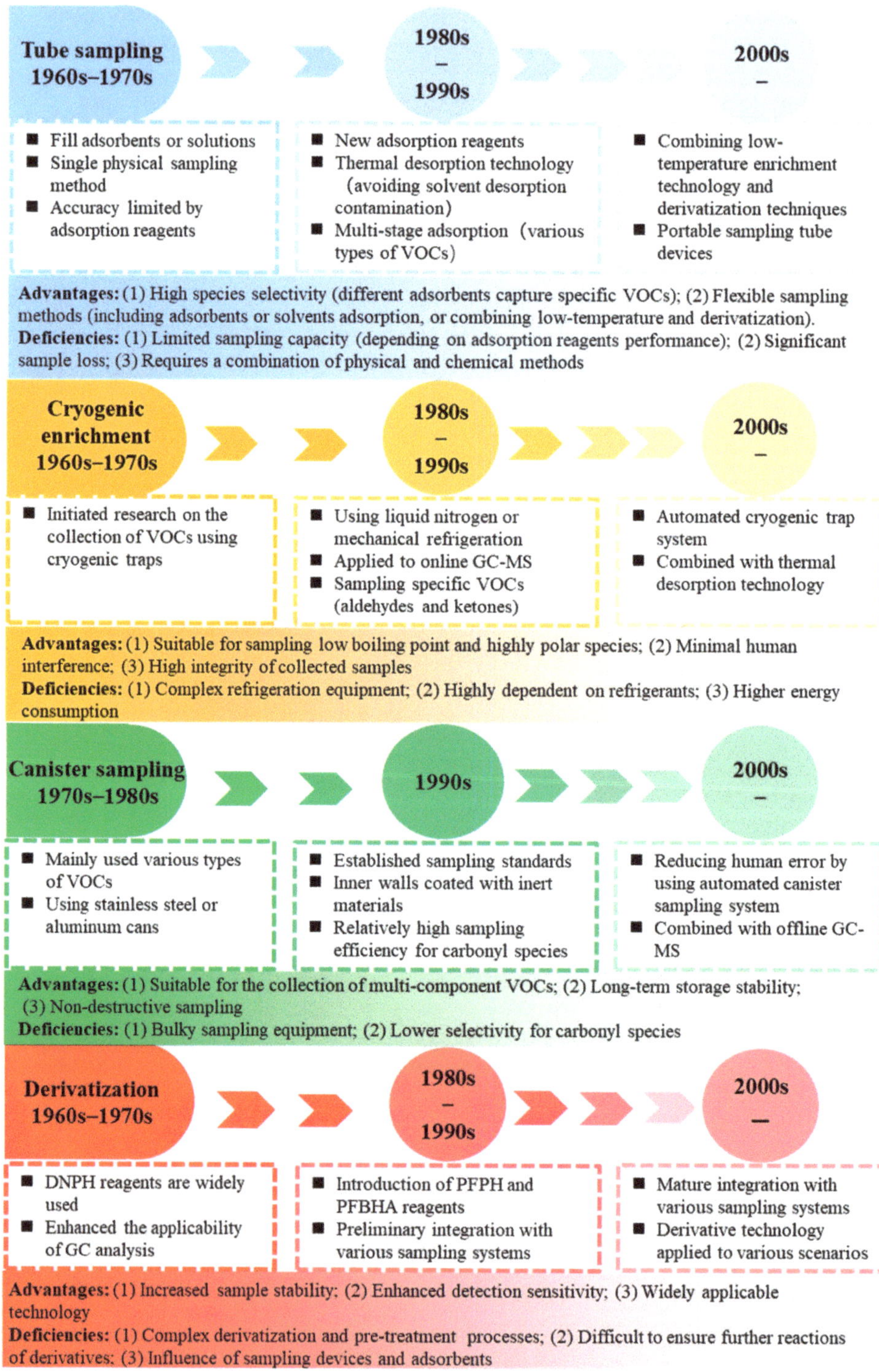


**Figure 3.** Development of main sampling methods for atmospheric carbonyl compounds.

### *3.1. Direct Sampling Methods*

Tube sampling techniques achieve efficient collection of carbonyl compounds by enriching them with solvents or adsorbents. Solvent absorption is simple and cost-effective but is only suitable for carbonyl compounds with low molecular weight and high concentration, such as formaldehyde. Adsorbents are more effective for aldehydes (>C5). The use of multi-layer adsorption tubes enhances the ability to sample C4 aldehydes effectively [30]. Under certain cryogenic conditions, low-boiling-point carbonyl compounds can also be adsorbed. Furthermore, highly polar carbonyl compounds may undergo irreversible adsorption in adsorption tubes [31]. Using cryogenic polytetrafluoroethylene (Teflon) sampling tubes can mitigate this issue [32], effectively improving the sampling efficiency of carbonyl compounds. However, different solvents or adsorbents are only suitable for capturing specific types of compounds, posing certain limitations in applications involving simultaneous sampling of multi-component carbonyl compounds. Currently, tube sampling methods often integrate physical collection with chemical derivatization, such as DNPH-coated silica gel or PFPH (pentafluorophenylhydrazine)-coated Tenax TA adsorbents, to collect a broader range of carbonyl compounds [33,34]. Solvent absorption or adsorbent collection alone is currently limited to formaldehyde sampling.

Canister sampling techniques use specially designed stainless steel or metal canisters to collect atmospheric samples. It is a standard VOC sampling method recommended by the U.S. Environmental Protection Agency (EPA) (TO-14A, TO-15), typically coupled with GC-MS for offline detection of various types of VOCs [35,36]. Its application to carbonyl compound sampling represents a natural extension of existing VOC analytical methods. Canister sampling is widely used due to its advantages of multi-component analysis, environmental friendliness, and repeatable sample injection. However, due to the high reactivity of carbonyl compounds, the storage stability of samples collected using canister sampling is relatively poor, making this method currently suitable only for a limited number of carbonyl compound species [37,38].

Cryogenic enrichment techniques create low-temperature environments to concentrate low-boiling-point atmospheric carbonyl compounds. These techniques are typically coupled with online GC-FID/MS for detecting VOCs, including carbonyl compounds. Currently, most applications utilize liquid nitrogen or other refrigerants to achieve ultra-low temperatures. Cryogenic trap-freezing techniques based on liquid nitrogen enhance the capture of polar carbonyl compounds [39]. When combined with thermal desorption, these methods improve collection efficiency and reduce losses of carbonyl compounds [32], enabling continuous monitoring of more than ten species. However, these techniques heavily rely on refrigerants, and the associated cryogenic traps are bulky and complex, limiting their use in remote areas. Mechanical refrigeration systems that do not require refrigerants are currently applicable to only a limited range of carbonyl compounds (C3–C6) [40]. In addition, the online spectroscopy and sensor-based monitoring technologies for formaldehyde sampling directly draw air through pipeline systems.

### *3.2. Indirect Sampling Methods*

Indirect sampling methods primarily employ derivatization techniques, which reduce the polarity and reactivity of target analytes to enhance the stability of carbonyl compound samples [41]. Following sampling with this method, carbonyl compound samples require pretreatment before analysis with relevant instruments, which can result in higher sample loss and increased time consumption [42]. Derivatization techniques are typically used for the offline detection of formaldehyde or combined with chromatography for analyzing multiple carbonyl compounds. The choice of derivatization reagent requires careful evaluation to ensure stable reaction products compatible with the selected separation and detection procedures [43]. Common derivatization reagents include DNPH, PFPH, and PFBHA.

DNPH is suitable for derivatization detection using LC, and the most commonly used sampling method involves directly passing gas into a device containing a DNPH solution or a cartridge with an acidic solid adsorbent. The device with the DNPH solution is suitable for the enrichment of formaldehyde alone [44], while the cartridge with the adsorbent is suitable for multiple species, with its adsorption performance dependent on the type of adsorbent. Silica cartridges with DNPH have a cleaner background and better breakthrough resistance compared with C18 [45], and they can remain stable at <4 °C for up to six months. However, the background levels of formaldehyde and acetone may increase with prolonged storage times and elevated temperatures. Studies have found that using a gas-phase coating method [46] can reduce the background concentration levels of carbonyl compounds. DNPH is prone to react with oxidants in the environment, such as $NO_X$ and $O_3$, generating interfering compounds. Attaching an IK column or a glass annular tube to the front of the sampling device can remove interference from $O_3$ and other photochemical oxidants on the derivatives [47,48]. Furthermore, unsaturated carbonyl compound derivatives can further react with DNPH, forming dimers and trimers, and silica gel adsorbents may promote this reaction, leading to uncertainties in concentration measurements [49]. Kahnt et al. [50] compared denuders coated with XAD-4 resin and those simultaneously coated with XAD-4 and DNPH and found that using XAD-4 with DNPH improved the collection efficiency of carbonyl compounds. In contrast, XAD-4 resin alone had lower collection efficiency for high-volatility carbonyl compounds, such as formaldehyde, acetaldehyde, acetone, and methacrolein, but was less affected by atmospheric oxidants, including $NO_2$ and $O_3$.

The five fluorine atoms in O-(2,3,4,5,6-pentafluorobenzyl) hydroxylamine (PFBHA) and PFPH make the carbonyl-PFPH/PFBHA derivatives more thermally stable and volatile than DNPH derivatives, making them suitable for GC [51]. PFPH is relatively inexpensive compared with PFBHA and exhibits better derivatization performance. The compatible sampling systems include adsorption tubes, denuders, impactors, and cartridges. The most common and easy-to-operate method is to use derivatizing reagents coated on solid adsorbent tubes. Coating methods include solid-phase adsorption, gas-phase coating, and liquid-phase coating [52]. The same GC-MS analytical technique requires adjustments in coating methods based on different detection needs and analytical objectives [53]. Common adsorbents include silica gel, XAD series, and Tenax series adsorbents. XAD-2 and XAD-7 are thermally unstable, while C18 silica gel has lower efficiency than Tenax adsorbents in desorbing PFBHA derivatives [54]. Tenax TA adsorbents show varying absorption efficiencies and collection rates for different species under different derivatizing reagent coatings. Tenax TA adsorbents with PFBHA require lower rates and collect fewer species, while Tenax TA adsorbents with PFPH are widely used and suitable for C1-C10 species [55]. The combination of the same adsorbent with different derivatizing reagents can also affect the derivatization reaction. Tenax GR adsorbents with PFBHA remain stable for only one day, while Tenax TA adsorbents with PFBHA can last for ten days. Additionally, silica gel cartridges coated with PFPH can effectively capture carbonyl compounds in the air, with a lower detection limit compared with the classic HPLC-UV/visible spectrometry method [56].

Solid-phase microextraction (SPME) is a sampling and enrichment technique first developed by Arthur and Pawliszyn (1990) [57]. SPME fibers can be directly derivatized [58] and exposed to air or headspace during sampling to extract and retain analytes. Headspace solid-phase microextraction (HS-SPME) avoids direct contact with complex sample matrices by extracting the compounds in the sample's headspace, while other SPME methods extracts compounds by direct immersion in the sample. The choice of method depends on the nature of the sample, the volatility of the target analytes, and the requirements for analytical sensitivity and selectivity. The advantages of SPME include high selectivity and sensitivity, the option for automation to minimize human error and solvent consumption, and increased analytical throughput due to fewer preparation steps and shorter extraction times [59]. However, due to the limited adsorption capacity of the fibers, competition between saturation effects and analytes may affect the performance and reproducibility of SPME. Fiber effects can alter the extraction efficiency [60], and cooling the fibers during extraction can improve the SPME efficiency [61]. Studies have found that the effectiveness of different derivatizing reagents on SPME fibers varies significantly. PFPH-coated SPME fibers remain stable for about 11 weeks, while PFBHA-coated SPME fibers remain stable for only about 3 days. Ketones react much more slowly than aldehydes with PFBHA-coated SPME fibers. Furthermore, the loading of PFPH-coated SPME fibers is at least five times greater than that of PFBHA, but the byproducts generated from PFPH derivatization are significantly more than those from PFBHA [62].

The sampling methods mentioned above lack the ability to differentiate between the gas phase and particle phase, making it difficult to simultaneously measure the gas- and particle-phase concentrations of carbonyl compounds. Traditional filter membrane sampling systems are a technique used to collect particulate matter and semi-volatile organic compounds (SVOCs) in the atmosphere. By combining with derivatization techniques, this system can facilitate the collection of carbonyl compounds. Particles are collected on quartz or PTFE filters, followed by sample extraction using solvent extraction, supercritical fluid extraction, or thermal desorption. The extracted samples are then analyzed using analytical techniques. Filtered samples typically require several hours or days of sampling, and compound loss may occur due to volatilization, gas-phase adsorption, and reactions during the collection process [9].

Collecting particle-phase samples on impact surfaces is an effective method for high-throughput sample collection. The thermally desorbed particle beam mass spectrometer (TDPB-MS) [63] focuses particles using an aerodynamic lens, then samples the cooled impact surface in a high-vacuum chamber. Williams et al. [64] proposed a new in situ instrument, the thermal desorption aerosol GC/MS-FID (TAG), collecting aerosols on impact surfaces, including both the gas and particle phases of carbonyl compounds. These methods offer high sample throughput, but they are not compatible with a range of different detectors. Hohaus et al. [65] developed an aerosol collection module (ACM) for sampling atmospheric aerosols, which allows derivatization during the collection process. This system is suitable for smoke chamber studies with limited sample volumes where large-volume collectors cannot be used. The ACM allows for the selection of detector types based on specific analytical capabilities, addressing different scientific questions without altering the sampling method. For example, it can be used with GC-MS to study isomers in samples or with PTR-MS for high-resolution detection of species.

The denuder sampling method has been proven effective for accurately measuring the gas-particle distribution of volatile organic compounds. Denuders coated with XAD-4 resin and PFBHA are an effective method for measuring the gas-particle distribution and partitioning of semi-volatile carbonyl compounds. Appropriate sampling flow rates can achieve satisfactory collection efficiency for mono-, di-, and aromatic carbonyl compounds, as well as carbonyl-containing furans and quinones. However, prolonged sampling times lead to a sharp decline in collection efficiency. Temime et al. [66] compared the absorption efficiency of denuders using adsorbents alone versus those using both adsorbents and derivatization reagents. They found that the PFBHA-coated denuder had a collection efficiency exceeding 90%, significantly higher than that of denuders using XAD-4 as the sole adsorbent. Compared with traditional filter sampling methods, the PFBHA-coated denuder–filter system reduces gas-phase adsorption onto the filter, thereby minimizing errors during the sampling process. Based on the denuder technique, Eichler et al. [67] proposed a novel modular injection system called "chemical analysis of aerosol online" (CHARON). It consists of a gas-phase denuder, an aerodynamic lens, and a thermal desorption device, aiming to provide online chemical characterization of semi-volatile submicron particles.

### *3.3. Applicability and Selection Principles of Sampling Methods*

Direct sampling methods for atmospheric carbonyl compounds do not require chemical reactions, thus minimizing human-induced interference; however, they face challenges in ensuring the stability of the sampled compounds. Among them, tube sampling methods are suitable for collecting strong volatile low-molecular-weight carbonyl compounds, and when combined with chemical derivatization, a wider range of carbonyl compound species can be collected. Canister sampling methods are used for offline sampling of various VOCs, including less reactive carbonyl compound species; however, the storage stability of samples containing highly reactive carbonyl compounds is relatively poor. The cryogenic enrichment technique requires the rapid and stable capture and release of target compounds, making it suitable for online enrichment of low-boiling and highly polar carbonyl compound species.

Samples collected via indirect sampling methods are more stable but require sample pre-treatment, leading to higher sample loss and time consumption. The selection of derivatization reagents, adsorbents, coating methods for reagents, and sampling devices must be carefully evaluated according to different detection requirements and analytical objectives. For example, DNPH solid-phase adsorbent cartridge and denuder sampling are commonly combined with LC, while PFPH and PFBHA are typically used with various adsorbents and sampling devices for sampling, followed by GC analysis. Filters, impactors, and denuders generally use PFPH and PFBHA for derivatization, coupled with GC or PTR-MS for detecting particle-phase carbonyls.

## 4. Analytical Methods for Atmospheric Formaldehyde

Analytical methods for formaldehyde are classified into online and offline techniques. In the 1970s, derivatization-based fluorescence and spectrophotometry methods were introduced for offline formaldehyde detection, followed by efforts to develop novel derivatization reagents and optimize chemical reaction conditions. During the 1980s and 1990s, techniques such as DOAS, LIF, and semiconductor and electrochemical sensors were the primary methods for online formaldehyde monitoring. In the early 21st century, CEAS, CRDS, and photochemical sensors were introduced.

### *4.1. Offline Analytical Methods for Formaldehyde*

Offline analytical methods for formaldehyde primarily include derivatization-based fluorescence and spectrophotometry methods. The fluorescence method is based on the Hantzsch reaction, quantifying formaldehyde by measuring the fluorescence intensity of the sample. Fluorescence intensity is influenced by reaction temperature, time, and reagents, with the choice of fluorescence reagents directly affecting detection sensitivity, selectivity, stability, and cost efficiency [68]. Dong et al. [69] designed and synthesized a fluorescence reagent with high selectivity for formaldehyde. This reagent operates under mild reaction conditions, exhibits vigorous chemical reactivity, and produces strong fluorescence intensity. Li et al. [70] developed a reagent applicable to both fluorescence and spectrophotometry methods. The fluorescence method offers a lower detection limit and higher sensitivity, while the spectrophotometry method is more economical and user-friendly.

The spectrophotometric method, also based on the Hantzsch reaction, detects formaldehyde by utilizing its light absorption characteristics at specific wavelengths. The results are significantly influenced by factors such as chromogenic temperature, reaction time, and chromogenic reagent [71]. By incorporating different reagents and optimizing analytical conditions, this method enables rapid detection of formaldehyde under various scenarios [23]. The phenol-reagent spectrophotometric method offers advantages such as low reagent cost and operational simplicity. Moreover, it demonstrates excellent overall performance in terms of detection limits, sensitivity, and anti-interference capabilities [44]. This method is widely used as an offline technique for detecting low concentrations of formaldehyde [72,73]. Ji et al. [74], following the experimental steps for formaldehyde detection specified in the national standard GB/T 18204.2—2014 [75], achieved low-concentration detection (0.061 μg) by optimizing the chromogenic reagent concentration, absorbent concentration, and chromogenic conditions. The acetylacetone-reagent method exhibits stable linearity and a wider detection limit range, making it suitable for detecting higher concentrations of formaldehyde with greater accuracy [72]. The AHMT-reagent method offers high selectivity and sensitivity [76], while spectrophotometric methods based on colorimetric probes demonstrate broad applicability [77]. Both approaches enable the accurate detection of formaldehyde concentrations in air. Ashraf et al. [78] developed a low-cost, highly sensitive, and selective offline analytical method for determining formaldehyde and acetaldehyde based on the DPD (N, N-diethyl-p-phenylenediamine) spectrophotometric method, achieving a resolution of at least 30 s.

### 4.2. Online Analytical Methods for Formaldehyde

Online analytical methods for formaldehyde primarily include spectroscopic and sensor-based methods (Table 1). Spectroscopy quantifies formaldehyde concentration based on the linear relationship between the light absorption or reflection intensity of formaldehyde or its derivatives and their concentrations. The detection limit is typically lower than that of sensor-based methods [24]. Spectroscopic techniques require a stable light source to minimize environmental interference and an extended optical path to achieve high sensitivity in field measurements [79,80], resulting in large and bulky equipment. Catoire et al. [81] demonstrated that the optical path length significantly affects TDLAS performance, with longer paths leading to lower detection limits and improved resolution. CRDS utilizes high-reflectivity mirrors, enabling light to reflect multiple times within the cavity to extend the effective optical path. However, this requires a large and complex cavity design [82,83]. Online application of fluorescence methods requires well-designed workflows for derivatization-based sample enrichment [84] and the ability to achieve low time resolution [85] to monitor formaldehyde in environments with significant concentration fluctuations [86].

Additionally, spectroscopic detection technologies have been used to detect dicarbonyl compounds, addressing the issue where PTR-MS is unable to effectively monitor acetaldehyde and methylglyoxal due to interference from similar fragment ions or isomers. The laser-induced phosphorescence of (methyl) GLyOxal spectrometry (LIPGLOS) method can be used for the simultaneous real-time quantification of acetaldehyde and methylglyoxal in the air, utilizing the unique phosphorescence lifetime of each molecule. At a single wavelength, acetaldehyde has a detection limit of 11 pptv in 5 min, and methylglyoxal has a detection limit of 243 pptv in 5 min [87]. The DOAS technique was first used for the direct measurement of acetaldehyde, with a detection sensitivity of 150 pptv over 2–15 min [88]. Improved technologies, such as long-path DOAS (LP-DOAS) [89] and passive multi-axis DOAS (MAX-DOAS) [90], have enhanced the sensitivity of acetaldehyde measurement, with detection limits of 100–200 ppt/10 min and 20 ppt/15 min, respectively. Light-emitting-diodes cavity-enhanced DOAS (LED-CE-DOAS) can also be used to measure acetaldehyde (detection limit, 28.5 ppt/min) and methylglyoxal (detection limit, 255 ppt/min) [91]. Another spectroscopic technique, incoherent broadband cavity-enhanced absorption spectroscopy (IBBCEAS), involves simultaneous detection across multiple wavelengths. When measuring acetaldehyde, it can achieve a detection limit of 29 ppt/min [92]. Compared with CRDS with wavelength-specific detection, IBBCEAS is better suited for humid environments and complex samples. Broadband cavity-enhanced absorption spectroscopy (BBCEAS) combined with a charge-coupled device (CCD) offers higher time resolution and lower detection limits (10 ppt/min), while coupling with a photomultiplier tube (PMT) provides a more cost-effective alternative [93].

Gas sensors can be categorized based on their detection principles into electrochemical, optical, and semiconductor gas sensors. These sensors are portable and exhibit high selectivity for formaldehyde but are susceptible to temperature and humidity fluctuations. Their performance is highly dependent on the sensor materials [94]. Electrochemical sensors quantify formaldehyde based on electrical signals generated by its adsorption onto sensor materials. However, these signals are prone to interference and lack stability [26]. Optical sensors rely on the optical properties of gases or their reaction products but are sensitive to dust interference [95,96]. Semiconductor gas sensors operate on the principle of changes in electrical properties caused by the adsorption of gas molecules onto semiconductor materials. Despite limitations such as a short linear range and susceptibility to interference, advancements in semiconductor materials and their diversity have made these sensors a research hot spot. Their advantages include fast response times, low cost, simple structures, long lifespan, miniaturization, and ease of operation [97]. Additionally, Warneke et al. [98] employed PTR-MS techniques to detect formaldehyde exclusively, enabling a rapid single 16 s data acquisition of formaldehyde in the troposphere and marine boundary layer from aircraft. The above technologies offer diverse solutions for real-time and accurate formaldehyde monitoring, ranging from portable environmental monitoring to high-precision industrial and airborne detection.

**Table 1.** Main online monitoring methods for atmospheric formaldehyde [1].

| Method | Principle | Optimization | LOD | TR | OP | Remarks | Ref. |
|---|---|---|---|---|---|---|---|
| DOAS | Spectral absorption characteristics | LP-DOAS | 0.45 ppb | 20 min | 5 km | Pathway located in the atmosphere; wide measurement range; no calibration required; poor time resolution | [99] |
| | | Xenon lamp as light source | 540–1200 ppt | 10–20 min | 4.9–8.2 km | | [79] |
| | | Tungsten lamp as light source and different detectors | 198–1500 ppt | 5–20 min | 5–15 km | | [100] |
| | | Low optical path | 3–4 ppt | -[2] | 426 m | | [101] |
| | | MAX-DOAS | 0.7–4.2 ppb | 5 min | 40 km | | [102] |
| FTIR | Infrared spectral absorption | Eight-mirror multi-reflection unit system | 4 ppb | 10 min | 1.08 km | Simultaneous analysis of multiple components; requires appropriate gas qualitative and quantitative prediction models | [103] |
| | | FTIR system equipment | 0.6 ppb | 5 min | 1 km | | [80] |
| | | Commercial FTIR | 1.05–1.1 ppm | 0.2–1 s | 3.2–10 m | | [104] |
| Hantzsch fluorimetry | Hantzsch reaction | Online analytical instrument based on Hantzsch reaction | 84 ppt | 120 s | - | Operate stably over the long term; significant differences between devices; requires independent design | [86] |
| | | | 0.05 ppb | 18 min | - | | [85] |
| LIF | Fluorescence signal intensity at a specific wavelength of the laser | Frequency-doubled tunable dye laser | 50 ppt | - | - | Laser as the light source; relatively complex calibration process; larger size; high measurement accuracy | [105] |
| | | No sample collection, water extraction, or further chemical treatment required | 10 ppt | 100 s | - | | [106] |
| | | No background detection and aerial surveying used | 36 ppt | 1 s | - | | [107] |
| | | Direct in situ detection in a white multi-channel cell | 51 ppt | 1 s | - | | [108] |
| | | Non-resonant-LIF (NR-LIF) | 261 ppt | 10 s | - | | [109] |

**Table 1.** *Cont.*

| Method | Principle | Optimization | LOD | TR | OP | Remarks | Ref. |
|---|---|---|---|---|---|---|---|
| TDLAS | Molecular absorption spectroscopy | Lead salt laser | 0.75 ppb/ 300 ppt | 3 min | 33.5 m/ 153 m | High accuracy; relatively compact setup; suitable for airborne and other fields | [110,111] |
| | | Difference frequency generation (DFG) laser | 222 ppt | 1 min | 100 m | | [112] |
| | | Quantum cascade laser (QCL) | 450 ppt/ 30–120 ppt | 1 min/ 1–10 s | 76 m/ 240 m | | [113,114] |
| | | Interband cascade laser (ICL) | 207 ppt/ 153 ppt | 90 s/ 10 s | 54.6 m/ 96 m | | [115,116] |
| CRDS | Vibrational absorption spectral characteristics | Pulsed CRDS | 300 ppb | - | - | Small sampling volume; unaffected by light source fluctuations. Interference by external environmental | [82] |
| | | OPO (optical parametric oscillator) light source | 112 ppt | 1 s | 300 m | | [83] |
| | | System based on CRDS | 1–2 ppb/ 3 ppb | ~s | - | | [117,118] |
| CEAS | Beer–Lambert law | IBBCEAS | 1.14 ppb | 30 s | 2.15 km | High sensitivity requires proper technological integration; complex system with poor stability. High stability and resolution be achieved by sacrificing optical path. | [119,120] |
| | | Mode-locked (ML)-CEAS | 3.3 ppb | - | - | | [118] |
| | | V-shaped CEAS | 15 ppt | 10 s | 1.97 km | | [121] |
| | | Pound–Drever–Hall (PDH) technique | 75 ppb | 1 s | 20 m | | [122] |
| Electrochemical sensor | Electrical signal generated by formaldehyde adsorption on the sensor material | | 20~50 ppb | 20 s to 30 min | Good repeatability; high resolution; high cost; poor stability | | [26] |
| Optical sensor | Optical properties | | 0.03~0.20 ppm | 15 min | High accuracy; affected by dust; expensive | | [95,96] |
| Semi-conductor sensor | Changes in electrical properties generated by the adsorption of gas molecules on semiconductor materials | | 50 ppm to 10 ppb | 8~131 s | Performance is related to the semiconductor material. | | [97] |

[1] LOD, limit of detection; TR, time resolution; OP, optical path; Ref., reference. [2] "-" means no data.

In summary, the results of offline formaldehyde detection are primarily influenced by chemical reagents and reaction conditions. By integrating various reagents and optimizing analytical conditions, rapid formaldehyde detection can be achieved for different scenarios. Spectroscopic online monitoring requires a stable light source to minimize environmental interference and an extended optical path to achieve high sensitivity. Most instruments are bulky, and optimized technologies such as CRDS and CRES feature complex systems. Gas sensors are portable and exhibit high selectivity for formaldehyde, but they are susceptible to environmental factors and highly dependent on the performance of sensor materials.

## 5. Simultaneous Analytical Methods for Multiple Atmospheric Carbonyl Compounds

Analytical methods for multiple carbonyl compounds are classified into online and offline techniques. In the 1970s and 1980s, HPLC-UV began to be used for offline detection of multiple carbonyl compounds. With advances in MS, HPLC-MS became a primary method in the 1990s, while canister-GC-MS enabled the detection of multiple VOC types. In the early 2000s, the development of various sampling methods led to significant advancements in GC-MS, and online monitoring technologies for multiple carbonyl compounds also began to emerge. By the 2010s, various online MS techniques were applied to the online monitoring of multiple carbonyl compounds (Figure 4).

### *5.1. Offline Analytical Methods for Multiple Carbonyl Compounds*

Offline analytical methods for multiple carbonyl compounds are suitable for simultaneous multi-point sampling and for studying regional distribution patterns. These methods allow ample time for analytical preparation, enabling thorough qualitative and quantitative detection of target carbonyl compounds. The most commonly used offline methods include DNPH-HPLC-UV, GC-MS, and HPLC-MS techniques (Table 2).

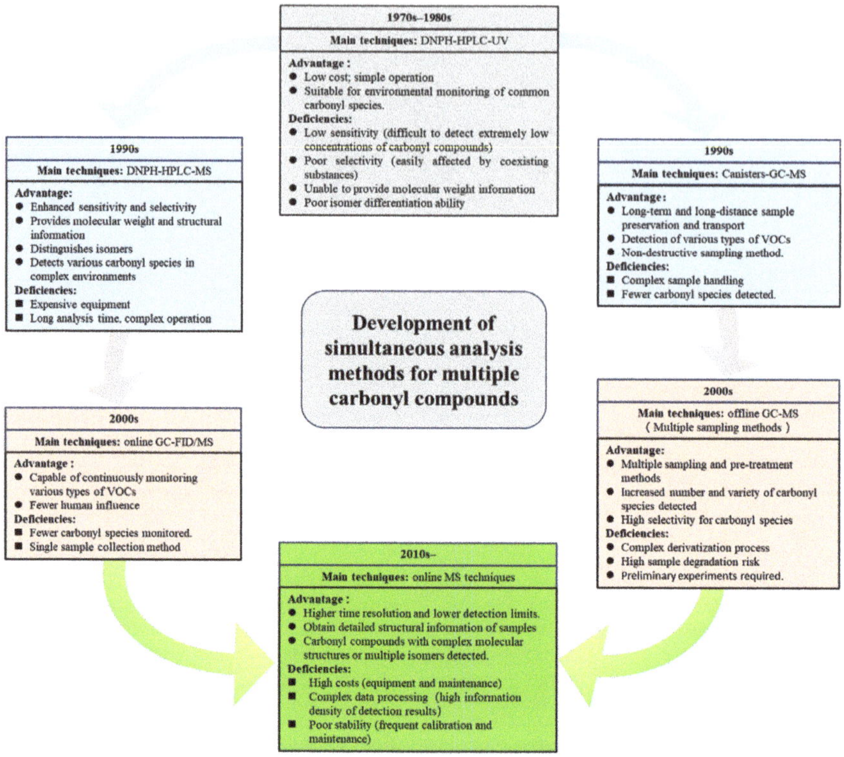


**Figure 4.** Development of simultaneous analytical methods for multiple atmospheric carbonyl compounds.

#### 5.1.1. DNPH-HPLC-UV

The DNPH-HPLC-UV technique is based on the reaction of carbonyl compounds with DNPH to form hydrazone derivatives, which are then eluted and analyzed using HPLC equipped with a UV detector [138]. This method's high sensitivity, selectivity, and ease of use have led to its widespread adoption globally. In China, the "Ambient air—Determination of aldehyde and ketone compounds—High performance liquid chromatography (HJ 683-2014)" standard [139] follows the main guidelines from the U.S. EPA's 1999 TO-11 method, making it suitable for detecting low-carbon atmospheric carbonyl compounds [42,140], although it has limitations in distinguishing between C3 [45] and C4 [141] carbonyls. The EPA's 2016 TO-11 revision also removed unsaturated aldehydes from its analyte list. Researchers have found that using appropriate gradient elution conditions, column temperature, or specific chromatographic columns can enhance HPLC's resolution of C3 and C4 carbonyls [142]. Possanzini et al. [143] demonstrated a secondary derivatization mechanism for alkenals (acrolein and crotonaldehyde) and optimized the mobile phase to separate these from other carbonyls. Wang et al. [131] selected appropriate columns and used a dual-gradient HPLC method with temperature and solvent control to address separation challenges for C3 (acrolein, acetone, and propanal) and C4 (crotonaldehyde, methacrolein, butanone, and butyraldehyde) compounds. Zhang et al. [144] improved separation of acrolein, acetone, and propanal using multiple mobile phases, while other researchers completed the detection of 13 OVOCs within 10 min by using specialized columns [145]. The 2020 Chinese standard, "Ambient air—Determination of aldehyde and ketone compounds—Solution absorption-High performance liquid chromatography (HJ 1154-2020)", builds on the 2014 standard by enriching carbonyl compounds via solution absorption, followed by liquid extraction and concentration for the detection of 16 different carbonyl compounds [146]. Although the above DNPH-HPLC-UV techniques provides stable detection for common carbonyls, the number of detectable species remains limited due to detector constraints.

**Table 2.** Main offline analytical methods for multiple atmospheric carbonyl compounds.

| Method | Optimization | LOD [1] | TR [2] | Species | Remarks | Ref. |
|---|---|---|---|---|---|---|
| Canisters GC-MS | Optimize instrument parameters and column temperature box heating program | 0.49~1.1 μg/m$^3$ | 30 min | 13 carbonyls | Suitable for the determination of low-concentration OVOCs (C4–C8) | [39,42] |
| | Combining Deans Switch center cutting technique | 0.009~0.134 nmol/mol | 1 h | 13 carbonyls | Improved separation performance of GC | [123,124] |
| Super-critical fluid chromatography (SFC)-GC-MS | Supercritical fluid as the mobile phase | - | 1 h | 57 carbonyls (13 aldehydes and 44 ketones) | Enhanced separation capability; addressing the limited peak capacity issue of GC-MS | [125] |
| Liquid coating PFPH-GC-MS | PFPH as derivatization reagent | 3.7–11.6 ng/tube | 1 h | 20 carbonyls | Improved stability of derivatives and increased sampling efficiency | [34] |
| | | 3–10 ng/tube | | C4-C11 Straight-chain aldehydes and some ketones | | [51] |
| | Multi-bed adsorbent and TD technique | 1.8~25.4 ng/tube | 1 h | 6 carbonyls | Achieved simultaneous detection of aldehydes, ketones, and aromatic hydrocarbons | [126] |
| | TD technique | 0.3 ppb/24 L | 1 h | 16 carbonyls | Low detection limit and simplified operation process | [52] |
| Gas coating PFPH-GC-MS | Optimal sampling flow rate and solvent extraction | 0.08~0.20 ppb/24 L | 1 h | 21 carbonyls | Low species background concentration level | [46] |
| GC-MS | Impinger containing PFBHA | 5~679 ppb | Several to dozens of hours | Aromatic aldehydes, quinones, di-carbonyls | Suitable for the detection and identification of various carbonyl compounds in smog | [127] |
| | Denuder coated with PFBHA and XAD-4 | 0.2–0.7 μg/m$^3$ | Several tens of minutes | 23 carbonyls | Suitable for gas-particle distribution research, with strong ability to distinguish isomers | [66] |
| GC-MS | Derivatization with PFBHA on SPME fiber | - | Several minutes | C4~C9 linear aldehydes | In situ derivatization | [128] |
| GC-MS/GC-FID | | 0.2–1.9 μg/m$^3$ | <30 min | 5 carbonyls | | [62] |
| Solid adsorption DNPH-HPLC-UV | Single-use micro-silica DNPH column and automated system | - | 2 h | 6 carbonyls | Achieved continuous operation of sampling and analysis, reducing manual intervention | [129] |
| | Ultra-HPLC (UHPLC) | 0.002~0.004 mg/L | 4 min | 13 carbonyls | Significantly shortened analysis time, addressing the high-throughput sample demand | [130] |
| | Temperature and binary solvent gradient | 0.04~0.19 μg/m$^3$ | 35 min | | Addressed the issue of difficulty in separating C3 and C4 components | [131] |
| | Acidic DNPH adsorption tube sampling, elution time controlled within 6 h | 0.16~0.71 μg/m$^3$ | 30 min | | From reaction kinetics perspective, addressed the issue of the impact of hydrazone derivatives of unsaturated aldehydes on quantification | [132] |
| | New type of chromatography column and column temperature regulation | 0.025~0.091 μg/m$^3$ | 40 min | 22 carbonyls | Improved chromatographic separation capability | [133] |
| DNPH-HPLC-UV | Solution absorption | 0.001~0.002 mg/m$^3$ | 40 min | 16 carbonyls | Liquid-phase sampling method | [134] |
| HPLC-ESI-TOF-MS | Denuder coated by XAD-4 and 2,4-DNPH | 0.9 ppt–0.3 ppb | 1 h | 12 carbonyls | In situ derivatization | [50] |
| DNPH-HPLC/ESI-MS/MS | ESI ion source | 0.4–9.4 ng/m$^3$ | 1 h | 32 carbonyls | Difficult to distinguish compounds with the same molecular weight or similar chemical properties | [135] |
| DNPH-HPLC/APCI-MS | APCI ion source | 0.005–0.049 μg/m$^3$ | 1 h | 30 carbonyls | Difficult to distinguish isomers with similar retention times | [136] |
| DNPH-UHPLC-MS/MS | MS/MS | 0.002~0.07 μg/m$^3$ | 1 h | 47 carbonyls | Capable of analyzing multiple types of OVOCs | [137] |

[1] The limit of detection for offline measurements depends on the sampling volume and specie. [2] TR here refers to the time required for the instrument to analyze the sample as the sampling time is significantly influenced by human factors.

### 5.1.2. Chromatography Tandem MS

Chromatography tandem MS is currently the most widely used method in laboratories for detecting multiple carbonyl compounds, primarily including LC-MS and GC-MS techniques. Optimizing this technique allows for the differentiation of various types of carbonyl compounds and enables the detection of a broader range of species, although it often involves complex procedures [147]. Studies have found that the ionization source in MS significantly influences detection performance. When using electrospray ionization (ESI) in combination with the single-ion monitoring (SIM) mode, the detection limit is minimized [148]. The use of atmospheric pressure photoionization (APPI) allows for the detection of carbonyls without the need for dopants, reducing the potential interference caused by dopants [149]. The HPLC-MS/MS (tandem MS) method has been shown to differentiate between aldehydes and ketones, linear and branched structures, and unsaturated and aromatic carbonyl compounds [46]. Fourier transform ion cyclotron resonance MS (FTICR-MS) with nano-electrospray ionization (nano-ESI) allows for direct sample injection without the need for derivatization [150]. Zhang et al. [136] optimized sampling and analysis processes, using DNPH-HPLC/APCI-MS (atmospheric pressure chemical ionization) to detect 30 OVOCs, including mono-carbonyl, di-carbonyl, and oxygen-containing and heterocyclic carbonyl. Chi et al. [135] made customized adjustments to ESI and MS/MS parameters for different carbonyl compounds, achieving the detection of 32 carbonyl species using DNPH-HPLC-ESI-MS/MS. Xu et al. [137] developed a novel UHPLC-MS/MS-based analytical method to identify and quantify 47 carbonyl compounds with carbon numbers ranging from 1 to 13, including 28 aliphatic saturated mono-carbonyls, 8 aromatic mono-carbonyls, 8 other unsaturated mono-carbonyls, and 3 di-carbonyl compounds.

The canister sampling GC-MS method is the standard offline method for measuring VOCs as specified by the Ministry of Ecology and Environment, typically measuring only a limited number of carbonyl compounds, such as acetone, acrolein, and 2-butanone [151]. Some researchers have achieved the measurement of low-concentration carbonyl compounds in the C4–C8 range by adjusting various parameters of instruments and the column temperature program [39,42]. The Deans Switch central cutting device enables the simultaneous measurement of other VOCs and 13 carbonyl compounds in a single analysis run [123,124]. The use of supercritical fluid as a mobile phase has addressed the issue of limited peak capacity in GC-MS [125]. The collection efficiency of carbonyl compounds can be greatly enhanced by using sampling methods such as sampling tubes, sampling columns, denuders, and SPME fibers, as well as optimizing the sample pretreatment and analysis processes. Zou et al. [152] collected atmospheric samples using a Tenax-TA sampling tube coated with PFBHA and determined 14 mono-carbonylic compounds and 2 di-carbonyl compounds in the atmosphere using solvent extraction coupled with GC-MS. After collecting carbonyl compounds with PFPH tubing, pretreatment is required, typically involving solvent desorption, and this process can be complex [51]. Li et al. [34] developed a GC/MS method for the simultaneous determination of 20 carbonyl compounds by altering the solvent extraction reagent and found that formaldehyde was not completely separated due to significant human factors. The TD technique can significantly simplify the operation process and enable automated desorption, concentration, and analysis [52]. Chien et al. [126] used liquid-coated PFPH tubing combined with TD-GC-MS and multiple-bed adsorbent tubes to achieve simultaneous detection of carbonyl compounds and aromatic hydrocarbons. Wang et al. [153] used multi-bed absorption tubes combined with two-stage TD and GC-MS techniques to detect 28 OVOC species in Hong Kong, including aldehydes, ketones, alcohols, acrylates, acetate esters, and ethers, although the process was highly complex. Kahnt et al. [50] used an XAD-4/DNPH denuder for in situ derivatization on the surface, reducing the sample preparation steps, and detected 12 carbonyl compounds using HPLC/ESI-TOF-MS techniques. Temime and Brice [66] employed an XAD-4/PFBHA denuder-filter sampling method, combined with GC-MS, to detect gas- and particle-phase arbonyl compounds, identifying 23 aldehydes and ketones, with a distinct advantage in the analysis of isomers. Bourdin and Desauziers [62] and Poole et al. [128] used SPME fibers for in situ PFPH derivatization, significantly improving the time resolution for aldehyde species detection while reducing the sample preparation steps.

Quinone compounds are organic compounds that contain one or more carbonyl groups (C=O), typically consisting of one or more aromatic rings. Until the 2010s, only a few studies had reported the concentration and gas-particle distribution of environmental quinones. Few reports have been made on the spectroscopic detection techniques for quinones. As early as the 1950s, Fuson et al. [154] conducted an infrared spectroscopic study on various types of ketones, benzoquinones, and naphthoquinones, investigating the influence of carbonyl (C=O) stretching frequencies. They found that the molecular structure, conjugation effects, substituent effects, and ring strain all affect spectral detection. Spectroscopic techniques, such as LP-FTIR, detect fewer species compared with GC-MS techniques [155]. Compared with spectroscopic detection techniques, GC-MS is the most commonly used method. Due to the large differences in the properties of quinone species, some studies choose to use derivatization followed by CI or EI ionization for GC-MS detection to enhance analytical sensitivity [156]. However, not all quinones can be effectively derivatized. During the derivatization and analysis processes, thermally unstable quinones may undergo transformation and/or degradation [157]. Some studies have found that for certain compounds, the GC-MS measurements of derivatized and non-derivatized quinones showed no significant differences [158]. HPLC-MS/MS does not require derivatization of thermally unstable, non-volatile, or highly polar compounds, offering certain advantages [159]. dos Santos et al. [160] used cold-fiber SPME-GC-MS (CF-SPME-GC-MS) to achieve higher detection sensitivity and lower detection limits. Four quinone compounds (2-methyl-1,4-naphthoquinone, 2,6-di-tert-butyl-1,4-benzoquinone, methyl-1,4-benzoquinone, and 2,3-dimethylanthraquinone) were detected for the first time in ambient air. Jo et al. [161] developed a quantitative chemical ionization technique based on GC-MS and GC-MS/MS, which can be used for the quantitative analysis of oxidized polycyclic aromatic hydrocarbons (OPAHs) across different concentration ranges and is suitable for detecting various types of OPAHs.

In summary, the main offline analytical methods for multiple carbonyl compounds include DNPH-HPLC-UV and chromatographic tandem MS techniques. DNPH-HPLC-UV is a straightforward technique, and by optimizing gradient elution conditions and column temperature or selecting the appropriate chromatographic column, it can enable the detection of more than ten common carbonyl compounds. The optimization of chromatographic tandem MS techniques is always accompanied by improvements in the complex sampling, pretreatment, and analytical processes, which demand high expertise from the researchers. By optimizing chromatographic tandem MS, it is possible to distinguish between various types of carbonyl compounds (e.g., aldehydes and ketones, straight-chain and branched structures, and unsaturated aromatics; mono-carbonyl, di-carbonyl, oxygenated, and heterocyclic carbonyls; and carbonyl compounds, alcohols, acrylic esters, acetate esters, ethers, and quinones).

### *5.2. Online Analytical Methods for Multiple Carbonyl Compounds*

Online analytical methods for multiple carbonyl compounds can provide higher time resolution, avoid measurement errors caused by target species' reactions or volatilization during the measurement process, and obtain complete and detailed time series. The most commonly used online methods include online MS techniques, such as GC-FID/MS and RTR-MS (Table 3).

**Table 3.** Main online analytical methods for multiple atmospheric carbonyl compounds.

| Method | Optimization | LOD | TR | Species | Remarks | Ref. |
|---|---|---|---|---|---|---|
| GC-FID/MS | Ultra-low-temperature pre-concentration | - | 10 min | C2–C3 hydrocarbons; C5–C12 hydrocarbons, halogenated hydrocarbons, and OVOCs | Detection limit lower than that of manual sampling detection | [39] |
| | Three-stage trap pre-concentration | <5 ppt | 30 min | MAC and MVK | Improved detection capability for low-concentration compounds | [162] |
| | Low-temperature empty tube cryo-concentration technique | 0.03~0.7 μg/m$^3$ | 1 h | 13 carbonyls | Addressed the irreversible loss in adsorption tube sampling; improved capture capability for polar aldehydes and ketones | [30] |
| | Refrigeration system without refrigerant | 0.008~ 0.026 ppb | 60 min | C3–C6 carbonyls | Suitable for remote areas or regions lacking liquid nitrogen supply | [40] |
| | Ultra-low-temperature freezing collection device and thermal desorption | 0.004~ 0.012 ppb | 1 h | 12 carbonyls | Species loss is reduced through cold trap water removal and PTFE tubing | [32] |
| GC-MS | Online SPME derivatization | 6–100 pptv | <20 min | 12 carbonyls | Direct sampling and derivatization | [59] |
| GC×GC-TOF/MS | Online TD combined with two-dimensional GC | 10,000 components/10 μg | - | Linear C5 to C9 aldehydes, C8 to C13 ketones, and some cyclic products such as furanone | In situ testing; used in conjunction with PTR-MS; complex data | [163] |
| PTR-IT-MS | Ion trap (IT) | ~100 ppb | 220 ms | Acetaldehyde, acetone, MVK, and propionaldehyde | Capable of detecting species overlooked by GC-MS low detection limits | [164,165] |
| PTR-TOF-MS | Time-of-flight (TOF) | ~ppt | - | Acrolein, acetone, 2-butanone, hexanone | Suitable for the differentiation of isomers | [166] |
| CIFI-TOF-MS | Chemical ionization focused integrated (CIFI) ionization source | 6~200 ppt | 40 μs | 12 carbonyls | Complex ion source design and instrument setup | [167] |

### 5.2.1. GC-FID/MS

GC-FID/MS can achieve online monitoring of various types of VOCs but has limited detection capability for carbonyl compounds. Traditional online GC-FID/MS systems mainly use a dual-channel sampling, dual-column separation, and dual-detector systems [41]. Different types of atmospheric species are split into two streams at the channel inlet, separately sampled and analyzed by different detectors. The FID detector is primarily used for low-carbon hydrocarbons (C2–C5, with lower boiling points), while hydrocarbons from C5 to C12, halogenated hydrocarbons, and carbonyl compounds are detected by the MS detector. Typically, up to 12 carbonyl compound species can be detected [168]. The availability of additional standard gases for carbonyl compounds can expand the range of target species identified by GC-FID/MS [32]. The entire analysis process takes about 1 h. To capture rapid variations or fluctuations of environmental pollutants, Apel et al. [169] introduced a portable, fast-response GC-MS capable of high-frequency monitoring of C2–C4 carbonyl compounds on board aircraft within 5 min. However, this technique is still based on direct injection measurement without enrichment or efficient desorption of species. It remains insufficiently sensitive for trace compounds, and the range of detected species is still limited.

Researchers often optimize sampling and pretreatment methods to improve the detection performance of online GC-FID/MS technique. Apel et al. [162] utilized a custom three-stage preconcentration system combined with MS (SIM), enabling stable online monitoring of low-concentration carbonyl compounds. Li et al. [30] applied a cryogenic empty-tube freezing preconcentration technique combined with a GC-FID/MS analysis system for automated online monitoring. This method allowed flexible selection of the sampling flow rate, sampling time, and trapping temperature, achieving stable online monitoring of 13 carbonyl compounds. However, the reliance on refrigerants limits the applicability of these cryogenic preconcentration techniques in remote areas. Wang et al. [40] developed a refrigerant-free GC-FID/MS system, which was validated to accurately detect C3–C6 carbonyl compounds through comparison with results from other technologies.

The analysis of carbonyl compounds in aerosols is particularly challenging. Typically, chemical reagents are used for derivatization prior to GC/MS analysis, with specially designed in situ online aerosol collection devices and TD techniques being the primary sampling methods [170]. Williams et al. [171] used a thermal desorption aerosol gas chromatography system (TAG) for real-time, in situ measurement of organic and inorganic compounds in atmospheric aerosols. A small portion of organic aerosol mass (10–40%) was analyzed, and alcohol species derived from isoprene were detected. Hamilton et al. [163] used direct thermal desorption combined with comprehensive two-dimensional GC×GC-TOF/MS to analyze oxygenated components in organic aerosol samples. They detected linear-chain aldehydes from C5 to C9, ketones from C8 to C13, and some cyclic products such as furanone.

#### 5.2.2. Online MS Techniques Without Chromatographic Separation

Online MS allows direct sample injection, minimizing potential human errors during sample collection and pretreatment [25], and provides high temporal resolution for a deeper understanding of chemical transformations. The development of ionization sources and mass analyzers is one of the driving forces behind advancements in MS and a key reason for its online application [172]. Electron ionization (EI) [173] is the most traditional form of hard ionization, widely used in the GC-MS technique. It produces numerous fragments, making accurate determination of molecular ion masses challenging [174]. Soft ionization methods primarily include chemical ionization (CI) [175], photoionization (PI) [176], and ESI [177]. These ionization methods offer high temporal resolution and enable the online detection of atmospheric carbonyl compounds [178].

PTR-MS is an online MS technique for detecting multiple carbonyl compounds without chromatographic separation. Its basic principle involves proton transfer reactions between target compounds in the environment and reagent ions, with the resulting product ions analyzed in a mass analyzer. Its detection limit and analysis time are typically lower than those of GC-FID/MS, but the method often requires extensive data analysis, heavily relying on machine learning and artificial intelligence [179]. Currently, PTR-MS primarily employs CI and can be equipped with various mass analyzers, including quadrupole (Q), ion trap (IT), and time-of-flight (TOF). PTR-Q-MS can detect compounds overlooked by GC-MS [180]. However, Q mass analyzers exhibit low transmission efficiency for heavier ions, resulting in reduced sensitivity for high-mass compounds [181]. PTR-IT-MS has been shown to analyze the entire mass spectrum almost simultaneously and distinguish ions with the same nominal mass [165], although it has a relatively high detection limit [164]. PTR-TOF-MS offers higher sensitivity and lower detection limits [166]. It has been successfully applied in airborne measurements [182] and outperforms various online spectroscopic methods in detecting glyoxal and methylglyoxal [183]. Advances in light sources have greatly driven the development of the PI technique. Synchrotron light sources can reduce or eliminate ion fragmentation [184], but both synchrotron and laser light sources [185] are complex, bulky, and expensive, limiting their application to laboratory-based fundamental research rather than online or field applications. The vacuum ultraviolet (VUV) PI-MS technique, characterized by minimal ion fragmentation, no need for sample pretreatment, and low polarity requirements for analytes [186,187], has been widely applied in the online monitoring of VOCs under various conditions [188]. Liu et al. [167] combined CI with VUV-PI-MS to develop a new method integrating a chemical ionization focusing ion source (CIFI) with TOF-MS, significantly improving ionization and ion transmission efficiency. CIFI-TOF-MS has been validated as a highly efficient and sensitive online monitoring technique, capable of real-time detection and analysis of 12 carbonyl compounds. The selected ion flow tube mass spectrometer (SIFT-MS) [189] can be used to detect formaldehyde and acetaldehyde and is capable of using multiple reagent ions simultaneously, with a detection limit for formaldehyde of less than 200 ppt/1 s.

In the analysis of particle-phase carbonyl compounds, the online use of TD can be affected by thermal decomposition during desorption and ion dissociation during the ionization process in the PTR-MS drift tube [190] and may not avoid surface collection artifacts [191]. Vogel et al. [192] employed a miniaturized, versatile aerosol concentration enrichment system (mVACES) to address the issue of particulate matter collection and applied it in the field. Additionally, several aerosol online collection devices, such as ACM [65] and CHARON [67], have been widely used in smoke chambers. Thompson et al. [193] utilized four different instruments to study the gas-particle distribution of carbonyl compounds, including FIGAERO (filter inlet for gases and aerosols)-ToF-CIMS, SV-TAG (semi-volatile thermal desorption aerosol gas chromatograph)/TD-PTR-MS, and $NO_3^-$-IMS (nitrate ion mobility spectrometry)-TOF-MS. They found that the mass spectrometry techniques with specific ion sources limited the range of measurable compounds and exhibited interference from different ion sources, making it difficult to recognize isomers.

In summary, the main online analytical methods for multiple carbonyl compounds include GC-FID/MS and other online MS techniques, such as PTR-MS and VUV-PI-MS. GC-FID/MS systems are primarily used for detecting various types of VOCs. By optimizing sampling and pretreatment methods, stable detection of certain carbonyl compounds can be achieved. Other online MS techniques can integrate diverse ionization sources, mass analyzers, and online sampling systems to enhance detection performance, achieving lower detection limits and higher temporal resolution.

## 6. Analytical Methods for Carbonyl Compounds in Atmospheric Photochemical Smog Chamber Simulation Studies

The smog chamber allows for precise control of added reagents and environmental conditions, gradually increasing the complexity of chemical reactions to simulate different environmental and chemical systems. Currently, the analysis of multi-functional products (including carbonyl compounds) formed during VOC photooxidation in the smog chamber is typically conducted using spectroscopic methods or independent MS. Spectroscopic methods are simple to operate and require no complex pre-treatment steps, but they can only detect species with specific optical properties. Klotz et al. [194] used the DOAS technique to detect photooxidation products of toluene, identifying benzaldehyde and methylphenol. Olariu [155] used FTIR to detect carbonyl compounds formed from reactions between OH radicals and phenol, ortho-, meta-, and para- methylphenol and first identified 1,2-dihydroxybenzene and 1,4-benzoquinone as new products of OH radical-induced phenol and methylphenol oxidation. Thalman et al. [183] demonstrated the differences in detecting di-carbonyls using seven different measurement techniques in smog chambers in the U.S. and Europe. They found that spectral detectors, such as BBCEAS, CE-DOAS, white-cell DOAS, FTIR, and LIP, were all affected by interference from water vapor, aerosols, or other gases in the environment, leading to poor stability.

PTR-MS is a powerful soft ionization technique that allows for the direct injection and quantification of photochemical oxidation products and precursor VOCs, providing additional insights into potential photochemical oxidation products [195]. Eichler et al. [67] used a CHARON-PTR-ToF-MS combination device to measure organic and ammonium components in submicron particles in real time, detecting carbonyl compound signals. Holzinger et al. [196] used TD-PTR-MS to detect products such as formaldehyde and formic acid during the reaction between ozone and terpenes and found that the desorption temperature of samples in the smog chamber was much lower than that of environmental samples, suggesting a significant discrepancy between the smog chamber experimental conditions and atmospheric conditions. Gkatzelis et al. [197] conducted a comparative study on the gas-particle partitioning of major biogenic oxidation products. In the experiment, different sampling inlets (ACM, CHARON, and TD units) were connected to the PTR-ToF-MS, and it was found that all three systems were capable of detecting aldehyde and ketone compounds. However, since PTR-MS provides molecular formulas rather than specific compound structures, there is a possibility that original chemical components may not be accurately identified due to isomeric compounds or fragments generated by thermal dissociation and ion fragmentation. Furthermore, as different thermal desorption temperatures and operational conditions affect ion fragmentation, the detection limits and time resolutions of each system also vary. TD has the lowest detection limit ($10^{-3}$ ng/m$^3$), while ACM has the highest detection limit (250 ng/m$^3$). CHARON provides real-time measurements (10 s), while the sampling times for ACM and TD are 120 min and 240 min, respectively.

GC-MS can effectively identify isomeric information. Yu et al. [127] employed impinger -PFBHA-GC-MS in both indoor Teflon bag reactors (TBRs) and outdoor Teflon smog chambers to analyze the carbonyl compounds produced from the reactions between six alkylbenzenes and OH radicals. They detected aromatic aldehydes, quinones, di-carbonyls, and other carbonyl products, with a particular focus on distinguishing various isomeric forms of di-carbonyl products. Kourtchev et al. [198,199] utilized Denuder-PFBHA-GC-MS to study the reactions of (E)-β-farnesene with $O_3$ and OH radicals in the gas phase, including the determination of rate coefficient and analysis of carbonyl products and distinguishing isomeric forms. These offline techniques separate the sampling and derivatization processes, resulting in significant time consumption. Borrás et al. [59] employed in situ dual derivatization SPME-GC-MS at EUPHORE to identify various compounds, including aldehydes, ketones, α-di-carbonyl compounds, hydroxy aldehydes, hydroxy ketones, and carboxylic acids. They applied this method to the ozonolysis of isoprene, tracking the formation of multi-functional oxygenated compounds. Spittler et al. [200] studied the dark reactions between $NO_3$ radicals and two monoterpenes—limonene and α-pinene—and their role in SOA formation in the EUPHORE photoreactor facility and a large volume laboratory photoreactor, where the main carbonyl products, pinonaldehyde and endolim, were successfully detected and quantified using long-path FTIR, HPLC-UV/Vis, and GC-FID. Hohaus et al. [65] connected the ACM to a GC-MS/FID system and successfully detected several carbonyl compounds in SOA during β-pinene ozonolysis, including acetone, bicyclo[3,1,1]hept-3-ene-2-one, myrtanal, myrtenol, 1-hydroxynopinone, 3-oxonopinone, and 3,7-dihydroxynopinone. Although these techniques have undergone optimization in sampling systems and pretreatment processes, their temporal resolution remains inferior to that of PTR-MS technique.

Combining isomeric information from GC-MS with high-time-resolution data from PTR-ToF MS can expand and improve insights obtained from photochemical oxidation experiments. Gómez Alvarez et al. [201] employed a method combining SPME with PFBHA and used PTR-MS, GC-FID, and FTIR instruments to detect di-carbonyl compounds during the photochemical oxidation of toluene and benzene. Finja Löher et al. [202] constructed a Teflon atmospheric simulation chamber, which was connected to a PTR-ToF-MS and an SPME-GC-MS system. Under NOx-free conditions, they detected the first-generation products of toluene oxidation and compared their results with those of Borrás [59]. They found that different smog chamber devices (the Teflon chamber and the EUPHORE chamber) affected the experimental results for acrolein.

## 7. Problems and Challenges

In the ongoing development and optimization of various commonly used sampling techniques and analytical methods, these approaches have revealed distinct and significant problems and face different challenges for further development and optimization. Table 4 summarizes the main problems and challenges of commonly used sampling techniques and analytical methods.

**Table 4.** Problems and challenges of analytical methods for atmospheric carbonyl compounds.

| Type | Method | Problem | Challenge |
|---|---|---|---|
| Sampling method | Tube sampling | Simultaneous sampling of multiple carbonyl species | Developing novel adsorbents or efficient solvents |
| | Canister sampling | | Designing canisters made from new materials |
| | Cryogenic enrichment | | Advancing cryogen-free enrichment techniques and optimizing cold trap devices |
| | Derivatization technique | Sample loss; time consumption | Designing derivatization reagents with immediate derivatization, high derivative stability, and good sensitivity and selectivity; efficient combination of sampling equipment and derivative materials |
| Online analytical methods | Spectroscopic | Poor stability; bulky volume; multiple program units for derivative fluorescence equipment | Research on optical path improvement technique and design of precision equipment |
| | Gas sensors | Significant environmental impact; limitations in sensor materials | Developing new sensor materials |
| | GC-FID/MS | Unstable instrument status; lack of standard gas; single capture technique | Instrument maintenance; development of new standard gas and sampling technique |
| | Other online MS | Poor stability of ion source; difficult to distinguish between isomers; complex mass spectrum | Ion source optimization; MS maintenance; combining big data technique |
| Offline analytical methods | Fluorescence and spectrophotometry methods | Multiple types of reagents; low time resolution | Optimization of analysis conditions; reagent selection |
| | DNPH-HPLC-UV | High blanks; sampling parameter settings; derivative products impacting on detection results; limited availability of standards | Standardization of sampling methods; exploration of derivatization mechanisms |
| | Chromatography tandem MS | Complex MS operations; multiple pre-experiments | Improving MS operation techniques; determining the instrument analysis conditions |
| Analytical methods for chamber | Spectroscopic | Poor stability | Complex environment detection |
| | PTR-MS | Difficult to distinguish between isomers | Use of multiple methods combined |
| | GC-MS | Poor time resolution | |
| | PTR-MS/GC-MS | Complex data processing and operation, significant influence of chamber itself and sampling system | Combining big data technique; conduct research on the impact of chamber itself and sampling system |

### *7.1. Sampling Methods*

Direct sampling methods for atmospheric carbonyl compounds have limitations in simultaneous sampling of multiple carbonyl species. Tube sampling requires chemical derivatization to collect a broader range of carbonyl compounds, whereas canister sampling and cryogenic enrichment techniques are typically coupled with specific chromatographic systems and are suitable for sampling various VOCs (limited carbonyl species). Challenges in direct sampling of carbonyl compounds include developing novel adsorbents or efficient solvents to reduce reliance on derivatization in tube sampling, designing canisters made from new materials to enhance sample stability, and advancing cryogen-free enrichment techniques alongside optimizing cold trap devices.

Indirect sampling methods are often associated with significant sample loss and high time consumption. Furthermore, the selection of derivatization reagents, adsorbents, coating methods for reagents, and sampling devices is complex, requiring careful adjustment to meet diverse detection requirements and analytical objectives. Designing derivatization reagents with immediate derivatization, high derivative stability, and good sensitivity and selectivity, as well as rapidly determining sampling schemes, remains a significant challenge.

### *7.2. Analytical Methods of Formaldehyde*

Fluorescence and spectrophotometric methods are limited to the detection of single species and exhibit low time resolution. The selection of reagents and optimization of analytical conditions often vary with environmental contexts, making multiple pre-experiments necessary to achieve accurate formaldehyde detection results.

Spectroscopic detection techniques are highly sensitive to environmental conditions, with overlapping or missing spectral signals being common issues. The development of spectroscopic instruments with stable selectivity is therefore critical. Formaldehyde monitoring devices based on chemical methods often involve multiple steps, requiring customized designs. Currently, high-precision spectrometers are typically bulky to meet the requirements for long optical paths, limiting their applicability for field monitoring. Although advanced CRDS and CRES achieve extended optical paths, their instruments are complex and prohibitively expensive. Gas sensors are greatly affected by the environment, and their performance is highly dependent on sensor materials. Developing new sensor materials is crucial.

### *7.3. Analytical Methods of Multiple Carbonyl Compounds*

#### 7.3.1. Offline Analytical Methods

The DNPH-HPLC-UV method suffers from significant sample loss during sampling and high background concentrations of carbonyl compounds. Additionally, there is no coordinated control between the sampling volume and the flow rate. The duration of sampling affects sample recovery, and derivatives may undergo isomerization, leading to the formation of isomers that compromise accurate quantification and result in lower reproducibility than expected. This method has limitations in the species it can detect, and the lack of standard derivatives restricts the identification of unknown compounds [203]. Therefore, optimizing sorbent tubes, studying the derivatization mechanism, developing new standard reference materials, and determining appropriate analytical conditions are crucial for improving the method.

Chromatography tandem MS provides structural information on unknown compounds but requires advanced experimental techniques. The MS may fail to detect peaks captured by other detectors, such as UV detectors, which can lead to underestimation [136]. Therefore, combining multiple detectors in experiments is essential. Different ion sources in MS require distinct operational adjustments for chromatographic coupling, and each ionization source has specific environmental requirements for sample analysis. Achieving optimal results often requires multiple pre-experiments to fine-tune experimental conditions. Currently, MS often uses SIM mode, which detects only a few selected ions, eliminating interference from other ions, but fails to provide a full spectrum, making the qualitative analysis of unknown compounds challenging. In contrast, the full scan mode enables the exploration of unknown compounds, which is crucial for detecting new carbonyl compounds, although it is challenging to operate.

#### 7.3.2. Online Analytical Methods

During continuous operation, GC-FID/MS equipment may experience changes in the retention times of certain compounds due to factors such as decreased column efficiency and variations in the heating rates of preprocessing devices. Consequently, it is necessary to conduct weekly retention time checks for all target compounds to ensure accurate identification. The lack of available standard gases limits the ability of GC to identify and quantify various carbonyl compounds, often resulting in fewer detectable compounds compared with offline analytical methods. Currently, online systems rely on cryogenic enrichment techniques for carbonyl compounds, whereas alternative sampling methods remain unsuitable for GC-based online analysis. As a result, the online GC-FID/MS technique faces challenges related to instrument maintenance and the development of new standard gases and sampling methods.

The performance of ion sources in online MS can be influenced by various factors, such as temperature and humidity. Mass spectra are often complex, and simplifying them typically requires the use of high-precision instruments, which demand regular maintenance and calibration and are extremely costly, making them difficult for commercial use. MS often encounters multiple isomers or fragment ions, making the correct identification and differentiation of compounds a technical challenge. A thorough analysis of sample information requires the combined use of GC-MS and PTR-MS, integrating the separation advantages of GC with the qualitative capabilities of PTR-MS. However, the operation and data processing are complex, requiring the integration of artificial intelligence and machine learning techniques. In the analysis of the gas-particle distribution of carbonyl compounds, MS often requires specific sampling systems, which can affect the analytical performance.

### *7.4. Analytical Methods in Smog Chamber Simulation Studies*

In situ detection techniques, such as CEAS, DOAS, FTIR, and LIP, are more suited for detection in simple environments due to the poor stability of their spectroscopic detectors under certain conditions. Online techniques such as PTR-MS may fail to accurately identify species composition due to isomers and fragmentation caused by thermal dissociation or ion dissociation. Offline techniques such as GC-MS provide better isomer differentiation capabilities but suffer from poor temporal resolution. The combination of PTR-MS and GC-MS integrates isomer information with the high temporal resolution, thereby expanding and refining the information obtained from photochemical oxidation experiments. However, the data processing and operation are complex, and the smog chamber itself may influence chemical reactions, leading to discrepancies compared with real atmospheric detection. Therefore, detecting carbonyl compounds in smog chambers faces challenges in both spectral detection in complex environments and the combined use of PTR-MS and GC-MS. This necessitates the continued development of new sampling systems and comparative studies with real atmospheric detection to characterize the effects of various smog chambers, materials, and other factors on the samples

## 8. Future Research Needs and Prospects

Future analytical technologies for atmospheric carbonyl compounds must meet the requirements of detecting more high-molecular-weight carbonyl species, adapting to diverse monitoring scenarios, and ensuring stable long-term operation. This will enable a more comprehensive understanding of the environmental behavior and distribution characteristics of carbonyl compounds. For offline detection, the focus should be on simplifying procedures and minimizing measurement errors caused by manual operations. For online monitoring, achieving a more stable operating environment, compact instrumentation, and the ability to detect a wider range of carbonyl compounds with high reliability is essential. For the detection in smog chambers, it is essential to achieve stable spectral detection and characterize the impact of various sampling system, as well as the shape, light source, and materials of the smog chamber, on the results of PTR-MS and GC-MS detection.

To meet the future demands of both offline detection and online monitoring technologies for atmospheric carbonyl compounds, research should prioritize the following five areas: ① Develop derivatization reagents and adsorbents with immediate reaction capabilities, high derivative stability, sensitivity, and selectivity to fundamentally improve the accuracy of offline detection results for carbonyl compounds. ② Achieve high levels of automation in sampling, pretreatment, and analysis, which are crucial for improving the temporal resolution of offline detection. ③ Conduct analytical research combining multiple detectors to extract more comprehensive information on carbonyl compounds in samples. ④ Develop additional standard gases for carbonyl compounds, stable light sources, and MS ion sources, along with designing precise detectors. These advancements will reduce the size of online monitoring instruments and ensure stable detection of carbonyl compounds across diverse environments. ⑤ Integrate artificial intelligence (AI) and machine learning (ML) technologies to process and analyze the extensive data generated by online monitoring instruments. ⑥ Continuously developing the sampling system for smog chambers and characterizing the differences in the shape, light source, and materials of various types of smog chambers are essential to minimize their impact on detection results.

## 9. Conclusions

Optimizing analytical methods for atmospheric carbonyl compounds has been an ongoing effort. Over the past few decades, researchers worldwide have refined sampling, pretreatment, and analysis processes to achieve adequate sensitivity, selectivity, temporal resolution, accuracy, and operability for both laboratory research and routine monitoring. Based on an extensive literature review and systematic analysis, the key conclusions are as follows:

(1) Direct sampling methods for atmospheric carbonyl compounds encompass tube sampling, canister sampling, and cryogenic enrichment techniques. These methods are effective for collecting low-reactivity, low-boiling-point, highly polar, and volatile low-molecular-weight carbonyl compounds. Indirect sampling methods primarily involve the use of derivatization techniques in combination with specific sampling devices to collect samples, suitable for carbonyl compound sampling in various environments and phases. The reagents, adsorbents, coating methods for the reagents, and sampling devices all need to be adjusted according to specific detection requirements and analytical objectives.

(2) Offline analytical methods for formaldehyde, including fluorescence and spectrophotometry, are highly sensitive to chemical reagents and reaction conditions. By combining different reagents and optimizing analytical conditions, these methods enable rapid detection of formaldehyde under varying scenarios. Online monitoring techniques for formaldehyde primarily include spectroscopic and sensor-based detection methods. Spectroscopic monitoring relies on stable light sources to reduce environmental interfer-

ence and extended optical paths to achieve high sensitivity. However, these instruments are frequently bulky and complex. In contrast, gas sensors are portable and exhibit high selectivity for formaldehyde, but their performance is susceptible to environmental factors and strongly dependent on sensor materials.

(3) Offline analytical methods for multiple carbonyl compounds primarily include DNPH-HPLC-UV and chromatographic tandem MS. The DNPH-HPLC-UV method is simple to operate and provides reliable detection of over ten common carbonyl compounds. Advances in chromatographic tandem MS have been accompanied by improvements in complex sampling, pretreatment, and analytical methods, facilitating the differentiation of various types of carbonyl compounds. Online analytical methods for multiple carbonyl compounds primarily include GC-FID/MS and other online MS techniques, such as PTR-MS and VUV-PI-MS. GC-FID/MS systems are primarily employed for VOC analysis and can reliably detect a limited range of carbonyl compounds. Other online MS techniques can integrate various ionization sources, mass analyzers and online sampling systems to enhance performance, enabling improved isomer differentiation, lower detection limits, and shorter analysis times.

(4) Analytical methods for carbonyl compounds in smog chamber simulation studies primarily include spectroscopy, PTR-MS, and GC-MS. Spectral detectors are suitable for detection in simple environments. PTR-MS offers high time resolution, and GC-MS has strong isomer identification capabilities. The combination of PTR-MS and GC-MS can expand and improve the information obtained from detected samples. However, the smog chamber itself and the connected sampling system influence the chemical reactions, which in turn affect the detection results.

(5) Current offline analytical methods for carbonyl compounds heavily rely on the optimization of sampling, pretreatment, and analytical methods, requiring extensive time and advanced experimental skills for repeated pre-experiments. Online monitoring methods face challenges including limited stability and species coverage. In smog chamber simulation studies, the detection of carbonyl compounds is heavily influenced by both the sampling system and the chamber itself. Future detection methods should focus on improving environmental adaptability and automation while emphasizing the integration of multiple techniques and the impact of the sampling system and the smog chamber itself on the detection in smog chamber simulation studies. This approach aims to enhance the sensitivity, selectivity, temporal resolution, accuracy, and practicality of analytical methods.

**Author Contributions:** Conceptualization, X.G. and X.Y.; methodology, X.G. and Y.Y.; validation, X.Z.; formal analysis, X.G.; investigation, X.G. and Y.N.; resources, H.L.; data curation, X.G. and X.Y.; writing—original draft preparation, X.G.; writing—review and editing, W.W. and H.L.; visualization, X.G.; supervision, J.L. and Y.L.; project administration, J.B.; funding acquisition, R.G. All authors have read and agreed to the published version of the manuscript.

**Funding:** This work was supported by the Ministry of Ecology and Environment of China under the project DQGG202121 and the Dongying Ecological and Environmental Bureau (2021DFKY-0779).

**Institutional Review Board Statement:** Not applicable.

**Informed Consent Statement:** Not applicable.

**Data Availability Statement:** No new data were created or analyzed in this study.

**Conflicts of Interest:** The authors declare no conflicts of interest.

## References


1. Atkinson, R. Atmospheric chemistry of VOCs and NOx. *Atmos. Environ.* **2000**, *34*, 2063–2101. [CrossRef]
2. Zhang, Y.; Xue, L.; Dong, C.; Wang, T.; Mellouki, A.; Zhang, Q.; Wang, W. Gaseous carbonyls in China's atmosphere: Tempo-spatial distributions, sources, photochemical formation, and impact on air quality. *Atmos. Environ.* **2019**, *214*, 116863. [CrossRef]
3. Chen, T.; Xue, L.; Zheng, P.; Zhang, Y.; Liu, Y.; Sun, J.; Han, G.; Li, H.; Zhang, X.; Li, Y.; et al. Volatile organic compounds and ozone air pollution in an oil production region in northern China. *Atmos. Chem. Phys.* **2020**, *20*, 7069–7086. [CrossRef]
4. Kroll, J.A.; Hansen, A.S.; Moller, K.H.; Axson, J.L.; Kjaergaard, H.G.; Vaida, V. Ultraviolet Spectroscopy of the Gas Phase Hydration of Methylglyoxal. *ACS Earth Space Chem.* **2017**, *1*, 345–352. [CrossRef]
5. Mellouki, A.; Wallington, T.J.; Chen, J. Atmospheric Chemistry of Oxygenated Volatile Organic Compounds: Impacts on Air Quality and Climate. *Chem. Rev.* **2015**, *115*, 3984–4014. [CrossRef]
6. Ling, Z.H.; Guo, H.; Chen, G.X.; Lam, S.H.M.; Fan, S.J. Formaldehyde and Acetaldehyde at Different Elevations in Mountainous Areas in Hong Kong. *Aerosol Air Qual. Res.* **2016**, *16*, 1868–1878. [CrossRef]
7. Wang, C.; Huang, X.F.; Han, Y.; Zhu, B.; He, L.Y. Sources and Potential Photochemical Roles of Formaldehyde in an Urban Atmosphere in South China. *J. Geophys. Res. Atmos.* **2017**, *122*, 2169–8996. [CrossRef]
8. Chou, C.C.K.; Tsai, C.Y.; Chang, C.C.; Lin, P.H.; Liu, S.C.; Zhu, T. Photochemical production of ozone in Beijing during the 2008 Olympic Games. *Atmos. Chem. Phys.* **2011**, *11*, 9825–9837. [CrossRef]
9. Turpin, B.J.; Saxena, P.; Andrews, E. Measuring and simulating particulate organics in the atmosphere: Problems and prospects. *Atmos. Environ.* **2000**, *34*, 2983–3013. [CrossRef]
10. Du, Z.; Mo, J.; Zhang, Y. Risk assessment of population inhalation exposure to volatile organic compounds and carbonyls in urban China. *Env. Int.* **2014**, *73*, 33–45. [CrossRef] [PubMed]
11. Delikhoon, M.; Fazlzadeh, M.; Sorooshian, A.; Baghani, A.N.; Golaki, M.; Ashournejad, Q.; Barkhordari, A. Characteristics and health effects of formaldehyde and acetaldehyde in an urban area in Iran. *Environ. Pollut.* **2018**, *242*, 938–951. [CrossRef] [PubMed]
12. Flowers, L.; Broder, M.W.; Forsyth, C. *Toxicological Review of Acetone*; US Environmental Protection Agency: Washington, DC, USA, 2003.
13. Li, L.; Jia, Q.; Zhang, Z. Advances in the study of crotonaldehyde toxicity. *Chin. J. Ind. Hyg. Occup. Dis.* **2016**, *43*, 758–761.
14. Atkinson, R.; Arey, J. Atmospheric Degradation of Volatile Organic Compounds. *Chem. Rev.* **2003**, *103*, 4605–4638. [CrossRef]
15. Zhang, K.; Duan, Y.; Huo, J.; Huang, L.; Wang, Y.; Fu, Q.; Wang, Y.; Li, L. Formation mechanism of HCHO pollution in the suburban Yangtze River Delta region, China: A box model study and policy implementations. *Atmos. Environ.* **2021**, *267*, 118755. [CrossRef]
16. Bao, J.M.; Li, H.; Wu, Z.H.; Zhang, X.; Zhang, H.; Li, Y.F.; Qian, J.; Chen, J.H.; Deng, L.Q. Atmospheric carbonyls in a heavy ozone pollution episode at a metropolis in Southwest China: Character istics, health risk assessment, sources analysis. *J. Environ. Sci.* **2022**, *113*, 40–54. [CrossRef] [PubMed]
17. Chai, W.; Wang, M.; Li, J.; Tang, G.; Zhang, G.; Chen, W. Pollution characteristics, sources, and photochemical roles of ambient carbonyl compounds in summer of Beijing, China. *Environ. Pollut.* **2023**, *336*, 122403. [CrossRef] [PubMed]
18. Qiu, X.H.; Wang, S.X.; Ying, Q.; Duan, L.; Xing, J.; Cao, J.Y.; Wu, D.; Li, X.X.; Xing, C.Z.; Yan, X.; et al. Importance of Wintertime Anthropogenic Glyoxal and Methylglyoxal Emissions in Beijing and Implications for Secondary Organic Aerosol Formation in Megacities. *Environ. Sci. Technol.* **2020**, *54*, 11809–11817. [CrossRef]
19. Doussin, J.-F.; Fuchs, H.; Kiendler-Scharr, A.; Seakins, P.; Wenger, J. *A Practical Guide to Atmospheric Simulation Chambers*, 1st ed.; Springer Nature Switzerland AG: Cham, Switzerland, 2023; pp. 1–339.
20. Liu, Q.; Gao, Y.; Huang, W.; Ling, Z.; Wang, Z.; Wang, X. Carbonyl compounds in the atmosphere: A review of abundance, source and their contributions to $O_3$ and SOA formation. *Atmos. Res.* **2022**, *274*, 106184. [CrossRef]
21. Tang, J.; Wang, X.; Sheng, G.; Fu, J. Advances in the analysis of formaldehyde and carbonyl compounds in the atmosphere. *Chin. J. Anal. Chem.* **2005**, *33*, 134–140.
22. Nier, A.O. Some reflections on the early days of mass spectrometry at the university of minnesota. *Int. J. Mass. Spectrom. Ion Process.* **1990**, *100*, 1–13. [CrossRef]
23. Nash, T. The colorimetric estimation of formaldehyde by means of the Hantzsch reaction. *Biochem. J.* **1953**, *55*, 416–421. [CrossRef] [PubMed]
24. Pinheiro, H.L.C.; de Andrade, M.V.; Pereira, P.A.P.; de Andrade, J.B. Spectrofluorimetric determination of formaldehyde in air after collection onto silica cartridges coated with Fluoral P. *Microchem. J.* **2004**, *78*, 15–20. [CrossRef]
25. Yuan, B.; Koss, A.R.; Warneke, C.; Coggon, M.; Sekimoto, K.; de Gouw, J.A. Proton-Transfer-Reaction Mass Spectrometry: Applications in Atmospheric Sciences. *Chem. Rev.* **2017**, *117*, 13187–13229. [CrossRef]
26. Chung, P.R.; Tzeng, C.T.; Ke, M.T.; Lee, C.Y. Formaldehyde Gas Sensors: A Review. *Sensors* **2013**, *13*, 4468–4484. [CrossRef]
27. Wu, T.; Hu, R.; Xie, P.; Wang, J.; Liu, W. Research progress of atmospheric HCHO spectroscopy detection technology. *J. Quantum Electron.* **2021**, *38*, 699.
28. Huang, M.; Ma, F.; Zhang, J.; Zhou, Y.; Wen, S.; Liu, Z. Research Progress of the Environmental Monitoring Analytical Methods of Aldehydes and Ketones Carbonyl Compounds. *Sichuan Environ.* **2023**, *42*, 300–305.

29. Xu, Y.; Hui, L.; Zheng, P.; Liu, G.; Yu, J.Z.; Wang, Z. Monitoring techniques of airborne carbonyl compounds: Principles, performance and challenges. *TrAC Trends Anal. Chem.* **2023**, *169*, 117395. [CrossRef]
30. Li, H.; Xiao, Q.; Han, C.; Liu, L.; Fan, X.; Chen, N.; Pan, Y. On-Line Monitoring of VOCs in Ambient Air Based on Cryo-Trapping-GC-MS-FID. *Chin. Environ. Monit.* **2017**, *33*, 100–110.
31. MEE. *Determination of Volatile Organic Compounds in Ambient Air—Adsorbent Tube Sampling—Thermal Desorption/Gas Chromatography-Mass Spectrometry: HJ 644-2013*; China Environmental Science Press: Beijing, China, 2013.
32. Hui, L.; Liu, X.; Tan, Q.; Feng, M.; An, J.; Qu, Y.; Zhang, Y.; Jiang, M. Characteristics, source apportionment and contribution of VOCs to ozone formation in Wuhan, Central China. *Atmos. Environ.* **2018**, *192*, 55–71. [CrossRef]
33. Druzik, C.M.; Grosjean, D.; Van Neste, A.; Parmar, S.S. Sampling of Atmospheric Carbonyls with Small DNPH-Coated C18 Cartridges and Liquid Chromatography Analysis with Diode Array Detection. *Int. J. Environ. Anal. Chem.* **1990**, *38*, 495–512. [CrossRef]
34. Li, J.; Feng, Y.L.; Xie, C.J.; Huang, J.; Yu, J.Z.; Feng, J.L.; Sheng, G.Y.; Fu, J.M.; Wu, M.H. Determination of gaseous carbonyl compounds by their pentafluorophenyl hydrazones with gas chromatography/mass spectrometry. *Anal. Chim. Acta.* **2009**, *635*, 84–93. [CrossRef] [PubMed]
35. U.S. Environmental Protection Agency (EPA). *Compendium of Methods for the Determination of Toxic Organic Compounds in Ambient Air. Second Edition. Compendium Method TO-14A: Determination of Volatile Organic Compounds (VOCs) in Ambient Air Using Specially Prepared Canisters with Subsequent Analysis by Gas Chromatography. EPA/625/R-96/010b.PU.S*; EPA: Cincinnati, OH, USA, 1999.
36. U.S. Environmental Protection Agency (EPA). *Method TO-15A: Determination of Volatile Organic Compounds (VOCs) in Air Collected in Specially Prepared Canisters and Analyzed by Gas Chromatography–Mass Spectrometry (GC-MS)*; U.S. EPA: Cincinnati, OH, USA, 2019.
37. Batterman, S.A.; Zhang, G.-Z.; Baumann, M. Analysis and stability of aldehydes and terpenes in electropolished canisters. *Atmos. Environ.* **1998**, *32*, 1647–1655. [CrossRef]
38. Kelly, T.J.; Holdren, M.W. Applicability of canisters for sample storage in the determination of hazardous air pollutants. *Atmos. Environ.* **1995**, *29*, 2595–2608. [CrossRef]
39. Xie, X.; Lu, X.; Zhang, S. Comparison on Monitoring Methods of Environmental Air Volatile Organic Compounds in Nanjing. *China Environ. Monit.* **2020**, *36*, 11–15.
40. Wang, M.; Zeng, L.; Lu, S.; Shao, M.; Liu, X.; Yu, X.; Chen, W.; Yuan, B.; Zhang, Q.; Hu, M.; et al. Development and validation of a cryogen-free automatic gas chromatograph system (GC-MS/FID) for online measurements of volatile organic compounds. *Anal. Methods* **2014**, *6*, 9424–9434. [CrossRef]
41. Li, X.; Lu, S.; Shao, M. Methods for Measurement of Oxygenated Volatile Organic Compounds in the Atmosphere. *Peking. Univ. Sci.* **2006**, *42*, 548–554.
42. Liu, L.; Dong, Y.; Sun, X. Comparison of HPLC and GC-MS for the Determination of Aldehydes and Ketones in Ambient Air. *Environ. Monit. Manag. Technol.* **2023**, *35*, 45–48.
43. Pal, R.; Kim, K.H. Experimental choices for the determination of carbonyl compounds in air. *J. Sep. Sci.* **2007**, *30*, 2708–2718. [CrossRef] [PubMed]
44. Han, C.; Han, B.; Xia, L. The comparison between the MBTH and Acetylacetone spectrophotometric methods in Determination of Formaldehyde in Indoor Air. *Energy Conserv. Environ. Prot.* **2020**, *06*, 58–59.
45. U.S. Environmental Protection Agency (EPA). *Compendium of Methods for the Determination of Toxic Organic Compounds in Ambient Air. Second Edition. Compendium Method TO-11A: Determination of Formaldehyde in Ambient Air Using Adsorbent Cartridge Followed by High Performance Liquid Chromatography (HPLC) [Active Sampling Methodology]. EPA/625/R-96/010b*; U.S. EPA: Cincinnati, OH, USA, 1999.
46. Pang, X.; Lewis, A.C.; Hamilton, J.F. Determination of airborne carbonyls via pentafluorophenylhydrazine derivatisation by GC–MS and its comparison with HPLC method. *Talanta* **2011**, *85*, 406–414. [CrossRef]
47. Arnts, R.R.; Tejada, S.B. 2,4-Dinitrophenylhydrazine-coated silica gel cartridge method for determination of formaldehyde in air: Identification of an ozone interference. *Environ. Sci. Technol.* **1989**, *23*, 1428–1430. [CrossRef]
48. Williams, E.L.; Grosjean, D. Removal of atmospheric oxidants with annular denuders. *Environ. Sci. Technol.* **1990**, *24*, 811–814. [CrossRef]
49. Ho, S.S.H.; Ho, K.F.; Liu, W.D.; Lee, S.C.; Dai, W.T.; Cao, J.J.; Ip, H.S.S. Unsuitability of using the DNPH-coated solid sorbent cartridge for determination of airborne unsaturated carbonyls. *Atmos. Environ.* **2011**, *45*, 261–265. [CrossRef]
50. Kahnt, A.; Iinuma, Y.; Böge, O.; Mutzel, A.; Herrmann, H. Denuder sampling techniques for the determination of gas-phase carbonyl compounds: A comparison and characterisation of in situ and ex situ derivatisation methods. *J. Chromatogr. B* **2011**, *879*, 1402–1411. [CrossRef] [PubMed]
51. Cecinato, A.; Di Palo, V.; Mabilia, R.; Possanzini, M. Pentafluorophenylhydrazine as a coating reagent for the HRGC-MS determination of semi-volatile carbonyl compounds in air. *Chromatographia* **2001**, *54*, 263–269. [CrossRef]
52. Ho, S.S.H.; Yu, J.Z. Determination of airborne carbonyls: Comparison of a thermal desorption/GC method with the standard DNPH/HPLC method. *Environ. Sci. Technol.* **2004**, *38*, 862–870. [CrossRef] [PubMed]

53. Apel, E.C.; Calvert, J.G.; Riemer, D.; Pos, W.; Zika, R.; Kleindienst, T.E.; Lonneman, W.A.; Fung, K.; Fujita, E.; Shepson, P.B.; et al. Measurements comparison of oxygenated volatile organic compounds at a rural site during the 1995 SOS Nashville Intensive. *J. Geophys. Res. Atmos.* **1998**, *103*, 22295–22316. [CrossRef]
54. Lomonaco, T.; Romani, A.; Ghimenti, S.; Biagini, D.; Bellagambi, F.G.; Onor, M.; Salvo, P.; Fuoco, R.; Di Francesco, F. Determination of carbonyl compounds in exhaled breath by on-sorbent derivatization coupled with thermal desorption and gas chromatography-tandem mass spectrometry. *J. Breath. Res.* **2018**, *12*, 046004. [CrossRef] [PubMed]
55. Ho, S.S.H.; Yu, J.Z. Feasibility of Collection and Analysis of Airborne Carbonyls by On-Sorbent Derivatization and Thermal Desorption. *Anal. Chem.* **2002**, *74*, 1232–1240. [CrossRef]
56. Cecinato, A.; Yassaa, N.; Di Palo, V.; Possanzini, M.; Cecinato, A. Observation of volatile and semi-volatile carbonyls in an Algerian urban environment using dinitrophenylhydrazine/silica-HPLC and pentafluorophenylhydrazine/silica-GC-MS. *J. Environ. Monit.* **2002**, *4*, 223–228. [CrossRef] [PubMed]
57. Arthur, C.L.; Pawliszyn, J. Solid phase microextraction with thermal desorption using fused silica optical fibers. *Anal. Chem.* **1990**, *62*, 2145–2148. [CrossRef]
58. Schmarr, H.G.; Sang, W.; Ganß, S.; Fischer, U.; Köpp, B.; Schulz, C.; Potouridis, T. Analysis of aldehydes via headspace SPME with on-fiber derivatization to their O-(2,3,4,5,6-pentafluorobenzyl)oxime derivatives and comprehensive 2D-GC-MS. *J. Sep. Sci.* **2008**, *31*, 3458–3465. [CrossRef] [PubMed]
59. Borrás, E.; Tortajada-Genaro, L.A.; Ródenas, M.; Vera, T.; Speak, T.; Seakins, P.; Shaw, M.D.; Lewis, A.C.; Muñoz, A. On-line solid phase microextraction derivatization for the sensitive determination of multi-oxygenated volatile compounds in air. *Atmos. Meas. Tech.* **2021**, *14*, 4989–4999. [CrossRef]
60. Tumbiolo, S.; Gal, J.-F.; Maria, P.-C.; Zerbinati, O. Determination of benzene, toluene, ethylbenzene and xylenes in air by solid phase micro-extraction/gas chromatography/mass spectrometry. *Anal. Bioanal. Chem.* **2004**, *380*, 824–830. [CrossRef] [PubMed]
61. Ghiasvand, A.R.; Hajipour, S.; Heidari, N. Cooling-assisted microextraction: Comparison of techniques and applications. *TrAC* **2016**, *77*, 54–65. [CrossRef]
62. Bourdin, D.; Desauziers, V. Development of SPME on-fiber derivatization for the sampling of formaldehyde and other carbonyl compounds in indoor air. *Anal. Bioanal. Chem.* **2014**, *406*, 317–328. [CrossRef] [PubMed]
63. Tobias, H.J.; Kooiman, P.M.; Docherty, K.S.; Ziemann, P.J. Real-Time Chemical Analysis of Organic Aerosols Using a Thermal Desorption Particle Beam Mass Spectrometer. *Aerosol Sci. Technol.* **2000**, *33*, 170–190. [CrossRef]
64. Williams, B.J.; Goldstein, A.H.; Kreisberg, N.M.; Hering, S.V. An in-situ instrument for speciated organic composition of atmospheric aerosols: T hermal desorption a erosol GC/MS-FID (TAG). *Aerosol Sci. Technol.* **2006**, *40*, 627–638. [CrossRef]
65. Hohaus, T.; Trimborn, D.; Kiendler-Scharr, A.; Gensch, I.; Laumer, W.; Kammer, B.; Andres, S.; Boudries, H.; Smith, K.A.; Worsnop, D. A new aerosol collector for quasi on-line analysis of particulate organic matter: The Aerosol Collection Module (ACM) and first applications with a GC/MS-FID. *Atmos. Meas. Tech.* **2010**, *3*, 1423–1436. [CrossRef]
66. Temime, B.; Healy, R.M.; Wenger, J.C. A denuder-filter sampling technique for the detection of gas and particle phase carbonyl compounds. *Environ. Sci. Technol.* **2007**, *41*, 6514–6520. [CrossRef]
67. Eichler, P.; Müller, M.; D'Anna, B.; Wisthaler, A. A novel inlet system for online chemical analysis of semi-volatile submicron particulate matter. *Atmos. Meas. Tech.* **2015**, *8*, 1353–1360. [CrossRef]
68. Vairavamurthy, A.; Roberts, J.M.; Newman, L. Methods for Determination of Low-Molecular-Weight Carbonyl-Compounds in the Atmosphere. *Atmos. Env. Part. A-Gen. Topics.* **1992**, *26*, 1965–1993. [CrossRef]
69. Dong, B.; Song, X.; Tang, Y.; Lin, W. A rapid and facile fluorimetric method for detecting formaldehyde. *Sens. Actuators B: Chemical.* **2016**, *222*, 325–330. [CrossRef]
70. Li, Q.; Sritharathikhun, P.; Motomizu, S. Development of Novel Reagent for Hantzsch Reaction for the Determination of Formaldehyde by Spectrophotometry and Fluorometry. *Anal. Sci.* **2007**, *23*, 413–417. [CrossRef] [PubMed]
71. Cheng, W.; Chen, J.; Chen, J.; Liu, B.; Song, G. Application of spectroscopic detection technology in formaldehyde concentration measurement. *J. Mudanjiang Univ.* **2010**, *19*, 131–132+136.
72. Huang, Y. Determination of formaldehyde in air by spectrophotometric method with phenol reagent and acetylacetone spectrophotometry. *Environ. Dev.* **2018**, *30*, 115–116.
73. Wang, D.; Jiang, Y.; Li, P.; Li, L.; Liu, Q.; Sun, Y. Comparison of phenol reagent spectrophotometry and acetylacetone spectrophotometry for determining formaldehyde content in air. *Chin. Med. Instrum. Inf.* **2017**, *23*, 13–15.
74. Ji, K.; Su, Y.; Xiao, Y.; Liang, K.; Wang, J.; Liu, P.; Zhang, K.; Zhao, L. Key factor of the determination of formaldehyde in low concentration by phenol reagent spectrophotometry. *Chem. Ind. Progr.* **2020**, *39*, 281–286.
75. *GB/T 18204.2-2014*; Examination Methods for Public Place. Part 2: Chemical Pollutant. Standards Press of China: Beijing, China, 2014.
76. Wang, Q. Influence factors of determination of formaldehyde concentration by AHMT spectrophotometry. *Chem. Eng.* **2018**, *32*, 34–37.
77. Wei, K.; Ma, L.; Ma, G.; Ji, C.; Yin, M. A two-step responsive colorimetric probe for fast detection of formaldehyde in weakly acidic environment. *Dye. Pigments.* **2019**, *165*, 294–300. [CrossRef]

78. Mohamed, A.A.; Mubarak, A.T.; Marestani, Z.M.H.; Fawy, K.F. Highly sensitive and selective catalytic determination of formaldehyde and acetaldehyde. *Talanta.* **2008**, *74*, 578–585. [CrossRef] [PubMed]
79. Cárdenas, L.M.; Brassington, D.J.; Allan, B.J.; Coe, H.; Alicke, B.; Platt, U.; Wilson, K.M.; Plane, J.M.C.; Penkett, S.A. Intercomparison of Formaldehyde Measurements in Clean and Polluted Atmospheres. *J. Atmos. Chem.* **2000**, *37*, 53–80. [CrossRef]
80. Hak, C.; Pundt, I.; Trick, S.; Kern, C.; Platt, U.; Dommen, J.; Ordóñez, C.; Prévôt, A.S.H.; Junkermann, W.; Astorga-Lloréns, C.; et al. Intercomparison of four different in-situ techniques for ambient formaldehyde measurements in urban air. *Atmos. Chem. Phys.* **2005**, *5*, 2881–2900. [CrossRef]
81. Catoire, V.; Bernard, F.; Mébarki, Y.; Mellouki, A.; Eyglunent, G.; Daële, V.; Robert, C. A tunable diode laser absorption spectrometer for formaldehyde atmospheric measurements validated by simulation chamber instrumentation. *J. Environ. Sci.* **2012**, *24*, 22–33. [CrossRef] [PubMed]
82. Kleine, D.; Mürtz, M.; Lauterbach, J.; Dahnke, H.; Urban, W.; Hering, P.; Kleinermanns, K. Atmospheric trace gas analysis with cavity ring-down spectroscopy. *Isr. J. Chemistry.* **2001**, *41*, 111–116. [CrossRef]
83. Peltola, J.; Vainio, M.; Ulvila, V.; Siltanen, M.; Metsälä, M.; Halonen, L. Off-axis re-entrant cavity ring-down spectroscopy with a mid-infrared continuous-wave optical parametric oscillator. *Appl. Phys. B* **2012**, *107*, 839–847. [CrossRef]
84. Motyka, K.; Mikuska, P. Continuous fluorescence determination of formaldehyde in air. *Anal. Chim. Acta.* **2004**, *518*, 51–57. [CrossRef]
85. Xu, R.; Li, X.; Dong, H.; Wu, Z.; Chen, S.; Fang, X.; Gao, J.; Guo, S.; Hu, M.; Li, D.; et al. Measurement of gaseous and particulate formaldehyde in the Yangtze River Delta, China. *Atmos. Environ.* **2020**, *224*, 117114. [CrossRef]
86. Jiang, J.; Zeng, L. The design and application of an online gaseous formaldehyde analyzer. *Acta Sci. Circumst.* **2018**, *38*, 3475–3481.
87. Henry, S.B.; Kammrath, A.; Keutsch, F.N. Quantification of gas-phase glyoxal and methylglyoxal via the Laser-Induced Phosphorescence of (methyl)GLyOxal Spectrometry (LIPGLOS) Method. *Atmos. Meas. Tech.* **2012**, *5*, 181–192. [CrossRef]
88. Volkamer, R.; Molina, L.T.; Molina, M.J.; Shirley, T.; Brune, W.H. DOAS measurement of glyoxal as an indicator for fast VOC chemistry in urban air. *Geophys. Res. Lett.* **2005**, *32*, L08806. [CrossRef]
89. MacDonald, S.M.; Oetjen, H.; Mahajan, A.S.; Whalley, L.K.; Edwards, P.M.; Heard, D.E.; Jones, C.E.; Plane, J.M.C. DOAS measurements of formaldehyde and glyoxal above a south-east Asian tropical rainforest. *Atmos. Chem. Phys.* **2012**, *12*, 5949–5962. [CrossRef]
90. Hong, Q.; Liu, C.; Hu, Q.; Zhang, Y.; Xing, C.; Ou, J.; Tan, W.; Liu, H.; Huang, X.; Wu, Z. Vertical distribution and temporal evolution of formaldehyde and glyoxal derived from MAX-DOAS observations: The indicative role of VOC sources. *J. Environ. Sci.* **2022**, *122*, 92–104. [CrossRef] [PubMed]
91. Thalman, R.; Volkamer, R. Inherent calibration of a blue LED-CE-DOAS instrument to measure iodine oxide, glyoxal, methyl glyoxal, nitrogen dioxide, water vapour and aerosol extinction in open cavity mode. *Atmos. Meas. Tech.* **2010**, *3*, 1797–1814. [CrossRef]
92. Washenfelder, R.A.; Langford, A.O.; Fuchs, H.; Brown, S.S. Measurement of glyoxal using an incoherent broadband cavity enhanced absorption spectrometer. *Atmos. Chem. Phys.* **2008**, *8*, 7779–7793. [CrossRef]
93. Flowerday, C.E.; Thalman, R.; Asplund, M.C.; Hansen, J.C.J.T. Broadband Cavity-Enhanced Absorption Spectroscopy (BBCEAS) Coupled with an Interferometer for On-Band and Off-Band Detection of Glyoxal. *Toxics* **2023**, *12*, 26. [CrossRef]
94. An, C.; Yan, Y.; Gao, X.; Yan, X.; Ji, Y.; Shang, F.; Li, J.; Tan, L.; Gao, R.; Bi, F.; et al. A Comparative Investigation of the Characteristics of Nocturnal Ozone Enhancement Events and Their Effects on Ground-Level Ozone and $PM_{2.5}$ in the Central City of the Yellow River Delta, China, in 2022 and 2023. *Atmosphere* **2024**, *15*, 475. [CrossRef]
95. Fang, B.; Yang, N.; Wang, C.; Zhao, W.; Zhou, H.; Zhang, W. Highly sensitive portable laser absorption spectroscopy formaldehyde sensor using compact spherical mirror multi-pass cell. *Sens. Actuators B Chem.* **2023**, *394*, 134379. [CrossRef]
96. Darder, M.d.M.; Bedoya, M.; Serrano, L.A.; Alba, M.Á.; Orellana, G. Fiberoptic colorimetric sensor for in situ measurements of airborne formaldehyde in workplace environments. *Sens. Actuators B Chem.* **2022**, *353*, 131099. [CrossRef]
97. Yuan, Z.; Yang, C.; Meng, F. Strategies for Improving the Sensing Performance of Semiconductor Gas Sensors for High-Performance Formaldehyde Detection: A Review. *Chemosensors.* **2021**, *9*, 179. [CrossRef]
98. Warneke, C.; Veres, P.; Holloway, J.S.; Stutz, J.; Tsai, C.; Alvarez, S.; Rappenglueck, B.; Fehsenfeld, F.C.; Graus, M.; Gilman, J.B.; et al. Airborne formaldehyde measurements using PTR-MS: Calibration, humidity dependence, inter-comparison and initial results. *Atmos. Meas. Tech.* **2011**, *4*, 2345–2358. [CrossRef]
99. Platt, U.; Perner, D.; Pätz, H.W. Simultaneous measurement of atmospheric $CH_2O$, $O_3$, and $NO_2$ by differential optical absorption. *J. Geophys. Res. Oceans* **1979**, *84*, 6329–6335. [CrossRef]
100. Stutz, J.; Platt, U. Improving long-path differential optical absorption spectroscopy with a quartz-fiber mode mixer. *Appl. Opt.* **1997**, *36*, 1105–1115. [CrossRef]
101. Grutter, M.; Flores, E.; Andraca-Ayala, G.; Báez, A. Formaldehyde levels in downtown Mexico City during 2003. *Atmos. Environ.* **2005**, *39*, 1027–1034. [CrossRef]
102. Heckel, A.; Richter, A.; Tarsu, T.; Wittrock, F.; Hak, C.; Pundt, I.; Junkermann, W.; Burrows, J.P. MAX-DOAS measurements of formaldehyde in the Po-Valley. *Atmos. Chem. Phys.* **2005**, *5*, 909–918. [CrossRef]
103. Tuazon, E.C.; Graham, R.A.; Winer, A.M.; Easton, R.R.; Pitts, J.N.; Hanst, P.L. A kilometer pathlength fourier-transform infrared system for the study of trace pollutants in ambient and synthetic atmospheres. *Atmos. Environ (1967)* **1978**, *12*, 865–875. [CrossRef]

104. Suarez-Bertoa, R.; Clairotte, M.; Arlitt, B.; Nakatani, S.; Hill, L.; Winkler, K.; Kaarsberg, C.; Knauf, T.; Zijlmans, R.; Boertien, H.J.F. Intercomparison of ethanol, formaldehyde and acetaldehyde measurements from a flex-fuel vehicle exhaust during the WLTC. *Fuel* **2017**, *203*, 330–340. [CrossRef]
105. Becker, K.H.; Schurath, U.; Tatarczyk, T. Fluorescence Determination of Low Formaldehyde Concentrations in Air by Dye Laser Excitation. *Appl. Opt.* **1975**, *14*, 310–313. [CrossRef]
106. Möhlmann, G.R. Formaldehyde Detection in Air by Laser-Induced Fluorescence. *Appl. Spectrosc.* **1985**, *39*, 98–101. [CrossRef]
107. Cazorla, M.; Wolfe, G.M.; Bailey, S.A.; Swanson, A.K.; Arkinson, H.L.; Hanisco, T.F. A new airborne laser-induced fluorescence instrument for in situ detection of formaldehyde throughout the troposphere and lower stratosphere. *Atmos. Meas. Tech.* **2015**, *8*, 541–552. [CrossRef]
108. Hottle, J.R.; Huisman, A.J.; Digangi, J.P.; Kammrath, A.; Galloway, M.M.; Coens, K.L.; Keutsch, F.N. A Laser Induced Fluorescence-Based Instrument for In-Situ Measurements of Atmospheric Formaldehyde. *Environ. Sci. Technol.* **2009**, *43*, 790–795. [CrossRef] [PubMed]
109. Clair, J.M.; Swanson, A.K.; Bailey, S.A.; Hanisco, T.F. CAFE: A new, improved nonresonant laser-induced fluorescence instrument for airborne in situ measurement of formaldehyde. *Atmos. Meas. Tech.* **2019**, *12*, 4581–4590. [CrossRef]
110. Harris, G.W.; Mackay, G.I.; Iguchi, T.; Mayne, L.K.; Schiff, H.I. Measurements of formaldehyde in the troposphere by tunable diode laser absorption spectroscopy. *J. Atmos. Chem.* **1989**, *8*, 119–137. [CrossRef]
111. Mackay, G.I.; Karecki, D.R.; Schiff, H.I. Tunable diode laser absorption measurements of $H_2O_2$ and HCHO during the Mauna Loa Observatory Photochemistry Experiment. *J. Geophys. Res. Atmos.* **1996**, *101*, 14721–14728. [CrossRef]
112. Richter, D.; Fried, A.; Wert, B.P.; Walega, J.G.; Tittel, F.K. Development of a tunable mid-IR difference frequency laser source for highly sensitive airborne trace gas detection. *Appl. Phys. B.* **2002**, *75*, 281–288. [CrossRef] [PubMed]
113. Herndon, S.C.; Zahniser, M.S.; Nelson, D.D.; Shorter, J.; McManus, J.B.; Jiménez, R.; Warneke, C.; de Gouw, J.A. Airborne measurements of HCHO and HCOOH during the New England Air Quality Study 2004 using a pulsed quantum cascade laser spectrometer. *J. Geophys. Res. Atmos.* **2007**, *112*, D10. [CrossRef]
114. McManus, J.B.; Zahniser, M.S.; Nelson, D.D. Dual quantum cascade laser trace gas instrument with astigmatic Herriott cell at high pass number. *Appl. Opt.* **2011**, *50*, A74–A85. [CrossRef] [PubMed]
115. Dong, L.; Yu, Y.; Li, C.; So, S.; Tittel, F.K. Ppb-level formaldehyde detection using a CW room-temperature interband cascade laser and a miniature dense pattern multipass gas cell. *Opt. Express.* **2015**, *23*, 19821–19830. [CrossRef] [PubMed]
116. Fang, B.; Yang, N.N.; Zhao, W.X.; Wang, C.H.; Zhang, W.J.; Song, W.; Venables, D.S.; Chen, W.D. Improved spherical mirror multipass-cell-based interband cascade laser spectrometer for detecting ambient formaldehyde at parts per trillion by volume levels. *Appl. Opt.* **2019**, *58*, 8743–8750. [CrossRef]
117. Rella, C.; Hoffnagle, J.; Kim-Hak, D. Quantification of atmospheric formaldehyde by near-infrared cavity ring-down spectroscopy. *EGU Gen. Assem. Conf. Abstr.* **2018**, *20*, 11101.
118. Hoffnagle, J.; Fleck, D.; Rella, C.; Kim-Hak, D. Quantification of atmospheric formaldehyde by infrared absorption spectroscopy. *EGU Gen. Assem. Conf. Abstr.* **2017**, *19*, 8972.
119. Liu, J.; Li, X.; Yang, Y.; Wang, H.; Wu, Y.; Lu, X.; Chen, M.; Hu, J.; Fan, X.; Zeng, L.; et al. An IBBCEAS system for atmospheric measurements of glyoxal and methylglyoxal in the presence of high $NO_2$ concentrations. *Atmos. Meas. Tech.* **2019**, *12*, 4439–4453. [CrossRef]
120. Liu, J.; Li, X.; Yang, Y.; Wang, H.; Kuang, C.; Zhu, Y.; Chen, M.; Hu, J.; Zeng, L.; Zhang, Y. Sensitive Detection of Ambient Formaldehyde by Incoherent Broadband Cavity Enhanced Absorption Spectroscopy. *Anal. Chem.* **2020**, *92*, 2697–2705. [CrossRef] [PubMed]
121. Gorrotxategi-Carbajo, P.; Fasci, E.; Ventrillard, I.; Carras, M.; Maisons, G.; Romanini, D. Optical-feedback cavity-enhanced absorption spectroscopy with a quantum-cascade laser yields the lowest formaldehyde detection limit. *Appl. Phys. B.* **2013**, *110*, 309–314. [CrossRef]
122. He, Q.; Zheng, C.; Lou, M.; Ye, W.; Wang, Y.; Tittel, F.K.J.O.E. Dual-feedback mid-infrared cavity-enhanced absorption spectroscopy for $H_2CO$ detection using a radio-frequency electrically-modulated interband cascade laser. *Optics* **2018**, *26*, 15436–15444. [CrossRef] [PubMed]
123. Zhou, Z.; Feng, S. Sampling by canister, combining with Agilent 5977B GC/MS to analysis 13 aldehyde and keystone volatile organic compounds in Air. *Environ. Chem.* **2019**, *38*, 697–700.
124. Zhou, Z.; Feng, S. Analysis of 104 volatile organic compounds in the atmosphere using canister sampling-atmospheric preconcentration instrument coupled with Agilent 5977B single quadrupole GC-MS system. *Environ. Chem.* **2018**, *37*, 1869–1872.
125. Cui, T.; Li, H.; Tang, S.; Liu, C.; Gao, X.; Li, Z.; Zhang, F.; Yang, G.; Jiang, W.; Liu, Z.; et al. Determination of Aldehydes and Ketones Compounds in Mainstream Cigarette Smoke by Supercritical Fluid Chromatography and Gas Chromatography—Mass Spectrometry. *J. Anal. Test.* **2018**, *37*, 766–771.
126. Chien, Y.-C.; Yin, K.-G.; Chien, Y.-C. Simultaneous determination of airborne carbonyls and aromatic hydrocarbons using mixed sorbent collection and thermal desorption-gas chromatography/mass spectrometric analysis. *J. Environ. Monit.* **2009**, *11*, 1013–1019. [CrossRef] [PubMed]
127. Yu, J.; Jeffries, H.E.; Sexton, K.G. Atmospheric photooxidation of alkylbenzenes—I. Carbonyl product analyses. *Atmos. Environ.* **1997**, *31*, 2261–2280. [CrossRef]

128. Poole, J.J.; Grandy, J.J.; Gómez-Ríos, G.A.; Gionfriddo, E.; Pawliszyn, J. Solid Phase Microextraction On-Fiber Derivatization Using a Stable, Portable, and Reusable Pentafluorophenyl Hydrazine Standard Gas Generating Vial. *Anal. Chem.* **2016**, *88*, 6859–6866. [CrossRef]
129. Aiello, M.; McLaren, R. Measurement of Airborne Carbonyls Using an Automated Sampling and Analysis System. *Environ. Sci. Technol.* **2009**, *43*, 8901–8907. [CrossRef] [PubMed]
130. Huo, Y.; Qiu, X.; Zhan, S. Determination of 13 aldehyde and ketone compounds in air using ultra-high performance liquid chromatography. *Environ. Chem.* **2013**, *32*, 1823–1825.
131. Wang, F.; Wang, T.; Fan, Z.; Wu, W.; Ma, W.; Bai, Z. Determination of Carbonyl Compounds in Ambient Air by Temperature/Solvent Double-Gradient High Performance Liquid Chromatography. *Anhui Agric. Bull.* **2015**, *21*, 23–25.
132. Si, L.; Xing, G.; Wang, C.; Tan, L.; Chen, Y.; Yu, J.; Liu, F.; Yuan, M. Determination of aldehydes and ketones in ambient air using adsorbent cartridge followed by high performance liquid chromatography. *Environ. Chem.* **2019**, *38*, 2222–2228.
133. Fan, Z.; Wang, F.; He, Y.; Ma, W. Simultaneous Determination of Multiple Carbonyl Compounds in Ambient Air. *Environ. Monit. Manag. Technol.* **2018**, *30*, 43–45.
134. Tao, M.; Yin, L.; Fu, X. Determination of 16 Aldehydes and Ketones in Ambient Air by Solution Absorption-High Performance Liquid Chromatography. *Sci. Technol. Innov.* **2024**, *2*, 107–110.
135. Chi, Y.; Feng, Y.; Wen, S.; Lu, H.; Yu, Z.; Zhang, W.; Sheng, G.; Fu, J. Determination of carbonyl compounds in the atmosphere by DNPH derivatization and LC-ESI-MS/MS detection. *Talanta* **2007**, *72*, 539–545. [CrossRef] [PubMed]
136. Zhang, X.; Kong, Y.; Cao, J.; Li, H.; Gao, R.; Zhang, Y.; Wang, K.; Li, Y.; Ren, Y.; Wang, W. A sensitive simultaneous detection approach for the determination of 30 atmospheric carbonyls by 2,4-dinitrophenylhydrazine derivatization with HPLC-MS technique and its preliminary application. *Chemosphere* **2022**, *303*, 134985. [CrossRef] [PubMed]
137. Xu, Y.; Feng, X.; Chen, Y.; Zheng, P.; Hui, L.; Chen, Y.; Yu, J.Z.; Wang, Z. Development of an enhanced method for atmospheric carbonyls and characterizing their roles in photochemistry in subtropical Hong Kong. *Sci. Total Environ.* **2023**, *896*, 165135. [CrossRef] [PubMed]
138. Fung, K.; Grosjean, D. Determination of nanogram amounts of carbonyls as 2,4-dinitrophenylhydrazones by high-performance liquid chromatography. *Anal. Chem.* **1981**, *53*, 168–171. [CrossRef]
139. MEE. *Ambient Air-Determination of Aldehyde and Ketone Compounds-High: HJ683-2014*; China Environmental Science Press: Beijing, China, 2020.
140. Guo, S.; Chen, M.; Tan, J. Seasonal and diurnal characteristics of atmospheric carbonyls in Nanning, China. *Atmos. Res.* **2016**, *169*, 46–53. [CrossRef]
141. Possanzini, M.; DiPalo, V. Determination of olefinic aldehydes and other volatile carbonyls in air samples by DNPH-coated cartridges and HPLC. *Chromatographia* **1995**, *40*, 134–138. [CrossRef]
142. Schulte-Ladbeck, R.; Lindahl, R.; Levin, J.-O.; Karst, U.; Schulte Ladbeck, R. Characterization of chemical interferences in the determination of unsaturated aldehydes using aromatic hydrazine reagents and liquid chromatography. *J. Environ. Monit.* **2001**, *3*, 306–310. [CrossRef] [PubMed]
143. Slemr, J. Determination of volatile carbonyl compounds in clean air. *Fresenius J. Anal. Chem.* **1991**, *340*, 672–677. [CrossRef]
144. Zhang, H.; Hou, H.; Pang, Y.; Jiang, X.; Tang, G. HPLC Determination of Major Carbonyl Compounds in Mainstream ofCigarette Smoke with Solvent Trapping. *Phys. Chem. Test. (Chem. Sect.)* **2013**, *49*, 19–22.
145. Liu, L.; Yuan, B.; Jin, Y. Analysis of 13 carbonyl compounds using ultra-high-performance liquid chromatography with a Carbonyl column. *Environ. Chem.* **2013**, *32*, 715–716.
146. MEE. *Determination of Aldehydes and Ketones in Ambient Air by Solution Absorption-High Performance Liquid Chromatography: HJ 1154-2020*; China Environmental Science Press: Beijing, China, 2020.
147. Castellani, F.; Antonucci, A.; Pindinello, I.; Protano, C.; Vitali, M. Determination of Carbonyl Compounds in Different Work Environments: Comparison between LC-UV/DAD and LC–MS/MS Detection Methods. *Int. J. Environ. Res. Public Health.* **2022**, *19*, 12052. [CrossRef] [PubMed]
148. Van den Bergh, V.; Coeckelberghs, H.; Vankerckhoven, H.; Compernolle, F.; Vinckier, C. Study of the carbonyl products of terpene/OH radical reactions: Detection of the 2,4-DNPH derivatives by HPLC-MS. *Anal. BioAnal. Chem.* **2004**, *379*, 484–494. [CrossRef] [PubMed]
149. van Leeuwen, S.M.; Hendriksen, L.; Karst, U. Determination of aldehydes and ketones using derivatization with 2,4-dinitrophenylhydrazine and liquid chromatography–atmospheric pressure photoionization-mass spectrometry. *J. Chromatogr. A* **2004**, *1058*, 107–112. [CrossRef]
150. Li, M.; Biswas, S.; Nantz, M.H.; Higashi, R.M.; Fu, X.-A. Preconcentration and Analysis of Trace Volatile Carbonyl Compounds. *Anal. Chem.* **2012**, *84*, 1288–1293. [CrossRef] [PubMed]
151. MEE. *Determination of 65 Volatile Organic Compounds in Ambient Air by Canister Sampling and Gas Chromatography-Mass Spectrometry: HJ 759–2023*; China Environmental Science Press: Beijing, China, 2023.
152. ZOU, T.; FENG, Y.; FU, Z.; MU, C.; ZHAI, J. Determination of mono- and dicarbonyls in the atmosphere using gas chromatography/mass spectrometry after PFBHA derivatization. *ASC* **2012**, *32*, 718–2724.

153. Fuson, N.; Josien, M.-L.; Shelton, E.M. An Infrared Spectroscopic Study of the Carbonyl Stretching Frequency in Some Groups of Ketones and Quinones. *J. Am. Chem. Soc.* **1954**, *76*, 2526–2533. [CrossRef]
154. Wang, F.; Ho, S.S.H.; Man, C.L.; Qu, L.; Wang, Z.; Ning, Z.; Ho, K.F. Characteristics and sources of oxygenated VOCs in Hong Kong: Implications for ozone formation. *Sci. Total Environ.* **2024**, *912*, 169156. [CrossRef] [PubMed]
155. Olariu, R.I.; Klotz, B.; Barnes, I.; Becker, K.H.; Mocanu, R. FT–IR study of the ring-retaining products from the reaction of OH radicals with phenol, o-, m-, and p-cresol. *Atmos. Environ.* **2002**, *36*, 3685–3697. [CrossRef]
156. Delgado-Saborit, J.M.; Alam, M.S.; Godri Pollitt, K.J.; Stark, C.; Harrison, R.M. Analysis of atmospheric concentrations of quinones and polycyclic aromatic hydrocarbons in vapour and particulate phases. *Atmos. Environ.* **2013**, *77*, 974–982. [CrossRef]
157. Chung, M.Y.; Lazaro, R.A.; Lim, D.; Jackson, J.; Lyon, J.; Rendulic, D.; Hasson, A.S. Aerosol-Borne Quinones and Reactive Oxygen Species Generation by Particulate Matter Extracts. *Environ. Sci. Technol.* **2006**, *40*, 4880–4886. [CrossRef] [PubMed]
158. Sousa, E.T.; Cardoso, M.P.; Silva, L.A.; de Andrade, J.B. Direct determination of quinones in fine atmospheric particulate matter by GC–MS. *Microchem. J.* **2015**, *118*, 26–31. [CrossRef]
159. Wnorowski, A.; Charland, J.-P. Profiling quinones in ambient air samples collected from the Athabasca region (Canada). *Chemosphere* **2017**, *189*, 55–66. [CrossRef]
160. Dos Santos, R.R.; Cardeal, Z.d.L.; Menezes, H.C. Phase distribution of polycyclic aromatic hydrocarbons and their oxygenated and nitrated derivatives in the ambient air of a Brazilian urban area. *Chemosphere* **2020**, *250*, 126223. [CrossRef] [PubMed]
161. Jo, J.; Lee, J.-Y.; Jang, K.-S.; Matsuki, A.; Natsagdorj, A.; Ahn, Y.-G. Development of Quantitative Chemical Ionization Using Gas Chromatography/Mass Spectrometry and Gas Chromatography/Tandem Mass Spectrometry for Ambient Nitro- and Oxy-PAHs and Its Applications. *Molecules* **2023**, *28*, 775. [CrossRef] [PubMed]
162. Apel, E.C.; Riemer, D.D.; Hills, A.; Baugh, W.; Orlando, J.; Faloona, I.; Tan, D.; Brune, W.; Lamb, B.; Westberg, H.; et al. Measurement and interpretation of isoprene fluxes and isoprene, methacrolein, and methyl vinyl ketone mixing ratios at the PROPHET site during the 1998 Intensive. *J. Geophys. Res. Atmos.* **2002**, *107*, 4034. [CrossRef]
163. Hamilton, J.F.; Webb, P.J.; Lewis, A.C.; Hopkins, J.R.; Smith, S.; Davy, P. Partially oxidised organic components in urban aerosol using GCXGC-TOF/MS. *Atmos. Chem. Phys.* **2004**, *4*, 1279–1290. [CrossRef]
164. Prazeller, P.; Palmer, P.T.; Boscaini, E.; Jobson, T.; Alexander, M. Proton transfer reaction ion trap mass spectrometer. *Rapid Commun. Mass Spectrom.* **2003**, *17*, 1593–1599. [CrossRef] [PubMed]
165. Warneke, C.; Rosén, S.; Lovejoy, E.R.; de Gouw, J.A.; Fall, R. Two additional advantages of proton-transfer ion trap mass spectrometry. *Rapid Commun. Mass Spectrom.* **2003**, *18*, 133–134. [CrossRef]
166. Graus, M.; Müller, M.; Hansel, A. High resolution PTR-TOF: Quantification and formula confirmation of VOC in real time. *J. Am. Soc. Mass Spectrom.* **2010**, *21*, 1037–1044. [CrossRef] [PubMed]
167. Liu, R.; Guo, Y.; Li, M.; Li, J.; Yang, D.; Hou, K. Development and Application of a Chemical Ionization Focusing Integrated Ionization Source TOFMS for Online Detection of OVOCs in the Atmosphere. *Molecules* **2023**, *28*, 6600. [CrossRef] [PubMed]
168. Song, M.; Li, X.; Yang, S.; Yu, X.; Zhou, S.; Yang, Y.; Chen, S.; Dong, H.; Liao, K.; Chen, Q.; et al. Spatiotemporal variation, sources, and secondary transformation potential of volatile organic compounds in Xi'an, China. *Atmos. Chem. Phys.* **2021**, *21*, 4939–4958. [CrossRef]
169. Apel, E.C.; Hills, A.J.; Lueb, R.; Zindel, S.; Eisele, S.; Riemer, D.D. A fast-GC/MS system to measure C2 to C4 carbonyls and methanol aboard aircraft. *J. Geophys. Res.* **2003**, *108*, 8794. [CrossRef]
170. Waterman, D.; Horsfield, B.; Leistner, F.; Hall, K.; Smith, S. Quantification of Polycyclic Aromatic Hydrocarbons in the NIST Standard Reference Material (SRM1649A) Urban Dust Using Thermal Desorption GC/MS. *Anal. Chem.* **2000**, *72*, 3563–3567. [CrossRef]
171. Williams, B.J.; Zhang, Y.; Zuo, X.; Martinez, R.E.; Walker, M.J.; Kreisberg, N.M.; Goldstein, A.H.; Docherty, K.S.; Jimenez, J.L. Organic and inorganic decomposition products from the thermal desorption of atmospheric particles. *Atmos. Meas. Tech.* **2016**, *9*, 1569–1586. [CrossRef]
172. Maher, S.; Jjunju, F.P.M.; Taylor, S. Colloquium: 100 years of mass spectrometry: Perspectives and future trends. *Rev. Mod. Phys.* **2015**, *87*, 113–135. [CrossRef]
173. Buiarelli, F.; Canepari, S.; Di Filippo, P.; Perrino, C.; Pomata, D.; Riccardi, C.; Speziale, R. Extraction and analysis of fungal spore biomarkers in atmospheric bioaerosol by HPLC–MS–MS and GC–MS. *Talanta* **2013**, *105*, 142–151. [CrossRef]
174. Mirsaleh-Kohan, N.; Robertson, W.D.; Compton, R.N. Electron ionization time-of-flight mass spectrometry: Historical review and current applications. *Mass Spectrom. Rev.* **2008**, *27*, 237–285. [CrossRef]
175. Bertram, T.H.; Kimmel, J.R.; Crisp, T.A.; Ryder, O.S.; Yatavelli, R.L.N.; Thornton, J.A.; Cubison, M.J.; Gonin, M.; Worsnop, D.R. A field-deployable, chemical ionization time-of-flight mass spectrometer. *Atmos. Meas. Tech.* **2011**, *4*, 1471–1479. [CrossRef]
176. Adam, T.; Zimmermann, R. Determination of single photon ionization cross sections for quantitative analysis of complex organic mixtures. *Anal. BioAnal. Chem.* **2007**, *389*, 1941–1951. [CrossRef] [PubMed]
177. Yamashita, M.; Fenn, J.B. Electrospray ion source. Another variation on the free-jet theme. *J. Phys. Chem.* **1984**, *88*, 4451–4459. [CrossRef]
178. Awad, H.; Khamis, M.M.; El-Aneed, A. Mass Spectrometry, Review of the Basics: Ionization. *Appl. Spectrosc. Rev.* **2014**, *50*, 158–175. [CrossRef]

179. Zhong, S.; Zhang, K.; Bagheri, M.; Burken, J.G.; Gu, A.; Li, B.; Ma, X.; Marrone, B.L.; Ren, Z.J.; Schrier, J.; et al. Machine Learning: New Ideas and Tools in Environmental Science and Engineering. *Environ. Sci. Technol.* **2021**, *55*, 12741–12754. [CrossRef]
180. Kim, S.; Karl, T.; Guenther, A.; Tyndall, G.; Orlando, J.; Harley, P.; Rasmussen, R.; Apel, E. Emissions and ambient distributions of Biogenic Volatile Organic Compounds (BVOC) in a ponderosa pine ecosystem: Interpretation of PTR-MS mass spectra. *Atmos. Chem. Phys.* **2010**, *10*, 1759–1771. [CrossRef]
181. Steinbacher, M.; Dommen, J.; Ammann, C.; Spirig, C.; Neftel, A.; Prevot, A.S.H. Performance characteristics of a proton-transfer-reaction mass spectrometer (PTR-MS) derived from laboratory and field measurements. *Int. J. Mass Spectrom.* **2004**, *239*, 117–128. [CrossRef]
182. Müller, M.; Mikoviny, T.; Feil, S.; Haidacher, S.; Hanel, G.; Hartungen, E.; Jordan, A.; Märk, L.; Mutschlechner, P.; Schottkowsky, R.; et al. A compact PTR-ToF-MS instrument for airborne measurements of volatile organic compounds at high spatiotemporal resolution. *Atmos. Meas. Tech.* **2014**, *7*, 3763–3772. [CrossRef]
183. Thalman, R.; Baeza-Romero, M.T.; Ball, S.M.; Borrás, E.; Daniels, M.J.S.; Goodall, I.C.A.; Henry, S.B.; Karl, T.; Keutsch, F.N.; Kim, S.; et al. Instrument intercomparison of glyoxal, methyl glyoxal and $NO_2$ under simulated atmospheric conditions. *Atmos. Meas. Tech.* **2015**, *8*, 1835–1862. [CrossRef]
184. Zhou, Z.; Guo, H.; Qi, F. Recent developments in synchrotron vacuum ultraviolet photoionization coupled to mass spectrometry. *TrAC Trends Anal. Chem.* **2011**, *30*, 1400–1409. [CrossRef]
185. Lipson, R.H.; Dimov, S.S.; Wang, P.; Shi, Y.J.; Mao, D.M.; Hu, X.K.; Vanstone, J. VACUUM ULTRAVIOLET AND EXTREME ULTRAVIOLET LASERS: PRINCIPLES, INSTRUMENTATION, AND APPLICATIONS. *Instrum. Sci. Technol.* **2000**, *28*, 85–118. [CrossRef]
186. Yang, B.; Zhang, H.; Shu, J.; Ma, P.; Zhang, P.; Huang, J.; Li, Z.; Xu, C. A VUV-excited and CHCl/HO-amplified ionization-coupled mass spectrometry for oxygenated organics analysis. *Anal. Chem.* **2017**, *90*, 1301–1308. [CrossRef] [PubMed]
187. Xie, Y.; Li, Q.; Hua, L.; Chen, P.; Hu, F.; Wan, N.; Li, H. Highly selective and sensitive online measurement of trace exhaled HCN by acetone-assisted negative photoionization time-of-flight mass spectrometry with in-source CID. *Anal. Chim. Acta.* **2020**, *1111*, 31–39. [CrossRef] [PubMed]
188. Dang, M.; Liu, R.; Dong, F.; Liu, B.; Hou, K. Vacuum ultraviolet photoionization on-line mass spectrometry: Instrumentation developments and applications. *TrAC Trends Anal. Chem.* **2022**, *149*, 116542. [CrossRef]
189. Zogka, A.G.; Romanias, M.N.; Thevenet, F. Formaldehyde and glyoxal measurement deploying a selected ion flow tube mass spectrometer (SIFT-MS). *Atmos. Meas. Tech.* **2022**, *15*, 2001–2019. [CrossRef]
190. Gkatzelis, G.I.; Tillmann, R.; Hohaus, T.; Müller, M.; Eichler, P.; Xu, K.-M.; Schlag, P.; Schmitt, S.H.; Wegener, R.; Kaminski, M.; et al. Comparison of three aerosol chemical characterization techniques utilizing PTR-ToF-MS: A study on freshly formed and aged biogenic SOA. *Atmos. Meas. Tech.* **2018**, *11*, 1481–1500. [CrossRef]
191. Timkovsky, J.; Dusek, U.; Henzing, J.S.; Kuipers, T.L.; Röckmann, T.; Holzinger, R. Offline thermal-desorption proton-transfer-reaction mass spectrometry to study composition of organic aerosol. *J. Aerosol Sci.* **2015**, *79*, 1–14. [CrossRef]
192. Vogel, A.L.; Äijälä, M.; Brüggemann, M.; Ehn, M.; Junninen, H.; Petäjä, T.; Worsnop, D.R.; Kulmala, M.; Williams, J.; Hoffmann, T. Online atmospheric pressure chemical ionization ion trap mass spectrometry (APCI-IT-MSn) for measuring organic acids in concentrated bulk aerosol—A laboratory and field study. *Atmos. Meas. Tech.* **2013**, *6*, 431–443. [CrossRef]
193. Thompson, S.L.; Yatavelli, R.L.N.; Stark, H.; Kimmel, J.R.; Krechmer, J.E.; Day, D.A.; Hu, W.; Isaacman-VanWertz, G.; Yee, L.; Goldstein, A.H.; et al. Field intercomparison of the gas/particle partitioning of oxygenated organics during the Southern Oxidant and Aerosol Study (SOAS) in 2013. *Aerosol Sci. Technol.* **2016**, *51*, 30–56. [CrossRef]
194. Klotz, B.; Sørensen, S.; Barnes, I.; Becker, K.H.; Etzkorn, T.; Volkamer, R.; Platt, U.; Wirtz, K.; Martín-Reviejo, M. Atmospheric oxidation of toluene in a large-volume outdoor photoreactor: In situ determination of ring-retaining product yields. *J. Phys. Chem. A* **1998**, *102*, 10289–10299. [CrossRef]
195. Romano, A.; Hanna, G.B. Identification and quantification of VOCs by proton transfer reaction time of flight mass spectrometry: An experimental workflow for the optimization of specificity, sensitivity, and accuracy. *J. Mass Spectrom.* **2018**, *53*, 287–295. [CrossRef] [PubMed]
196. Holzinger, R.; Williams, J.; Herrmann, F.; Lelieveld, J.; Donahue, N.M.; Röckmann, T. Aerosol analysis using a Thermal-Desorption Proton-Transfer-Reaction Mass Spectrometer (TD-PTR-MS): A new approach to study processing of organic aerosols. *Atmos. Chem. Phys.* **2010**, *10*, 2257–2267. [CrossRef]
197. Gkatzelis, G.I.; Hohaus, T.; Tillmann, R.; Gensch, I.; Müller, M.; Eichler, P.; Xu, K.-M.; Schlag, P.; Schmitt, S.H.; Yu, Z.; et al. Gas-to-particle partitioning of major biogenic oxidation products: A study on freshly formed and aged biogenic SOA. *Atmos. Chem. Phys.* **2018**, *18*, 12969–12989. [CrossRef]
198. Kourtchev, I.; Bejan, I.; Sodeau, J.R.; Wenger, J.C. Gas-phase reaction of (E)-β-farnesene with ozone: Rate coefficient and carbonyl products. *Atmos. Environ.* **2009**, *43*, 3182–3190. [CrossRef]
199. Kourtchev, I.; Bejan, I.; Sodeau, J.R.; Wenger, J.C. Gas phase reaction of OH radicals with (E)-β-farnesene at 296±2 K: Rate coefficient and carbonyl products. *Atmos. Environ.* **2012**, *46*, 338–345. [CrossRef]
200. Spittler, M.; Barnes, I.; Bejan, I.; Brockmann, K.J.; Benter, T.; Wirtz, K. Reactions of $NO_3$ radicals with limonene and α-pinene: Product and SOA formation. *Atmos. Environ.* **2006**, *40*, 116–127. [CrossRef]

201. Gómez Alvarez, E.; Viidanoja, J.; Muñoz, A.; Wirtz, K.; Hjorth, J. Experimental Confirmation of the Dicarbonyl Route in the Photo-oxidation of Toluene and Benzene. *Environ. Sci. Technol.* **2007**, *41*, 8362–8369. [CrossRef] [PubMed]
202. Löher, F.; Borrás, E.; Muñoz, A.; Nölscher, A.C. Characterization of a new Teflon chamber and on-line analysis of isomeric multifunctional photooxidation products. *Atmos. Meas. Tech.* **2024**, *17*, 4553–4579. [CrossRef]
203. Komenda, M.; Schaub, A.; Koppmann, R. Description and characterization of an on-line system for long-term measurements of isoprene, methyl vinyl ketone, and methacrolein in ambient air. *J. Chromatogr. A* **2003**, *995*, 185–201. [CrossRef] [PubMed]

MDPI AG
Grosspeteranlage 5
4052 Basel
Switzerland
Tel.: +41 61 683 77 34

*Atmosphere* Editorial Office
E-mail: atmosphere@mdpi.com
www.mdpi.com/journal/atmosphere

www.ingramcontent.com/pod-product-compliance
Lightning Source LLC
Chambersburg PA
CBHW061847210926
51850CB00006B/996
* 9 7 8 3 7 2 5 8 6 3 3 0 3 *